Simulation-based Labs for Circuit Analysis

Discovering Circuits with Multisim Live and Tinkercad

RIVER PUBLISHERS SERIES ELECTRONIC MATERIALS, CIRCUITS AND DEVICES

The "River Publishers Series in Electronic Materials, Circuits and Devices" is a series of comprehensive academic and professional books which focus on theory and applications of advanced electronic materials, circuits and devices. This includes analog and digital integrated circuits, memory technologies, system-on-chip and processor design. Also theory and modeling of devices, performance and reliability of electron and ion integrated circuit devices and interconnects, insulators, metals, organic materials, micro-plasmas, semiconductors, quantum-effect structures, vacuum devices, and emerging materials. The series also includes books on electronic design automation and design methodology, as well as computer aided design tools.

Books published in the series include research monographs, edited volumes, handbooks and text-books. The books provide professionals, researchers, educators, and advanced students in the field with an invaluable insight into the latest research and developments.

Topics covered in this series include:

- Analog Integrated Circuits
- Data Converters
- Digital Integrated Circuits
- Electronic Design Automation
- Insulators
- Integrated circuit devices
- Interconnects
- Memory Design
- MEMS
- Nanoelectronics
- Organic materials
- Power ICs
- Processor Architectures
- Quantum-effect structures
- Semiconductors
- Sensors and actuators
- System-on-Chip
- Vacuum devices

For a list of other books in this series, visit www.riverpublishers.com

Simulation-based Labs for Circuit Analysis

Discovering Circuits with Multisim Live and Tinkercad

Massimo Mitolo

Electrical Technology Department, Irvine Valley College,
Irvine, California, USA

River Publishers

Routledge
Taylor & Francis Group
NEW YORK AND LONDON

Published 2024 by River Publishers
River Publishers
Alsbjergvej 10, 9260 Gistrup, Denmark
www.riverpublishers.com

Distributed exclusively by Routledge
605 Third Avenue, New York, NY 10017, USA
4 Park Square, Milton Park, Abingdon, Oxon OX14 4RN

Simulation-based Labs for Circuit Analysis / by Massimo Mitolo.

Routledge is an imprint of the Taylor & Francis Group, an informa business

ISBN 978-87-7004-081-5 (hardback)
ISBN 978-87-7004-181-2 (paperback)
ISBN 978-10-4011-995-2 (online)
ISBN 978-10-0351-533-3 (master ebook)

As the electrical scholars mature, they should delve into circuits,
seeking the elusive truths they hold.
Their wisdom, passed on as professors,
empowers students to handle the contemporary challenges.
This journey unveils the profound essence of their craft.

Massimo Mitolo

Contents

Preface

Welcome to *Simulation-based Labs for Circuit Analysis.*

Circuit analysis is a fundamental aspect of electrical and electronic engineering, laying the foundation for various applications in modern technology. This book aims to provide students and enthusiasts with a comprehensive and interactive learning experience in this field.

With the advent of simulation software, a new dimension of learning has emerged, offering a virtual laboratory environment that replicates real-world circuit experiments with remarkable precision and flexibility. This book is designed to bridge the gap between theoretical knowledge and real-world applications by harnessing the power of advanced simulation platforms like Multisim Live and Tinkercad.

Multisim Live, developed by National Instruments, and Tinkercad, developed by Autodesk, are free web-based platforms. They are prominent simulation tools that have revolutionized the way circuits are studied and understood. These platforms enable users to create and simulate circuits, observe voltage and current behaviors, and analyze the impact of different components and configurations, all within the confines of a digital workspace. The simulations offer virtual "breadboards" to prototype and test circuits. They provide integrated options to design, analyze, and customize circuits with realistic component models. Devices can be combined, parameters swept, and signals probed, all through an intuitive software interface.

This book leverages the capabilities of Multisim Live and Tinkercad to empower readers with interactive, hands-on experiences, making circuit analysis an engaging and enriching journey.

The textbook outlines simulation-centric labs that parallel the theory covered in key electrical engineering courses. Students construct circuits, take measurements, generate plots, and document results, developing proficiency translating concepts to practical implementation. The ability to quickly iterate designs provides insights into cause–effect relationships and circuit behavior.

The key features of *Simulation-based Labs for Circuit Analysis* are:

Comprehensive circuit simulations: Through solved problems, readers will explore a wide range of circuits, both in DC and AC. The simulations will cover various circuit elements, including resistors, capacitors, inductors, and operational amplifiers. Each chapter includes *prelabs*, summarizing the basic theory, and *lab procedures* to be performed by the readers.

Real-time visualizations: Multisim Live and Tinkercad offer dynamic visualizations, allowing users to witness the behavior of voltage and current in real time. This visual feedback enhances comprehension and aids in grasping complex concepts.

Practical problem-solving: By immersing themselves in hands-on circuit simulations, readers will develop problem-solving skills critical for engineering disciplines. The book includes challenging exercises and practical applications that reinforce theoretical knowledge.

Experiment replication: With Multisim Live and Tinkercad, readers can recreate experiments multiple times, promoting a deeper understanding of circuit characteristics and behavior in various scenarios.

As the author of this book, I am enthusiastic about the potential that simulation-based labs offer in elevating the learning experience for aspiring engineers, technicians, and circuit enthusiasts. With Multisim Live and Tinkercad, students can explore, innovate, and experiment without constraints, fostering a profound appreciation for circuit analysis in the digital age. This lab manual aims to teach students to judiciously utilize these advantages in conjunction with analytical skills to become adept circuit designers.

I hope this book serves as a valuable resource for students, educators, and professionals alike, empowering them to unlock the limitless possibilities of circuit analysis through immersive simulation-based learning.

Happy Simulating!

Prof. Massimo Mitolo

List of Figures

List of Tables

List of Abbreviations

BW	Bandwidth
CCCS	Current-controlled current source
LED	Light emitting diode
rms	Root mean square
SR	Slew rate
VAR	Volt-amps reactive
VCVS	Voltage-controlled voltage source

Part I

DC Circuits

1

Fundamental Circuit Conventions

1.1 Student Learning Outcomes

- Correctly identify and label the passive and active convention for current direction in electrical components such as resistors, voltage sources, and current sources.
- Analyze a given DC circuit and determine the sign of power consumption or delivery based on the passive sign convention and the direction of current flow.
- Measure voltages and currents in DC circuits built in Multisim using virtual multimeters. Compare measured values to calculated values.

1.2 Prelab

A two-terminal component, for example a resistor, is a "lumped" element that interacts with a circuit exclusively through its two terminals. The lumped abstraction is only valid if the electric potential difference across the terminals can be univocally determined and does not depend on the path taken for its calculation[1]. This condition ensures that the lumped element assumption remains valid and that the behavior of the component is solely determined by the voltage and current at its terminals, without any influence from time-varying magnetic fields. It ensures that the component's response is purely electrical and that there are no electromagnetic interactions that could affect the potential difference across its terminals or violate the principle of charge conservation.

[1]This condition holds true for a two-terminal omponent under the lumped abstraction if, for any closed path (∂S) that bounds the surface (S) encompassing the component, there is no time-varying magnetic flux (Φ_B) passing through it for all times. In formulae: $\frac{\partial \Phi_B}{\partial t} = 0, \forall t$.

Additionally, the current entering one terminal must be equal to the current leaving the other terminal. However, this assumption holds true only when there is no charge accumulation or depletion within the component over time[2]. This is only true in the absence of charge build-up or depletion within the component over time.

The instantaneous electric power, denoted as $p(t)$, represents the rate at which energy is absorbed or generated by a component. It is defined as the time derivative of the energy, $W(t)$, with respect to time:

$$p(t) = \frac{dW(t)}{dt} = \frac{dW}{dq}\frac{dq}{dt} = v(t) \cdot i(t). \tag{1.1}$$

The unit of power is Watt, which is equivalent to the product of voltage (V) and current (A). In other words: $[V \cdot A] = \left[\frac{J}{C} \cdot \frac{C}{s}\right] = [W]$.

A passive element cannot provide net energy, that is, it cannot give more energy than that received.

The component is passive if the absorbed energy is greater than, or equal to, for all time.

A passive element is incapable of providing net energy to a circuit. This means that it cannot deliver more energy than it receives. For a passive component, the absorbed energy, $W\text{abs}(t)$, is always greater than or equal to zero for all times:

$$W_{\text{abs}}(t) = \int_{-\infty}^{t} p_{\text{abs}}(t')\, dt' \geq 0, \ \forall t. \tag{1.2}$$

On the other hand, an active component can supply net energy to the circuit. It allows for the possibility of negative absorbed energy, where there exists at least one instant of time, denoted as t_0, such that $W\text{abs}(t_0) < 0$. This implies that energy can be extracted from the component.

1.2.1 Passive and active sign conventions

The *passive sign* convention is a convention used to assign the polarities of voltage and current in passive two-terminal components, such as resistors. According to this convention, the polarity of the voltage (*V*) across the component is assigned in the opposite direction of the current (*I*) circulating through it (Figure 1.1(a)). This means that the terminal where the current enters the resistor is considered the *positive* terminal, which is at a higher

---

[2]The current density $\bar{J}$ is solenoidal, that is, $i = \oiint_s \vec{J} \cdot \vec{u}_n dS = 0, \forall S$.

Figure 1.1 (a) Passive and (b) active sign conventions.

potential than the *negative* terminal. In this convention, the current flows from a higher potential to a lower potential.

On the other hand, the *active sign* convention is applied to active two-terminal components, like a battery. In this convention, the polarity of the voltage (V) across the component is assigned in the same direction as the current (I) circulating through it (Figure 1.1(b)). When a current flows through a voltage source, it travels from a lower to a higher potential, resulting in a *voltage rise*. The terminal from which the current leaves the active component is defined as the *positive* terminal.

These sign conventions not only determine the polarity of voltage and current but also play a role in defining the sign of electric power in a circuit. If the passive sign convention is applied to a component, the absorbed power of that component is defined as the product of voltage and current, $P_{abs} = v \cdot i$. On the other hand, if the active sign convention is applied, the absorbed power is defined as the negative product of voltage and current, $P_{abs} = -v \cdot i$. In this context, if the absorbed power is negative, it indicates that the component is supplying power to the circuit.

These sign conventions provide a consistent framework for analyzing and interpreting the behavior of components and circuits, allowing us to determine the direction and magnitude of voltage, current, and power in a standardized manner.

1.2.2 Problem 1

In the circuit depicted in Figure 1.2, voltage and current measurements are made in accordance with the passive and active sign conventions. To analyze the circuit and determine the absorbed power of the two-terminal components, perform the following tasks:

1. Using Multisim simulation, measure the voltages and currents in the circuit, ensuring that the measurements are taken following the appropriate sign conventions.

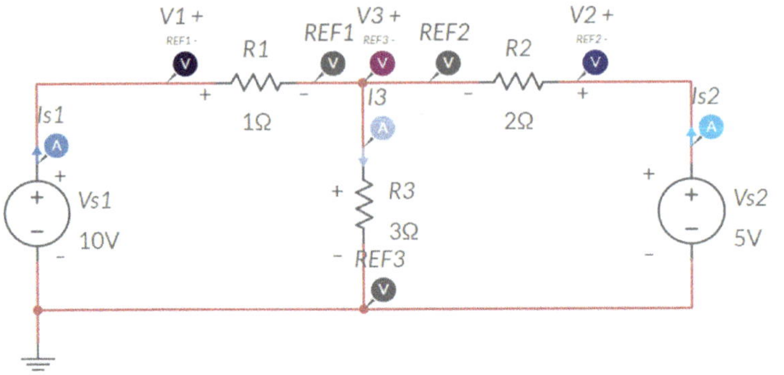

Figure 1.2 Circuit for voltage and current measurements in accordance with passive and active sign conventions.

2. For each two-terminal component in the circuit, calculate the absorbed power using the relevant voltage and current measurements. Apply the passive or active sign convention, depending on the nature of the component.

1.2.2.1 Solutions to Problem 1

1. Figure 1.3 shows a circuit simulation in Multisim with voltage and current measurements that follow the passive and active sign conventions.

When taking voltage or current measurements, negative readings imply a reversal in the polarity or direction relative to the assumed positive reference. For voltage measurements, a negative reading means that the actual voltage

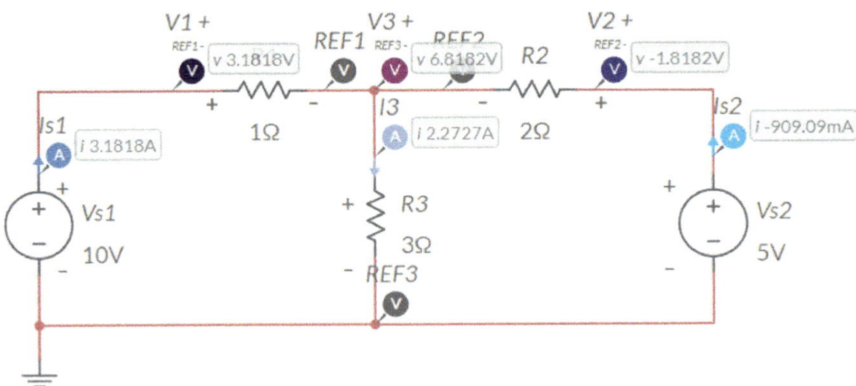

Figure 1.3 Multisim-simulated circuit with measurements of voltages and currents.

Table 1.1 Summary of readings

Component	P_{abs} (W)
V_{s1}	$-V_{s1}I_{s1} = -31.81$
V_{s2}	$-V_{s2}I_{s2} = 4.54$
R_1	$V_1I_1 = 10.11$
R_2	$V_2I_2 = 1.65$
R_3	$V_3I_3 = 15.4$

has the opposite polarity compared to the assumed positive polarity used for analysis. Similarly, a negative current measurement indicates that the true current direction is opposite to the assumed positive current direction.

2. Table 1.1 lists the absorbed power calculations for each two-terminal component in the circuit, expressed in units of Watts. The power absorbed P_{abs} is found by multiplying the voltage across and current through each element.

The voltage source V_{s1} absorbs -31.81 W of power, indicating it is delivering power to the circuit and acting as a true source. In contrast, the battery V_{s2} absorbs 4.5 W, meaning it is dissipating power and behaving as a load.

Overall, the total power absorbed by all the circuit components sums to $-31.81 + 4.54 + 10.11 + 1.65 + 15.4 \simeq 0$ W. This verifies the principle of conservation of power, where the power provided by the source V_{s1} equals the total absorbed power.

1.2.3 Problem 2

In the circuit depicted in Figure 1.4, voltage and current measurements are made in accordance with the passive and active sign conventions. To analyze the circuit and determine the absorbed power of the two-terminal components, perform the following tasks:

1. Using Multisim simulation, measure the voltages and currents in the circuit, ensuring that the measurements are taken following the appropriate sign conventions.
2. For each two-terminal component in the circuit, calculate the absorbed power using the relevant voltage and current measurements. Apply the passive or active sign convention, depending on the nature of the component.

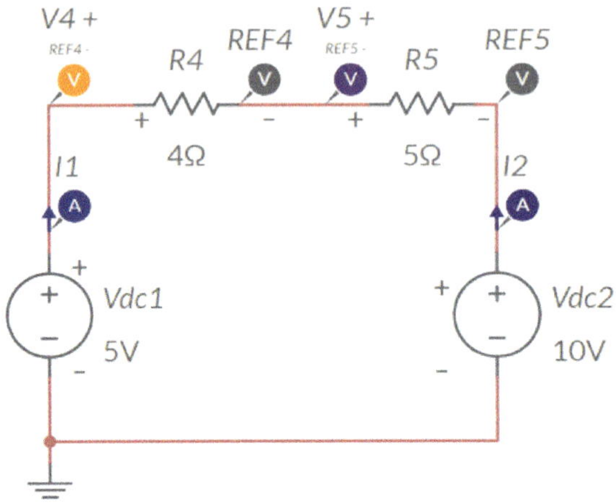

Figure 1.4 Circuit for voltage and current measurements in accordance with passive and active sign conventions.

1.2.3.1 Solutions to Problem 2

1. Figure 1.5 shows a circuit simulation in Multisim with voltage and current measurements that follow the passive and active sign conventions.

For voltage measurements, a negative reading means that the actual voltage has the opposite polarity compared to the assumed positive polarity used for analysis. Similarly, a negative current measurement indicates that the true current direction is opposite to the assumed positive current direction.

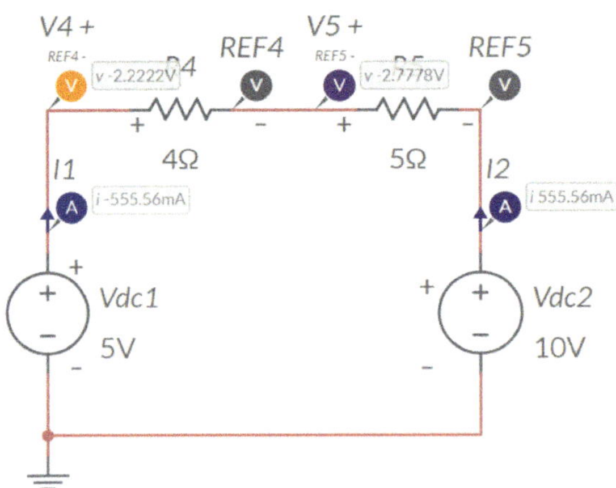

Figure 1.5 Multisim-simulated circuit with measurements of voltages and currents.

Table 1.2 Summary of readings

Component	P_{abs} (W)
V_{dc1}	$-V_{dc1}I_1 = 2.8$
V_{dc2}	$-V_{dc2}I_2 = -5.6$
R_4	$V_4I_1 = 1.23$
R_5	$V_2I_2 = 1.54$

2. Table 1.2 lists the absorbed power calculations for each two-terminal component in the circuit, expressed in units of Watts. The power absorbed P_{abs} is found by multiplying the voltage across and current through each element.

The voltage source V_{dc1} absorbs 2.8 W of power, indicating it is dissipating power and behaving as a load. In contrast, the battery V_{dc2} absorbs −5.6 W, meaning it is delivering power to the circuit and acting as a true source.

Table 1.2 also shows that the power provided by the source V_{dc2} equals the sum of the power absorbed by each component in compliance with the principle of conservation of power.

1.3 Lab Procedure 1

In the circuit depicted in Figure 1.6, voltage and current measurements are made in accordance with the passive and active sign conventions. To analyze the circuit and determine the absorbed power of the two-terminal components, perform the following tasks:

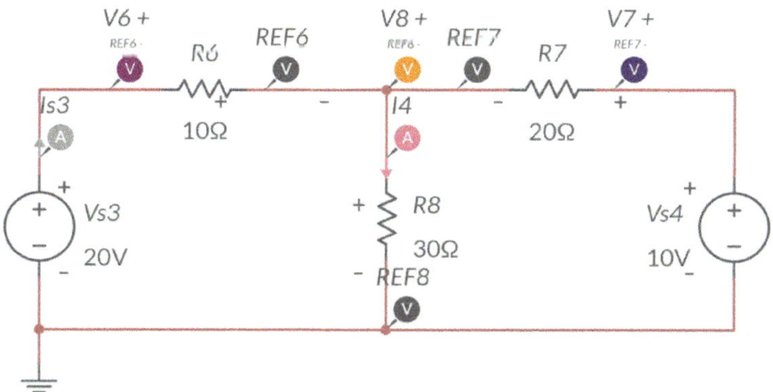

Figure 1.6 Circuit for voltage and current measurements in accordance with passive and active sign conventions.

Table 1.3 Summary of readings

Component	P_{abs} (W)
V_{s3}	
V_{s4}	
R_6	
R_7	
R_8	

1. Using Multisim simulation, measure the voltages and currents in the circuit, ensuring that the measurements are taken following the appropriate sign conventions.
2. For each two-terminal component in the circuit, calculate the absorbed power using the relevant voltage and current measurements. Apply the passive or active sign convention, depending on the nature of the component. Record results in Table 1.3

1.4 Lab Procedure 2

In the circuit depicted in Figure 1.7, voltage and current measurements are made in accordance with the passive and active sign conventions. To

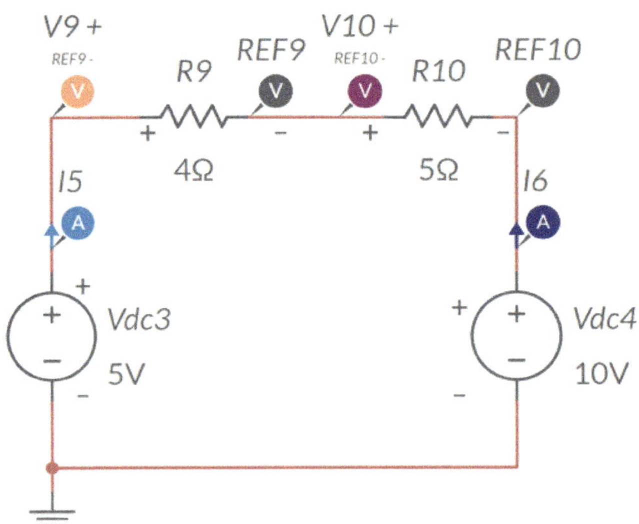

Figure 1.7 Circuit for voltage and current measurements in accordance with passive and active sign conventions.

Table 1.4 Summary of readings

Component	P_{abs} (W)
V_{dc3}	
V_{dc4}	
R_9	
R_{10}	

analyze the circuit and determine the absorbed power of the two-terminal components, perform the following tasks:

1. Using Multisim simulation, measure the voltages and currents in the circuit, ensuring that the measurements are taken following the appropriate sign conventions.
2. For each two-terminal component in the circuit, calculate the absorbed power using the relevant voltage and current measurements. Apply the passive or active sign convention, depending on the nature of the component. Record results in Table 1.4.

2

Ohm's Law

2.1 Student Learning Outcomes

- Construct resistor circuits in Multisim and Tinkercad, and simulate currents and voltages. Verify agreement between simulated and calculated values using Ohm's law.
- Understand the concept of the I-V characteristic curve and its significance in describing the behavior of a resistor.
- Interpret the slope of the I-V curve in terms of the resistance and its relationship with the physical properties of the resistor, as observed in the Multisim simulations.

2.2 Prelab

When applying a potential difference V across two points on the surface of an object of a given material and shape and measuring the resulting current I, the observed relationship between V and I follows Ohm's law. According to Ohm's law, the potential difference V is proportional to the circulating current I through a constant and positive quantity known as *resistance* R, expressed in Ohms (Ω). Mathematically, this relationship can be represented as follows:

$$V = I \cdot R. \tag{2.1}$$

In practice, the resistance of a given object can vary depending on factors such as the material and shape of the object, the specific points of potential difference application, and the area and shape of the voltage electrodes. However, to make resistance independent of these parameters, a cylindrical object with length l much greater than the square root of its cross-sectional area S

can be considered. Experiments have shown that, under this assumption, the resistance *R* can be calculated using the following formula:

$$R = \rho\frac{l}{S}, \tag{2.2}$$

where ρ is the resistivity of the material measured in units of $\Omega{\cdot}\mathrm{m}$ or $\Omega\frac{\mathrm{mm}^2}{\mathrm{m}}$.

Additionally, Ohm's law can be expressed in terms of conductance *G*, which is the reciprocal of resistance. Conductance *G* is measured in Siemens (S), and the relationship between current *I* and voltage *V* can be written as follows:

$$I = G \cdot V. \tag{2.3}$$

The linear correlation between current and voltage in a resistor is illustrated in Figure 2.1.

The equation of the straight line in this plot is given by

$$I = \frac{1}{R}V, \tag{2.4}$$

where the slope of the line, represented by the symbol m, is equal to $\frac{1}{R}$.

It is evident that as the resistance *R* increases, the slope decreases, and vice versa.

In the context of resistive two-terminals, the *short-circuit* and the *open-circuit* can also be considered. The short-circuit is an ideal component characterized by zero resistance and zero voltage across its terminals, regardless of the current circulating through it. On the other hand, the open-circuit is an ideal component with infinite resistance and zero current flowing through it, irrespective of the voltage applied across its terminals.

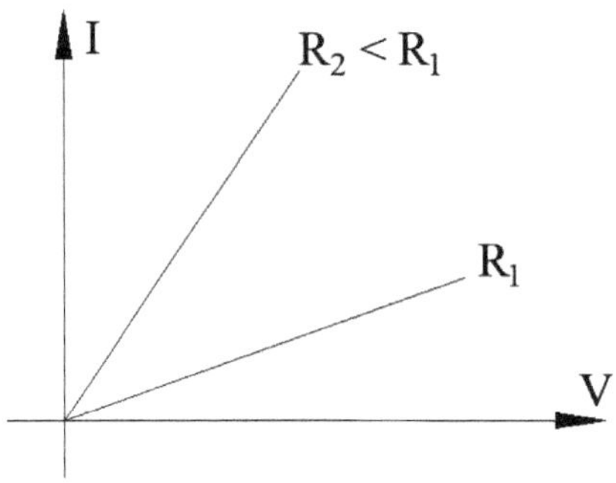

Figure 2.1 Linear correlation between current and voltage in a resistor.

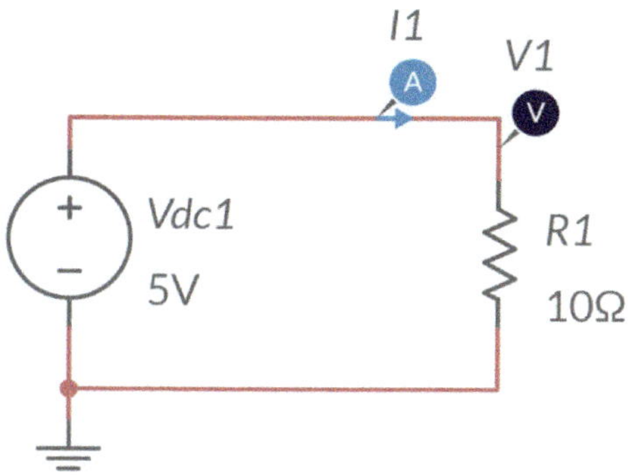

Figure 2.2 Circuit for voltage and current measurements in accordance with passive sign convention.

2.2.1 Problem 1

In the circuit depicted in Figure 2.2, voltage and current measurements are to be made in accordance with the passive sign convention. Perform the following tasks:

1. Use Multisim simulation to measure the current I_1 passing through the 10-Ω resistor R_1 and the voltage V_1 across it for the following values of supply voltage V_{dc1}: 5, 10, 15, 20, and 25 V.
2. Calculate the ratio $R_1 = V_1/I_1$ for each value of supply voltage V_{dc1} by dividing the measured voltage V_1 by the measured current I_1.
3. Determine the slope m for all the readings of V_1 and I_1. Determine the slope m for each set of readings of V_1 and I_1.
4. Repeat the above procedure for $R_1 = 20\ \Omega$ using the same set of supply voltages V_{dc1}. Measure the current I_1 passing through the 20-Ω resistor R_1 and the voltage V_1 across it, calculate the ratio $R_1 = V_1/I_1$, and determine the slope m for each reading.
5. Determine the slope of the straight line when $R_1 = 0$ and $R_1 = \infty$.

2.2.1.1 Solutions to Problem 1

1. The circuit has been simulated with Tinkercad (Figure 2.3).

The readings of voltage and current for the circuit simulation in Tinkercad are recorded in Table 2.1.

2. For a resistor value of $R_1 = 10\ \Omega$, the calculated ratios $R_1 = V_1/I_1$ are provided. From the readings, it can be observed that the ratio remains constant at 10 Ω for all supply voltage values.

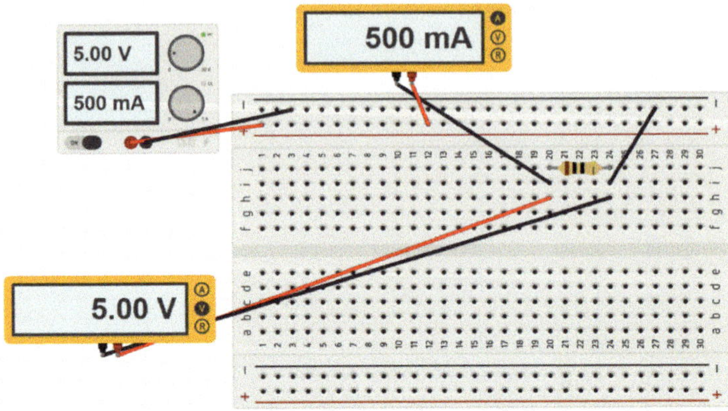

Figure 2.3 Simulated circuit with Tinkercad.

Table 2.1 Summary of readings and calculations

	$R_1 = 10\ \Omega$			
Supply voltage V_{dc1}	**I_1 (A)**	**V_1 (V)**	**$R_1 = V_1 / I_1$ (Ω)**	**m (S)**
5 V	0.5	5	10	0.1
10 V	1	10	10	0.1
15 V	1.5	15	10	0.1
20 V	2	20	10	0.1
25 V	2.5	25	10	0.1

3. The calculated values of the slope m are provided in Table 2.1. The slope m is consistently equal to 0.1 S, indicating a linear relationship between voltage and current in the resistor.

$$m = \frac{\Delta I}{\Delta V} = \frac{1}{R} = 0.1 \text{ S}. \tag{2.5}$$

This suggests that the resistor follows Ohm's law, as the slope represents the inverse of the resistance.

4. For the case of $R_1 = 20\ \Omega$, the measurements and calculations are recorded in Table 2.2.

Similar to the previous scenario, $R_2 = V_2/I_2$ remains constant and equal to 20 Ω for all supply voltage values. The slope m is also found to be 0.05 S, which is lower than the slope for $R_1 = 10\ \Omega$. This indicates that the line representing the relationship between voltage and current is less steep for $R_2 = 20\ \Omega$ compared to $R_1 = 10\ \Omega$ (Figure 2.4).

Table 2.2 Summary of readings and calculations

	R_1 = 20 Ω			
Supply voltage V_{dc1}	I_2 **(A)**	V_2 **(V)**	$R_2 = V_2/I_2$ **(Ω)**	m **(S)**
5 V	0.25	5	20	0.05
10 V	0.5	10	20	0.05
15 V	0.75	15	20	0.05
20 V	1	20	20	0.05
25 V	1.25	25	20	0.05

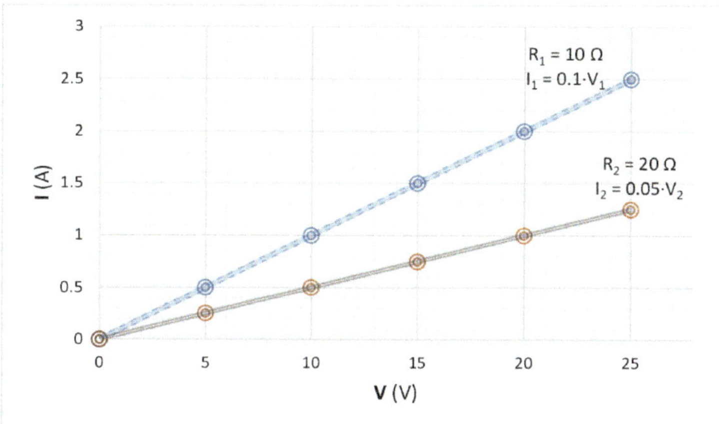

Figure 2.4 *I-V* characteristic graph for different values of resistance.

5. As R_1 approaches zero (short-circuit), a large current flows through the circuit. In this case, the current line on the *I-V* characteristic graph becomes almost vertical, approaching a 90° angle with the x-axis. The slope of the line, m, tends toward infinity (∞) as the change in voltage (ΔV) becomes negligible compared to the significant change in current (ΔI). On the other hand, as R_1 approaches infinity (open-circuit), the flow of current is reduced to almost zero. In this scenario, the current line on the *I-V* characteristic graph becomes nearly horizontal, approaching a 0° angle with the x-axis. The slope of the line, m, tends toward zero as the change in current (ΔI) becomes negligible compared to the change in voltage (ΔV). These observations align with the behavior of short-circuits and open-circuits, where a short-circuit allows for a large current flow with negligible voltage drop, while an open-circuit impedes the flow of current, resulting in negligible current with little to no voltage drop across the component.

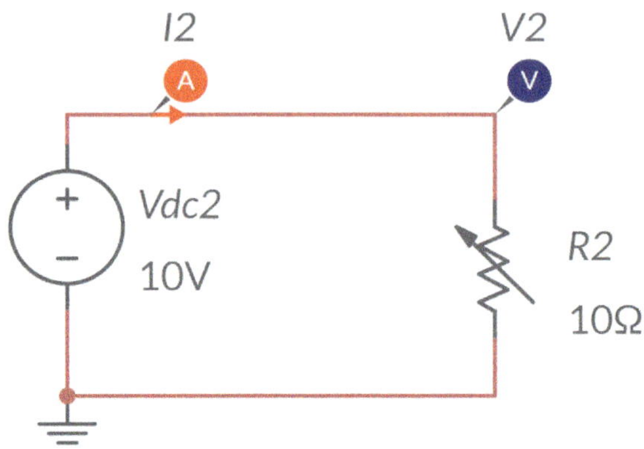

Figure 2.5 Circuit with variable resistor.

2.2.2 Problem 2

In the circuit of Figure 2.5, the supply voltage V_{dc2} is held constant, whereas the resistance R_2 varies ranging between 10 and 100 Ω in increments of 10 Ω. To analyze the circuit, perform the following tasks:

1. Measure the current I_2 passing through each value of R_2 by configuring the circuit in Multisim accordingly. Record the values of I_2 in Table 2.3.
2. Repeat the measurement steps mentioned above, but this time set the supply voltage to V_{dc2} = 20 V. Record the new values of I_2 in Table 2.3.
3. Create a plot that shows the relationship between the values of the current I_2 and the resistance R_2 for both supply voltage values (V_{dc2} = 10 V and V_{dc2} = 20 V).

Table 2.3 Summary of readings

	V_{dc2} = 10 V	V_{dc2} = 20 V
R_2 (Ω)	I_2 (A)	I_2 (A)
10	1.00	2.00
20	0.50	1.00
30	0.33	0.67
40	0.25	0.50
50	0.20	0.40
60	0.17	0.33
70	0.14	0.29
80	0.13	0.25
90	0.11	0.22
100	0.10	0.20

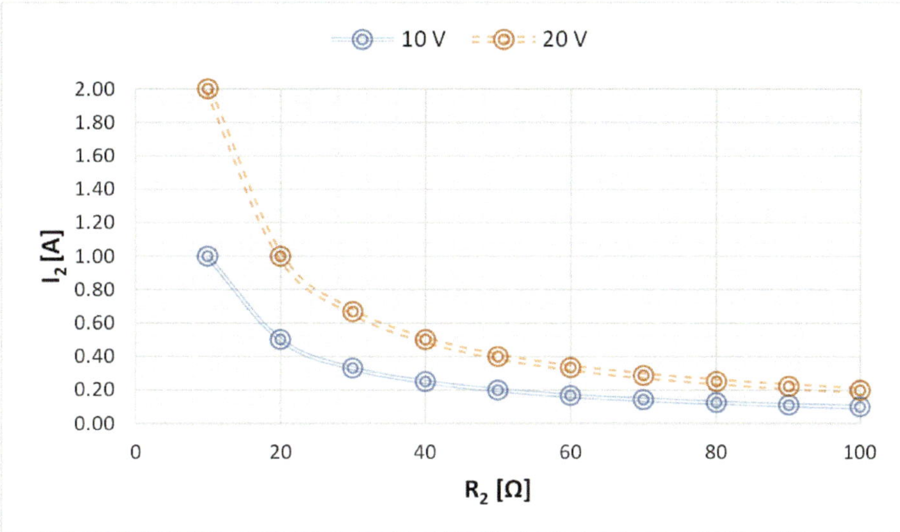

Figure 2.6 Plot of I_2 as a function of R_2 for both supply voltages V_{dc2} = 10 V and V_{dc2} = 20 V.

2.2.2.1 Solutions to Problem 2

1. The measurements of the current I_2 corresponding to each value of resistance R_2 in the simulation circuit of Figure 2.5 are recorded in Table 2.3.
2. The readings of I_2 in correspondence with each value of R_2 and for V_{dc2} = 20 V are also recorded in Table 2.3.
3. The plot of the values of $I_2 = \frac{V_{dc2}}{R_2}$ as a function of R_2 for both supply voltages V_{dc2} = 10 V and V_{dc2} = 20 V is shown in Figure 2.6.

The graph of $I_2 = \frac{V_{dc2}}{R_2}$ as a function of values of R_2, where V_{dc2} is the constant of proportionality, expresses an inverse variation relationship between the two quantities; as R_2 increases, I_2 decreases, and vice versa.

The graph illustrates the inverse variation relationship between I_2 and R_2, where an increase in R_2 leads to a decrease in I_2, and vice versa.

2.3 Lab Procedure 1

In the circuit depicted in Figure 2.7, voltage and current measurements are to be made in accordance with the passive sign convention using Multisim. The following tasks are to be performed:

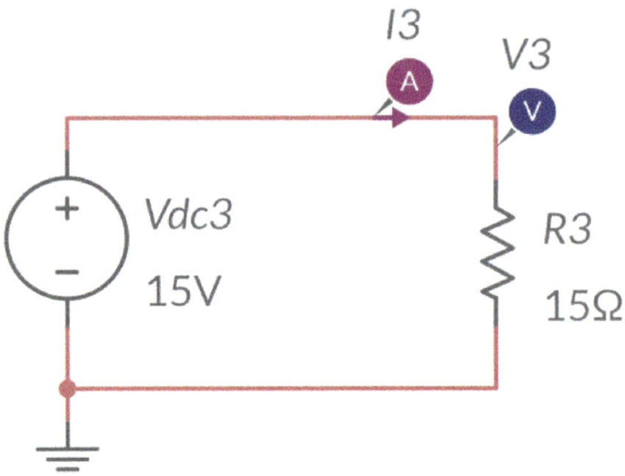

Figure 2.7 Circuit for voltage and current measurements in accordance with passive sign convention.

Table 2.4 Summary of readings and calculations

	$R_3 = 15\ \Omega$			
Supply voltage V_{dc3}	I_3 (A)	V_3 (V)	$R_3 = V_3 / I_3$ (Ω)	m (S)
5 V				
10 V				
15 V				
20 V				
25 V				

1. With Multisim simulation, measure the current I_3 passing through the resistor R_3 and the voltage V_3 across it for the following values of supply voltage V_{dc3}: 5, 10, 15, 20, and 25 V. Record the measured values in Table 2.4.
2. Calculate the ratio $R_3 = V_3/I_3$ for each of the values of V_{dc3}; record the results in Table 2.4.
3. Determine the slope m for each reading of V_3 and I_3; record the results in Table 2.4.

2.4 Lab Procedure 2

In the circuit of Figure 2.8, the supply voltage V_{dc4} is held constant, whereas the resistance R_4 varies ranging between 1 and 10 Ω in increments of 10 Ω. To analyze the circuit, perform the following tasks:

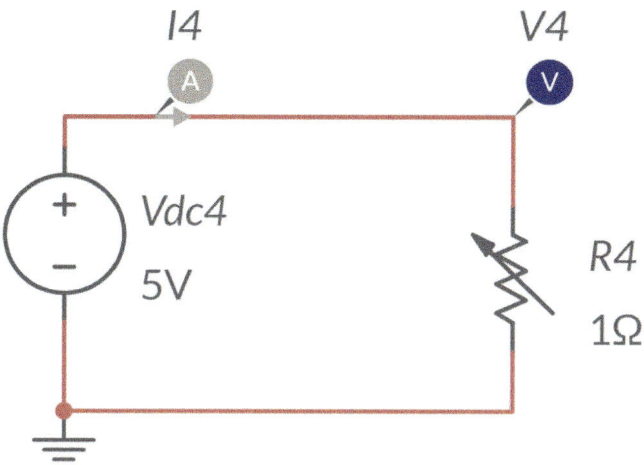

Figure 2.8 Circuit with variable resistor.

Table 2.5 Summary of readings

	$V_{dc4} = 5$ V	$V_{dc2} = 10$ V
R_4 (Ω)	I_4 (A)	I_4 (A)
1		
2		
3		
4		
5		
6		
7		
8		
9		
10		

1. Measure the current I_4 passing through each value of R_4 by configuring the circuit in Multisim accordingly. Record the values of I_4 in Table 2.5.
2. Repeat the measurement steps mentioned above, but this time set the supply voltage to $V_{dc4} = 10$ V. Record the new values of I_4 in Table 2.5.
3. Create a plot (Figure 2.9) that shows the relationship between the values of the current I_4 and the resistance R_4 for both supply voltage values ($V_{dc4} = 5$ V and $V_{dc2} = 10$ V).

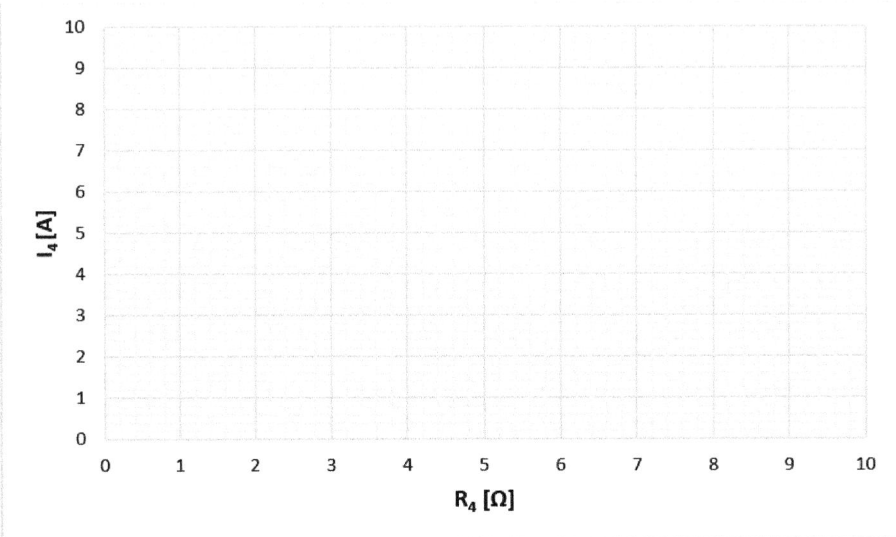

Figure 2.9 Plot of the current I_4 as a function of the resistance R_4 for V_{dc4} = 5 V and V_{dc4} = 10 V.

3

Power in DC Circuits

3.1 Student Learning Outcomes

- Use Multisim to measure voltages and currents to calculate power dissipated in a resistor circuit.
- Construct a Multisim circuit with resistors and sources. Apply Tellegen's theorem to verify total power supplied equals total power dissipated.
- Build resistor circuits in Multisim. Vary source voltage and resistance values. Plot power dissipation versus voltage and resistance.

3.2 Prelab

The instantaneous electric power $p(t)$ represents the time rate of change of energy $W(t)$ absorbed or generated by an electrical component:

$$p\left(t\right) = \frac{dW\left(t\right)}{dt} = \frac{dW}{dq}\frac{dq}{dt} = v\left(t\right)\cdot i\left(t\right), \tag{3.1}$$

where q is the charge in Coulombs. The unit of power is Watt, derived from: $[\mathrm{V}\cdot\mathrm{A}] = \left[\frac{\mathrm{J}}{\mathrm{C}}\cdot\frac{\mathrm{C}}{\mathrm{s}}\right] = [\mathrm{W}]$.

For a resistive element, Ohm's law allows alternate expressions for power:

$$p\left(t\right) = \frac{v^2(t)}{R} = i^2\left(t\right)R, \tag{3.2}$$

where $v(t)$ is the voltage across the resistance R, and $i(t)$ is the current flowing through it.

3.2.1 Tellegen's theorem

Tellegen's theorem, named after Bernard Tellegen, is a powerful result relating to conservation of power in electrical networks. It builds upon Kirchhoff's current and voltage laws. For a single network satisfying Kirchhoff's laws, the total power absorbed must equal the total power supplied at any instant:

$$\sum P_{\text{abs}} = \sum P_{\text{supplied}}. \tag{3.3}$$

Or alternatively, the theorem can be stated as follows: "the sum of the powers absorbed by each branch at any instant of time is zero" (the power absorbed by a true generator is negative).

Tellegen's theorem extends the principle of conservation of energy to two separate networks, which have a common linear graph (i.e., they share the same set of branches), but may otherwise be different. It states that if we have two networks, denoted as A and B, then the sum of the product of the branch currents (i_k^A) in network A and the corresponding branch voltages (v_k^B) in network B is always equal to zero:

$$\sum_{k=1}^{b} i_k^A v_k^B = 0, \tag{3.4}$$

where b is the number of branches of the networks. That is, the sum of cross-products of currents and voltages between the networks is zero.

This equation holds true as long as Kirchhoff's laws, which are the fundamental laws governing electric circuits, are satisfied in both networks A and B. Essentially, Tellegen's theorem states that the sum of the instantaneous "virtual" powers exchanged between the two networks is always balanced and does not result in net energy gain or loss.

3.2.2 Problem 1

The two networks A and B in Figure 3.1 have a common linear graph but are otherwise different. Perform the following tasks:

1. Construct the circuits shown in Figure 3.1 in Multisim. Measure voltages across each element and currents through each element of both networks A and B. Ensure that the probes are placed in compliance with passive and active sign conventions.

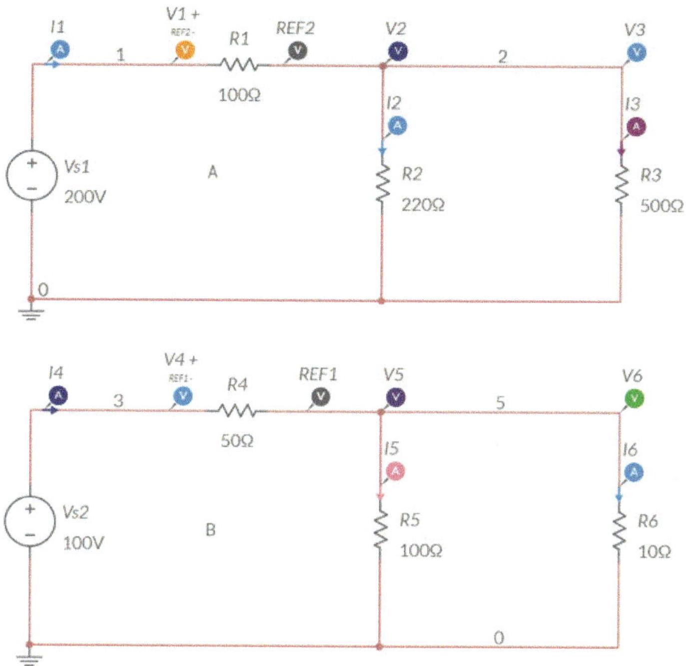

Figure 3.1 Networks A and B have a common linear graph but are otherwise different.

2. Determine the power absorbed by each element in networks A and B. Verify if Tellegen's theorem is satisfied.

3.2.2.1 Solutions to Problem 1

1. To measure with Multisim values of voltage and current, we can conduct a simulation of the circuit (Figure 3.2).

2. The powers absorbed by each element in each network, according to passive and active sign conventions, are recorded in Tables A and B.

The results confirm the principle of conservation of power, as the sum of the power absorbed by each branch equals the power supplied by the generator.

To verify if Tellegen's theorem is satisfied, we need to calculate the sum of the product of branch currents (i_k^A) in network A and the corresponding branch voltages (v_k^B) in network B (Table 3.1).

The resulting values demonstrate that the net "virtual" power is indeed zero, thus confirming Tellegen's theorem.

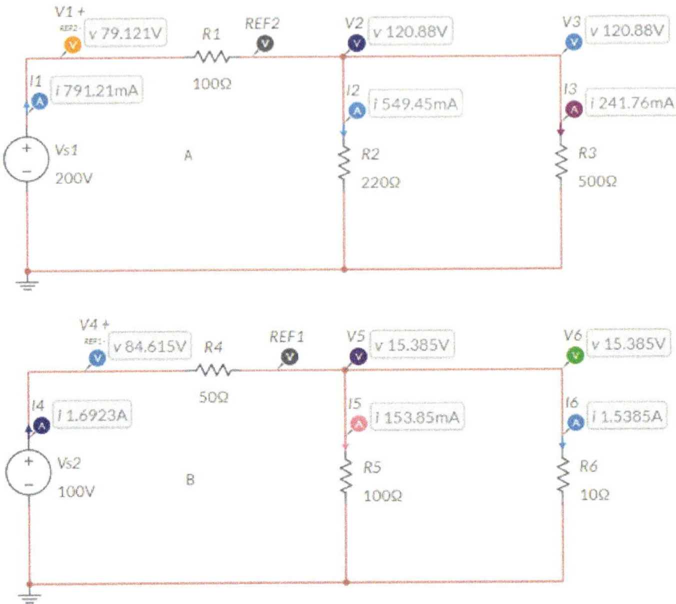

Figure 3.2 Multisim simulation with measurements of voltages and currents in the two networks.

Table A Summary of calculations

Component in **A**	P_{abs} (W)
V_{s1}	$-V_{s1}I_{s1} = -158.2$
R_1	$V_1 I_1 = 62.6$
R_2	$V_2 I_2 = 66.4$
R_3	$V_3 I_3 = 29.2$
$P_{supplied}$	158.2
$P_{absorbed}$	158.2

Table B Summary of calculations

Component in **B**	P_{abs} (W)
V_{s2}	$-V_{s1}I_{s1} = -169.2$
R_4	$V_4 I_4 = 143.2$
R_5	$V_5 I_5 = 2.4$
R_6	$V_6 I_6 = 23.6$
$P_{supplied}$	169.2
$P_{absorbed}$	169.2

Table 3.1 Summary of calculations

Component in A	Component in B	i^A (A)	v^B (V)	$\sum_{k=1}^{b} i_k^A v_k^B$ (W)
Source v_{s1}	Source v_{s2}	0.79121	100	−79.121
Resistor R_1	Resistor R_4	0.79121	84.615	66.948
Resistor R_2	Resistor R_5	0.54945	15.385	8.453
Resistor R_3	Resistor R_6	0.24176	15.385	3.719

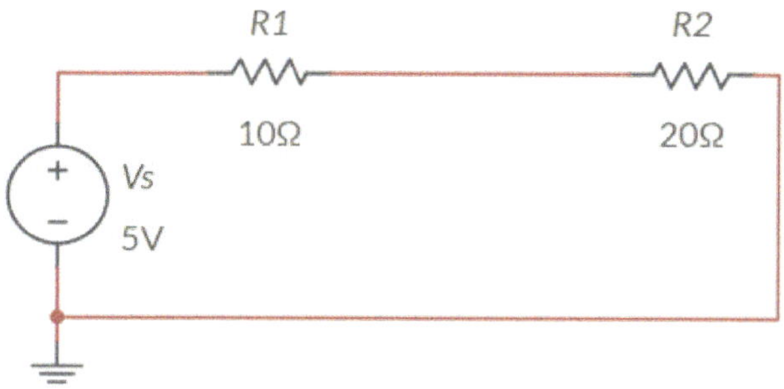

Figure 3.3 Resistive circuit for the indirect measurement of power.

3.2.3 Problem 2

Construct the circuit in Figure 3.3 in Multisim. Add ammeters to measure branch currents and voltmeters to measure voltage drops in compliance with passive and active sign conventions. Perform the following tasks:

1. In Multisim simulation, measure the total current I_T flowing in the circuit. Verify that the measured value matches the calculated values.
2. In Multisim, measure the voltage drops V_1 and V_2 across R_1 and R_2, respectively. Verify that the measured values match the calculated values.
3. Based on the measurements, calculate the values P_1 and P_2 of the power dissipated through R_1 and R_2, respectively.
4. Calculate the value P_s of the power supplied by the source V_s. Verify that $P_s = P_1 + P_2$.
5. Repeat all the measurements and calculations above with a supply voltage $V_s = 10$ V.

3.2.3.1 Solutions to Problem 2

1. Figure 3.4 shows the Multisim simulation results, including measured values for the total current I_T and voltage drops V_1 and V_2.

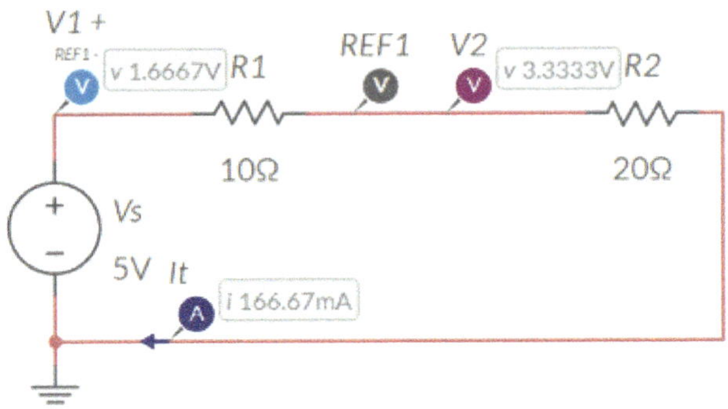

Figure 3.4 Multisim simulation showing measurements of voltages and current.

The total current I_t circulating in the circuit can be calculated using Ohm's law as follows:

$$I_t = \frac{V_s}{(R_1 + R_2)} = 166.67\,\text{mA}. \tag{3.5}$$

This matches the simulated value in Figure 3.4.

2. Similarly, the voltage drops V_1 and V_2 can be calculated using Ohm's law:

$$V_1 = I_t R_1 = 1.67\ \text{V} \tag{3.6}$$

$$V_2 = I_t R_2 = 3.33\ \text{V}. \tag{3.7}$$

The calculations align with the Multisim measurements (Figure 3.4).

3. The powers P_1 and P_2 dissipated in each resistor R_1 and R_2 can be calculated with the following formulae:

$$P_1 = I_t^2 R_1 = 277.8\,\text{mW} \tag{3.8}$$

$$P_2 = I_t^2 R_2 = 555.6\,\text{mW}. \tag{3.9}$$

4. The total supplied power P_s can be calculated with the following formula:

$$P_s = V_s I_t = 833.4\,\text{mW}. \tag{3.10}$$

This satisfies $P_s = P_1 + P_2$, confirming conservation of power.

5. Table 3.2 summarizes the measurements and calculations for V_s = 10 V, further validating the analysis.

Table 3.2 Summary of readings and calculations

V_s (V)	I_T (mA)	V_1 (V)	V_2 (V)	$P_1 = {I_T}^2 R_1$ (mW)	$P_2 = {I_T}^2 R_2$ (mW)	$P_s = V_s I_T$ (mW)	$P_s = P_1 + P_2$ (mW)
10	333.33	3.33	6.66	1,111	2,222	3,333	3,333

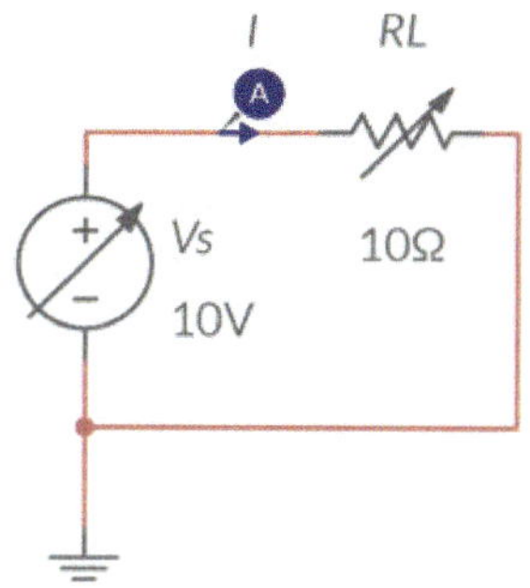

Figure 3.5 Resistive circuit for indirect power measurements.

3.2.4 Problem 3

Simulate the circuit in Figure 3.5. Add ammeters to measure branch currents and voltmeters to measure voltage drops in compliance with passive and active sign conventions. Perform the following tasks:

1. In Tinkercad, set the voltage source V_s to 10 V and insert a variable resistor R_L that can be swept from 1 to 10 Ω in 1-Ω increments. For each value of R_L, measure the branch current I_L and calculate the absorbed power P_L. Plot the power P_L versus resistance R_L on a graph.
2. In Multisim, with the fixed value of the resistance R_L = 10 Ω, measure the current I_L circulating in the circuit and calculate the associated dissipated power P_L for values of V_s ranging from 1 to 10 V, with increments of 1 V. Plot the power P_L as a function of the voltage V_s.

3.2.4.1 Solutions to Problem 3

1. The simulation of the circuit in Tinkercad is provided in Figure 3.6.

The absorbed power P_L can be calculated as follows:

$$P_L = V_s I_L. \tag{3.11}$$

Current readings and power calculations are provided in Table 3.3.

The plot of the power P_L as a function of the resistance R_L is provided in Figure 3.7.

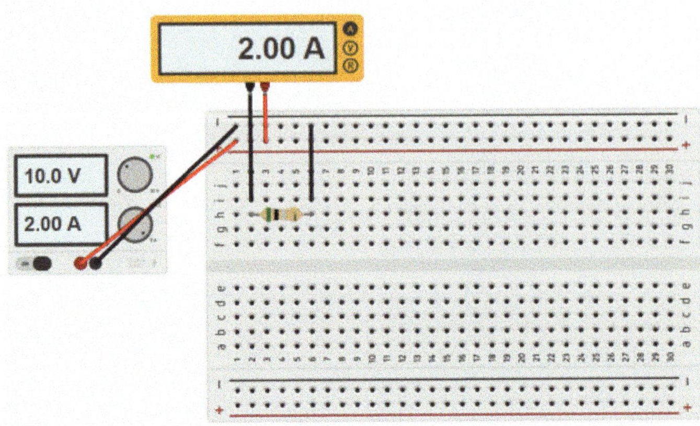

Figure 3.6 Tinkercad simulation of circuit for indirect power measurements.

Table 3.3 Summary of readings and calculations

V_s = 10 V		
R_L (Ω)	I_L (A)	$P_L = V_s \cdot I_L$ (W)
1	10.0	100.0
2	5.0	50.0
3	3.3	33.3
4	2.5	25.0
5	2.0	20.0
6	1.7	16.7
7	1.4	14.3
8	1.3	12.5
9	1.1	11.1
10	1.0	10.0

According to Ohm's law, the current I_L through the resistor R_L is inversely proportional to the resistance value, that is, $I_L = V_s/R_L$. For a fixed supply voltage V_s, increasing R_L causes a proportional decrease in I_L. However, the voltage drop V_s across R_L remains constant at $V_s = I_L \cdot R_L$, regardless of the value of R_L. Since V_s is constant while I_L decreases with increasing R_L, their product P_L must also decrease as R_L increases.

2. The simulation of the circuit in Multisim is provided in Figure 3.8.

The absorbed power P_L can be calculated as follows:

$$P_L = \frac{V_s^2}{R_L}. \tag{3.12}$$

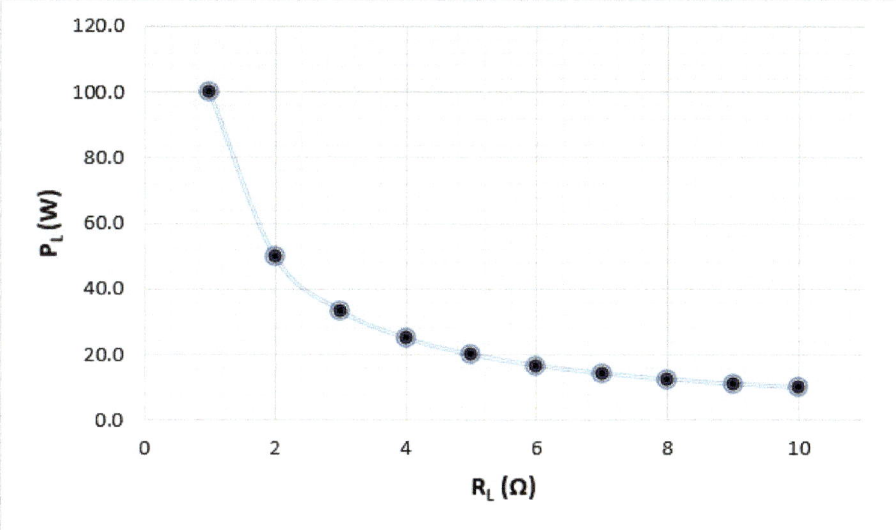

Figure 3.7 Plot of P_L as a function of the resistance R_L.

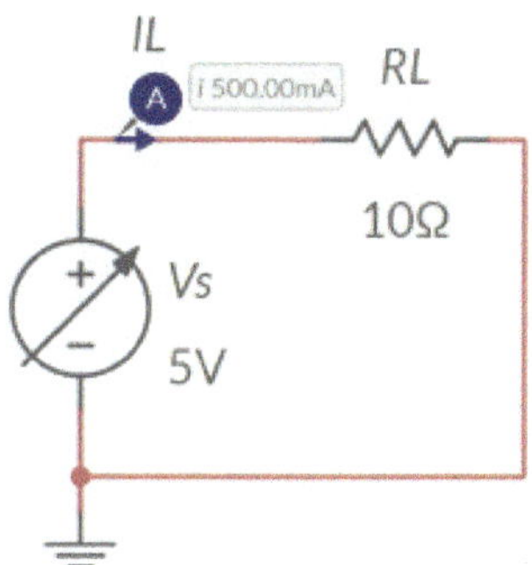

Figure 3.8 Multisim simulation.

Readings and calculations are provided in Table 3.4.

The plot of the power P_L as a function of the voltage V_s is shown in Figure 3.9.

According to Ohm's law, for a fixed resistance R_L, the current I_L through the resistor is directly proportional to the supply voltage V_s, such that I_L = V_s/R_L. Thus, increasing V_s, which is the voltage drop across the resistor, causes a proportional increase in I_L. The power dissipated in the resistor is given by $V_s \cdot I_L$, and since both V_s and I_L increase with increasing supply voltage V_s, their product P_L also increases with V_s.

Table 3.4 Summary of readings and calculations

$R_L = 10\ \Omega$		
V_s (V)	I_L (A)	$P_L = V_s \cdot I_L$ (W)
1	0.1	0.1
2	0.2	0.4
3	0.3	0.9
4	0.4	1.6
5	0.5	2.5
6	0.6	3.6
7	0.7	4.9
8	0.8	6.4
9	0.9	8.1
10	1.0	10.0

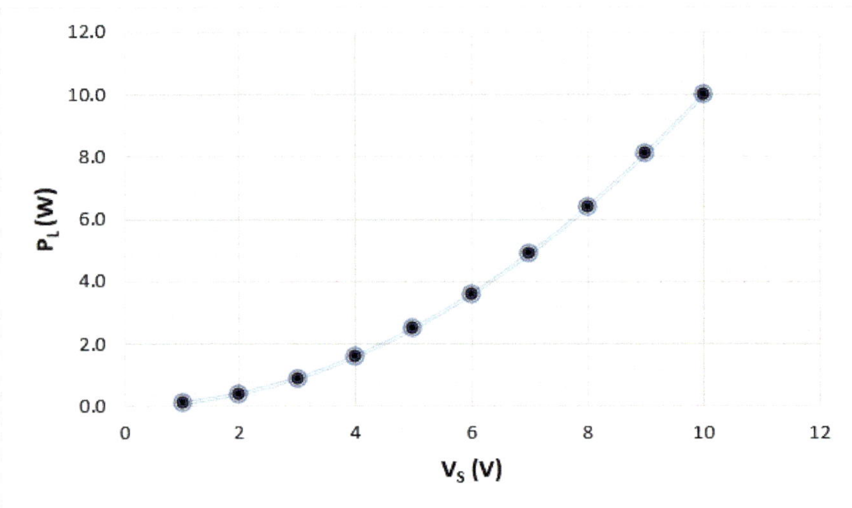

Figure 3.9 Plot of P_L as a function of the voltage V_s.

3.3 Lab Procedure 1

The two networks A and B in Figure 3.10 have a common linear graph but are otherwise different. Perform the following tasks:

1. Construct the circuits shown in Figure 3.10 in Multisim. Measure voltages across each element and currents current through each element of both networks A and B. Ensure that the probes are placed in compliance with passive and active sign conventions.

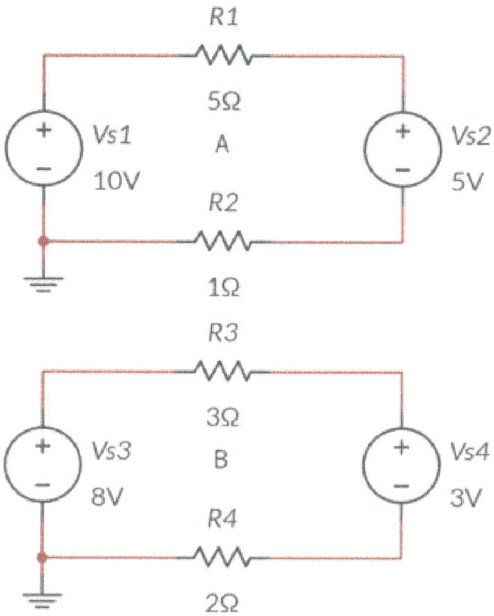

Figure 3.10 Networks A and B have a common linear graph but are otherwise different.

Table A Summary of calculations

Components in **A**	P_{abs} (W)
V_{s1}	
V_{s2}	
R_1	
R_2	
$P_{supplied}$	
$P_{absorbed}$	

Table B Summary of calculations

Components in **B**	P_{abs} (W)
V_{s3}	
V_{s4}	
R_3	
R_4	
$P_{supplied}$	
$P_{absorbed}$	

2. Determine the power absorbed by each element in networks A and B. Verify if Tellegen's theorem is satisfied.

3.4 Lab Procedure 2

Construct the circuit in Figure 3.11 in Multisim. Add ammeters to measure branch currents and voltmeters to measure voltage drops in compliance with passive and active sign conventions. Perform the following tasks:

1. In Multisim, measure the total current I_T flowing in the circuit. Verify that the measured value matches the calculated value. Report values in Table 3.5.
2. In Multisim, measure the voltage drops V_5 and V_6 across R_5 and R_6, respectively. Verify that the measured values match the calculated values. Report values in Table 3.5.
3. Based on the measurements, calculate the values P_5 and P_6 of the power dissipated through R_5 and R_6, respectively.
4. Calculate the value P_s of the power supplied by the source V_s. Verify that $P_s = P_1 + P_2$.

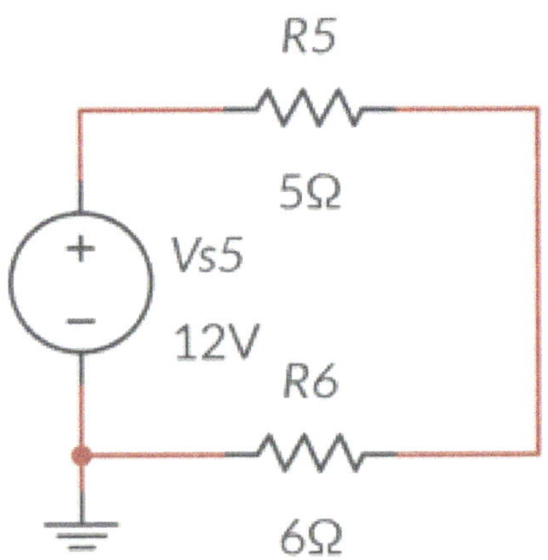

Figure 3.11 Indirect measurement of power.

Table 3.5 Summary of reading and calculations

I_T (mA)	V_5 (V)	V_6 (V)	P_5 (mW)	P_6 (mW)	P_s (mW)

4

Series and Parallel Resistors in DC Circuits

4.1 Student Learning Outcomes

- In Multisim, construct series and parallel resistor circuits. Take resistance measurements using Ohm's law.
- Calculate the total equivalent resistance of series and parallel circuits using analytical expressions. Compare calculated theoretical values with measured resistances in Multisim.
- In Multisim, construct voltage and current divider circuits. Take voltage and current measurements and compare to calculated values.

4.2 Prelab

When resistors are connected in *series* (Figure 4.1(a)), the circulating current through one resistor is the same as that of all the other resistors. In series connections, the resistors have only one terminal in common with each other. The total resistance R_S of a series connection of n resistors can be calculated

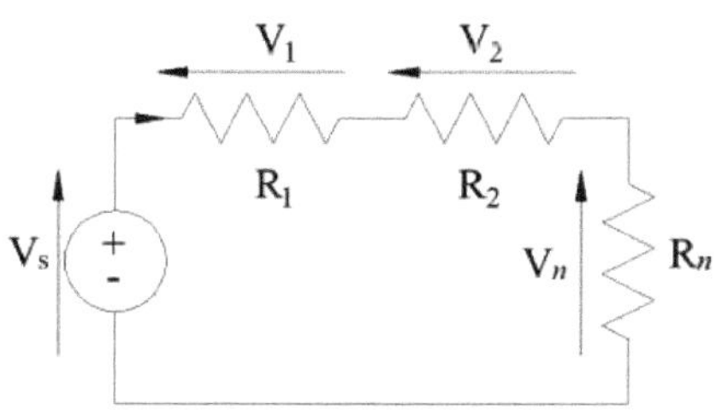

(a) resistors are connected in series

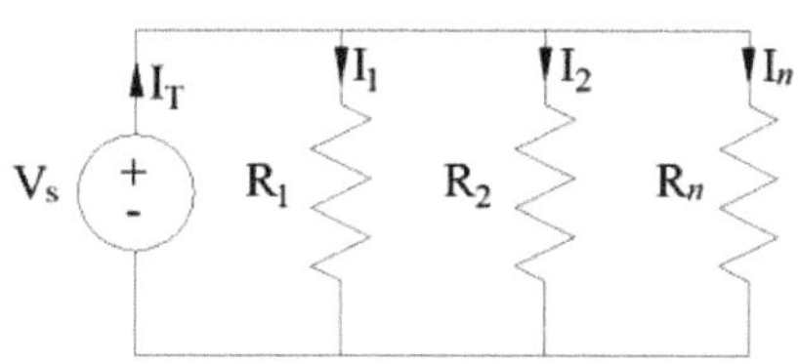

(b) resistors are connected in parallel

Figure 4.1 Series and parallel resistive combinations.

by summing up the individual resistance values:

$$R_S = \sum_{k=1}^{n} R_k. \tag{4.1}$$

On the other hand, when resistors are connected in *parallel* (Figure 4.1(b)), the voltage across one resistor is the same as that of all the other resistors. In parallel connections, the resistors have both terminals in common with each other. The total resistance R_P of a parallel connection of n resistors is calculated using the reciprocal of the sum of the reciprocals of the individual resistance values:

$$\frac{1}{R_P} = \sum_{k=1}^{n} \frac{1}{R_k}. \tag{4.2}$$

The parallel resistance R_P is always less than the smallest branch resistance.

If all the n resistors in a parallel connection are equal to R, the total resistance R_P can be simplified to:

$$R_P = \frac{R}{n}. \tag{4.3}$$

For a parallel connection of two resistors R_1 and R_2, the total resistance R_P can be calculated using the following formula:

$$R_P = \frac{R_1 R_2}{R_1 + R_2}. \tag{4.4}$$

4.2.1 Voltage divider

In a series configuration, where each resistor has the same current flowing through it, the voltage drops across the resistors are proportional to their resistance values. The voltage divider rule, as described in eqn (4.5), allows us to calculate the potential difference V_n across any one of the resistors in the series.

$$V_n = V_s \frac{R_n}{R_1 + R_2 + \cdots + R_n}. \tag{4.5}$$

The voltage V_s represents the total source voltage, and R_n corresponds to the resistance value of that specific resistor. The denominator of eqn (4.5) represents the sum of the resistances in the series connection.

4.2.2 Current divider

In a parallel configuration, the same voltage is across each resistor; therefore, the branch currents are inversely proportional to the resistance values. The *current divider* rule, as described in eqn (4.6), allows us to calculate the current I_n flowing through any one of the resistors in the parallel configuration.

$$I_n = I_T \frac{R_P}{R_n}. \tag{4.6}$$

R_n is the value of that specific resistance, and R_P is the value of the resistance of the parallel combination.

R_P/R_n is less than 1, causing the branch current I_n to be a fraction of the total current I_T.

4.2.3 Problem 1

In the circuit depicted in Figure 4.2, voltage and current measurements are to be made in accordance with the passive and active sign conventions. Perform the following tasks:

1. Simulate the circuit with Tinkercad. Vary the values of the voltage generator V_s from 1 to 10 V in increments of 1 V. For each value of V_s, run a simulation and record the current I.
2. Using Ohm's law, calculate the total resistance $R_S = \frac{V_s}{I}$ for each set of measurements. Verify that the calculated R_S values match the expected series resistance formula.

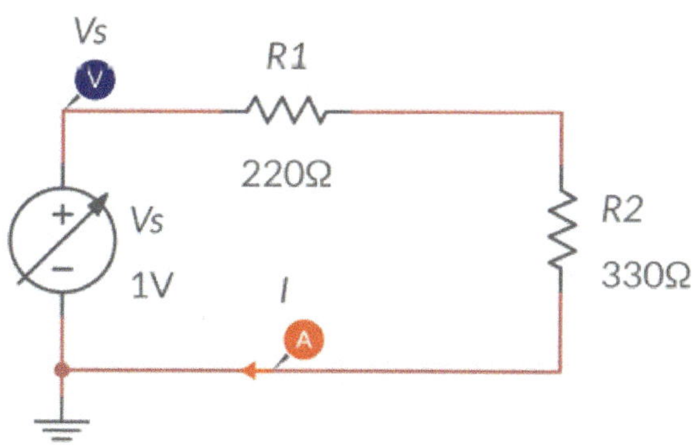

Figure 4.2 Series combination circuit.

3. After sweeping V_s, plot the *I*-*V* characteristic curve with current *I* on the *y*-axis and voltage V_s on the *x*-axis to visualize the relationship between current and voltage.

4.2.3.1 Solutions to Problem 1

1. Figure 4.3 shows the circuit schematic constructed in Tinkercad for simulation and analysis. The circuit contains a variable voltage source V_s and an ammeter to measure the current *I* according to the passive sign convention.

The readings of the current are shown in Table 4.1 for each value of V_s.

2. Table 4.1 shows the calculated total resistance R_S obtained at each simulated voltage and measured current data point.

The individual R_S values are seen to fluctuate slightly around the expected theoretical value of 550 Ω, calculated from the known resistances R_1 = 220 Ω and R_2 = 330 Ω. The small variation in calculated R_S is likely due to simulation accuracy limits. Overall, the simulation R_S values validate the total series resistance determined analytically.

3. Figure 4.4 shows the resulting *I*-*V* characteristic curve for the series resistor circuit, obtained by sweeping V_s and measuring *I* in Tinkercad.

The plot reveals a linear relationship between the current *I* and the supply voltage V_s, evidenced by an *R*-squared value of 1 for the linear fit. This

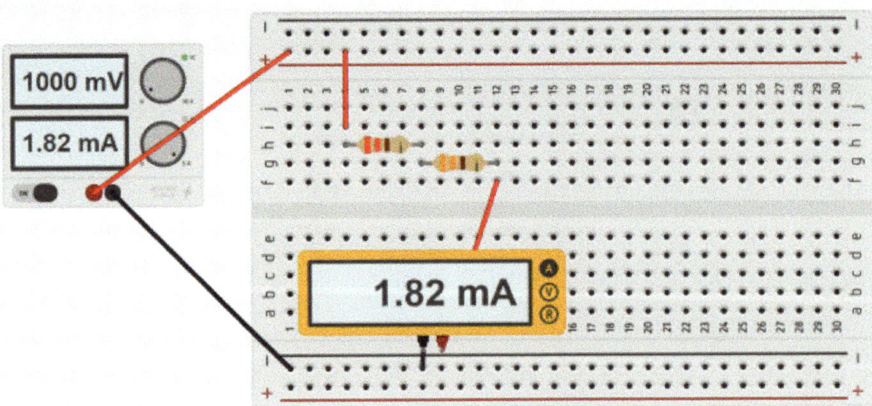

Figure 4.3 Series combination circuit constructed in Tinkercad for simulation and analysis.

Table 4.1 Summary of readings and calculations

V_s (V)	I (mA)	$R_S = V_s/I$ (Ω)
1	1.82	549.4
2	3.63	550.9
3	5.45	550.4
4	7.27	550.2
5	9.09	550.0
6	10.90	550.4
7	12.72	550.3
8	14.54	550.2
9	16.36	550.1
10	18.18	550.0

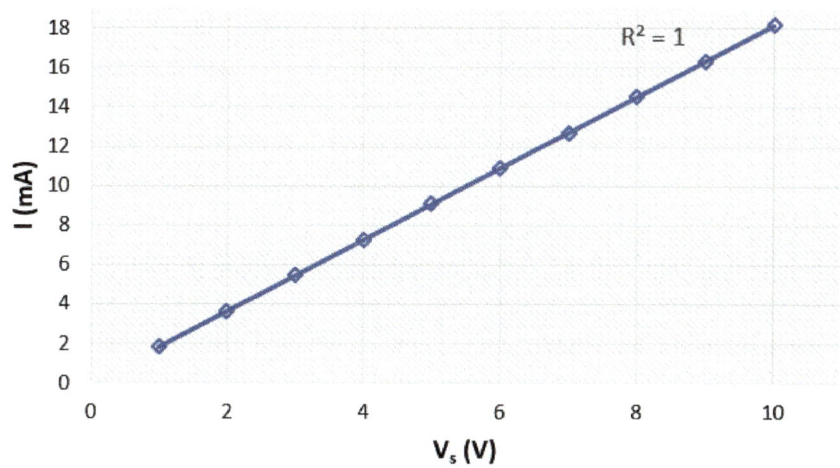

Figure 4.4 Plot of the I-V characteristic of the series circuit.

verifies that the series circuit follows Ohm's law, with the current proportional to the applied voltage across a constant series combination resistance.

4.2.4 Problem 2

Set up the circuit in Tinkercad according to Figure 4.5. Voltage and current measurements are to be made in accordance with the passive and active sign conventions. Perform the following tasks:

1. Vary the values of the voltage generator V_{s2} from 1 to 10 V in increments of 1 V. Measure the total current I_t supplied by the source.

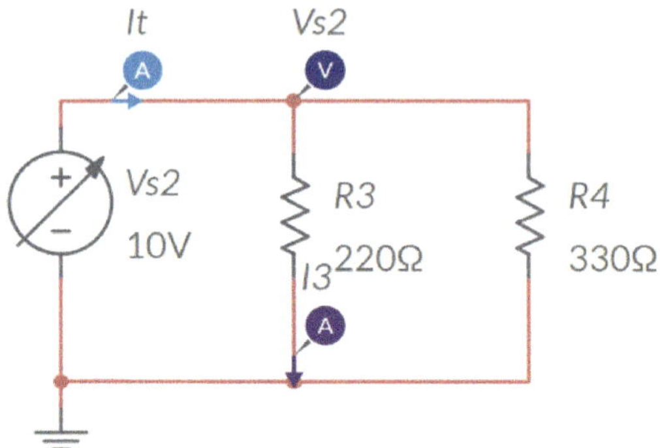

Figure 4.5 Parallel combination circuit.

2. For each set of V_{s2} and I_t, calculate the total resistance R_P using Ohm's law. Verify it matches the theoretical value from the parallel resistance formula.
3. After sweeping V_{s2}, plot the *I*-*V* characteristic curve with current I_t on the *y*-axis and voltage V_{s2} on the *x*-axis to visualize the relationship between current and voltage. Fit a trendline.
4. For all data points, measure the current I_3 circulating through *R*. Verify that I_3 matches the expected current divider formula.

4.2.4.1 Solutions to Problem 2

1. The circuit contains a variable voltage source V_{s2} and two ammeters to measure currents according to active and passive sign conventions (Figure 4.6).

The readings of the current I_t are shown in Table 4.1 for each value of V_{s2}.

2. The total resistance R_P can be calculated for each value of V_{s2} and measured current as follows:

$$R_P = \frac{V_{s2}}{I_t}. \tag{4.7}$$

Table 4.2 shows the values of R_P.

Calculated values of R_P in Table 4.2 are consistent with the parallel resistance formula R_P = 132 Ω.

The small variations in R_P are likely due to simulation accuracy limits. Overall, the simulation R_P values validate the parallel resistance determined analytically.

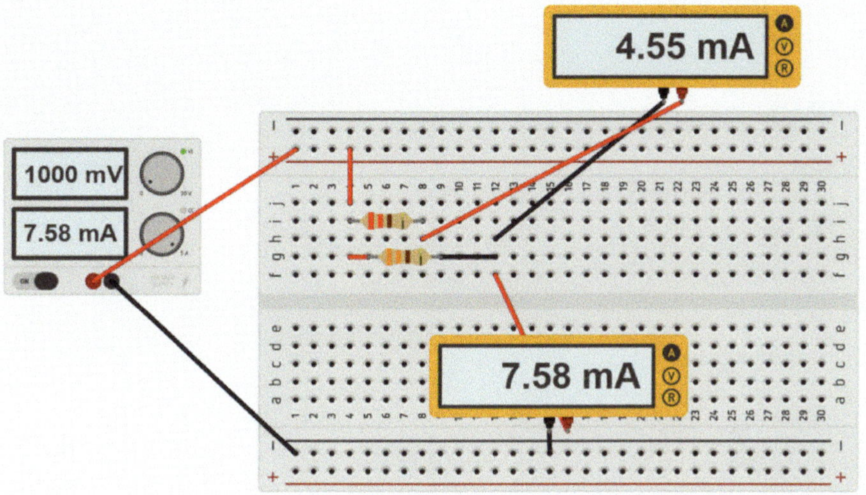

Figure 4.6 Parallel combination circuit constructed in Tinkercad for simulation and analysis.

Table 4.2 Summary of readings and calculations

V_{s2} (V)	I_t (mA)	R_P (Ω)	I_3 (mA)
1	7.58	131.9	4.5
2	15.2	131.6	9.1
3	22.7	132.2	13.6
4	30.3	132.0	18.2
5	37.9	131.9	22.7
6	45.5	131.9	27.3
7	53.0	132.1	31.8
8	60.6	132.0	36.4
9	68.2	132.0	40.9
10	75.8	131.9	45.5

3. The plot of the $I_t - V_{s2}$ characteristic of the parallel circuit is given in Figure 4.7. Figure 4.7 shows the resulting *I*-*V* characteristic curve for the parallel resistor circuit, obtained by sweeping V_{s2} and measuring I_t in Tinkercad.

The plot reveals a linear relationship between the current I_t and the supply voltage V_{s2}, evidenced by an *R*-squared value of 1 for the linear fit. This verifies that the parallel circuit follows Ohm's law, with the current proportional to the applied voltage across a constant parallel combination resistance.

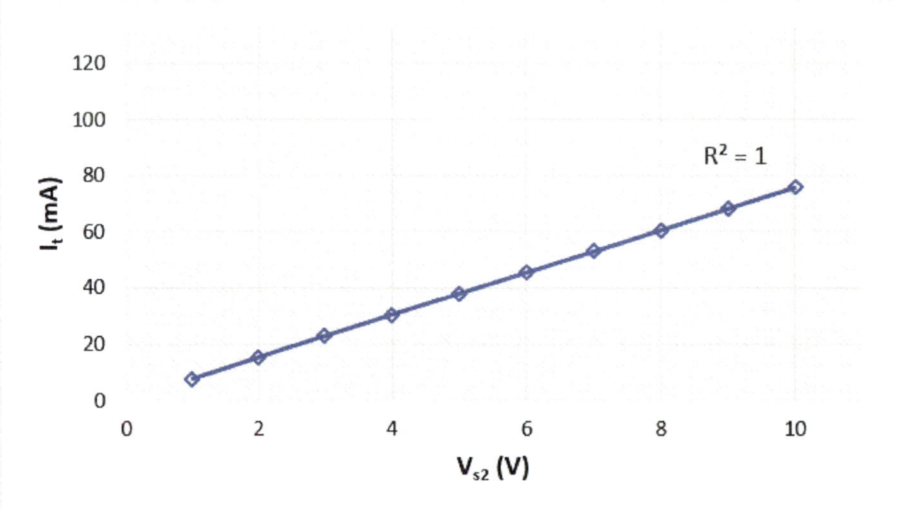

Figure 4.7 Plot of the *I-V* characteristic of the parallel circuit.

3. The measured values of the current I_3 circulating through R_3 for each value of V_{s2} are listed in Table 4.2. Readings are consistent with the current divider formula.

4.2.5 Problem 3

Set up the circuit in Multisim according to Figure 4.8, ensuring that voltage and current measurements are made in compliance with the passive and active sign conventions. Perform the following tasks:

1. With the switch SW open, measure in Multisim the current I_t supplied by the source and the voltage V_1 across the resistor R_1. Verify that the readings align with the theoretical expectations.
2. With the switch SW closed, calculate the parallel resistance R_P using the formula for the total resistance in a parallel configuration. In Multisim, verify that the calculated value aligns with the measurements.
3. With the switch SW closed, measure the currents I_t, I_1, and I_2 in Multisim. Verify that the measured currents align with the theoretical expectations based on the circuit configuration.

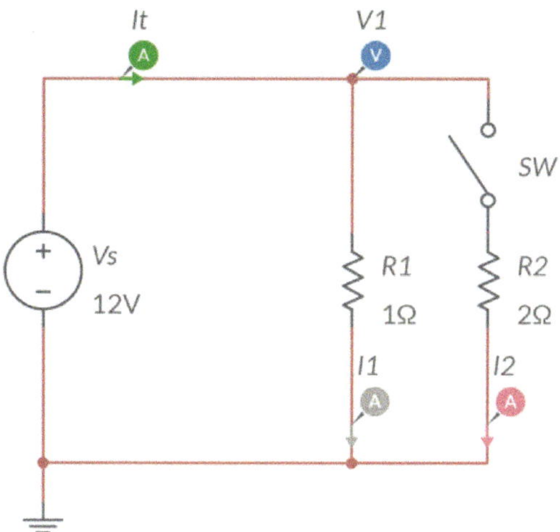

Figure 4.8 Parallel combination circuit.

Table 4.3 Summary of readings

SW	I_t (A)	V_1 (V)
Open	12	12

4.2.6 Solutions to Problem 3

1. Readings of current I_t and the voltage V_1 are provided in Table 4.3.

The reading of the current I_t and the voltage V_s are supported by the Ohm's law $I_t = \frac{V_1}{R_1}$.

The measured current I_t and the voltage V_1 align with the theoretical expectations based on Ohm's law, where $I_t = \frac{V_1}{R_1}$. This confirms that the readings are supported by the relationship between current and voltage in a resistor.

2. With the switch SW closed, the parallel resistance R_P is given by: Closing the switch SW connects the resistors R_1 and R_2 in parallel. The total parallel resistance R_P can be calculated using the formula for two resistors:

$$R_P = \frac{R_1 \cdot R_2}{R_1 + R_2} = 0.66\ \Omega. \tag{4.8}$$

To measure in Multisim the value of the parallel resistance, we can conduct a simulation of the circuit (Figure 4.9).

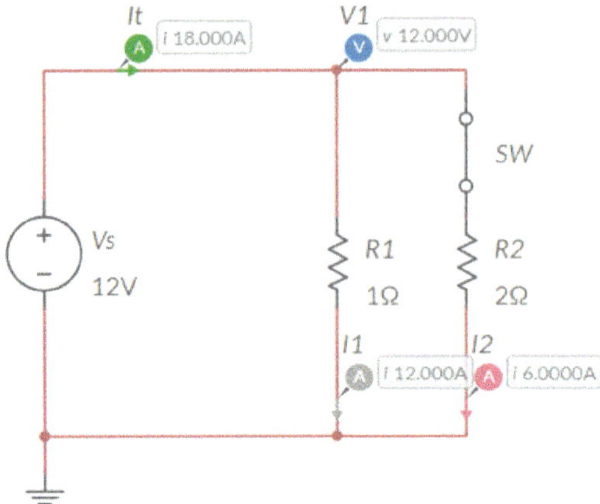

Figure 4.9 Multisim simulation providing values of voltage and current in the resistive parallel circuit.

Table 4.4 Summary of readings and calculations

SW	I_t (A)	I_1 (A)	I_2 (A)
Closed	18	12	6

We measure the values of the voltage across the parallel combination and the current supplied by the source.

The parallel resistance can be determined by applying Ohm's law dividing the measured values of voltage and current.

$$R_P = \frac{V_s}{I_t} = 0.66\ \Omega. \tag{4.9}$$

The calculated value of R_P matches the Multisim simulated value, confirming its accuracy.

1. With the switch SW closed, the measured currents I_t, I_1, and I_2 in Multisim (Figure 4.9) are listed in Table 4.4.

The measured currents in Multisim align with the theoretical expectations:

$$I_t = \frac{V_s}{R_P} = 18.2\ \text{A} \tag{4.10}$$

$$I_1 = I_t \frac{R_2}{R_1 + R_2} = 12.1\ \text{A} \tag{4.11}$$

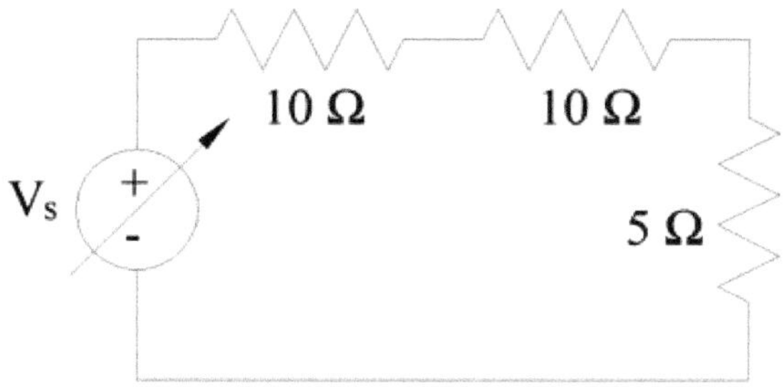

Figure 4.10 Resistive series combination circuit.

$$I_2 = I_t \frac{R_1}{R_1 + R_2} = 6.1 \text{ A}. \tag{4.12}$$

4.3 Lab Procedure 1

In the circuit depicted in Figure 4.10, voltage and current measurements are to be made in accordance with the passive and active sign conventions. Perform the following tasks:

1. Simulate the circuit with Tinkercad. Vary the values of the voltage generator V_s from 1 to 10 V in increments of 1 V. For each value of V_s, run a simulation and record the current *I*. Record readings in Table 4.5.
2. Using Ohm's law, calculate the total resistance $R_S = \frac{V_s}{I}$ for each set of measurements. Verify that the calculated R_S values match the expected series resistance formula. Record readings in Table 4.5.
3. After sweeping V_s, plot the *I*-*V* characteristic curve with current *I* on the *y*-axis and voltage V_s on the *x*-axis to visualize the relationship between current and voltage.

4.4 Lab Procedure 2

Set up the circuit in Tinkercad according to Figure 4.11. Voltage and current measurements are to be made in accordance with the passive and active sign conventions. Perform the following tasks:

1. Vary the values of the voltage generator V_s from 1 to 10 V in increments of 1 V. Measure the total current I_t supplied by the source. Record readings in Table 4.6.
2. For each set of V_s and I_t, calculate the total resistance R_P using Ohm's law. Verify whether it matches the theoretical value from the parallel resistance formula.

Table 4.5 Summary of readings and calculations

V_s (V)	I (mA)	$R_S = V_s/I$ (Ω)
1		
2		
3		
4		
5		
6		
7		
8		
9		
10		

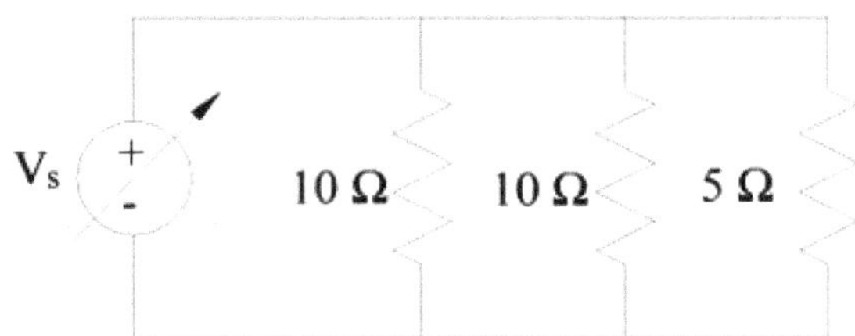

Figure 4.11 Resistive parallel combination circuit.

3. After sweeping V_s, plot the I-V characteristic curve with current I_t on the y-axis and voltage V_s on the x-axis to visualize the relationship between current and voltage. Fit a trendline.
4. For all data points, measure the current I_5 circulating through the 5-Ω resistor. Verify that I_5 matches the theoretical value from the current divider formula.

Table 4.6 Summary of readings and calculations

V_s (V)	I_t (mA)	R_P (Ω)	I_5 (mA)
1			
2			
3			
4			
5			
6			
7			
8			
9			
10			

5

Variable Resistors in DC Circuits

5.1 Student Learning Outcomes

- Discuss practical applications of rheostats and potentiometers for adjustable resistive control of current and voltage.
- In Tinkercad, construct circuits using rheostats. Take current and voltage measurements by appropriate connection of multimeters, following active and passive conventions.
- In Tinkercad, construct circuits using potentiometers. Take current and voltage measurements by appropriate connection of multimeters, following active and passive conventions.

5.2 Prelab

An adjustable resistor is a three-terminal component whose electrical resistance can be adjusted by sliding a contact over the resistive element.

According to IEC 60050 "*International Electrotechnical Vocabulary,*" a *rheostat* is an *adjustable resistor the resistance of which can be adjusted*

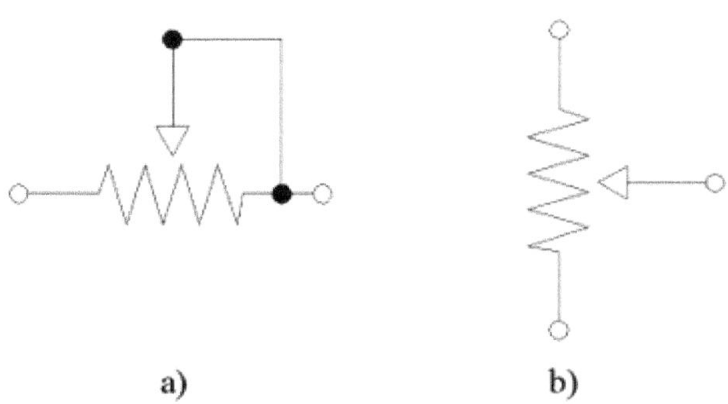

Figure 5.1 (a) Rheostat. (b) Potentiometer.

without interruption of electric current (Figure 5.1(a)). In the same standard, the *potentiometer* is defined as an *adjustable resistor having two outer and one or more intermediate terminals, the latter being fixed or sliding. A potentiometer allows fractions of the voltage between the outer terminals to be obtained* (Figure 5.1(b)).

A rheostat has two terminals and acts as a variable series resistor. Adjusting its wiper position changes the overall resistance without interrupting current flow. This provides adjustable current control.

A potentiometer acts as a voltage divider, with the wiper tapping off a fraction of the input voltage determined by the fraction of total resistance between the wiper and one end terminal. This allows adjustable voltage division. In general, the potentiometer's performance will be less affected by the connected load when the load resistance is significantly larger than the potentiometer's resistance. Therefore, a higher range of loads can be connected to the potentiometer without significantly affecting its performance when the load resistance is much larger than the potentiometer's resistance.

The total end-to-end resistance between the two outer terminals of a potentiometer or rheostat remains constant regardless of the wiper position. This is because the overall resistive element has a fixed resistance value (Figure 5.2).

When an ohmmeter is connected between one outer terminal and the wiper of a potentiometer or rheostat, the resistance reading depends on the wiper position. However, the full range of the variable resistor can be verified by turning the knob fully in both directions. For example, consider a 300-Ω variable resistor (Figure 5.3).

With the wiper fully counterclockwise, it contacts the same terminal as the ohmmeter. This shorts out the entire resistive element, giving a reading of 0 Ω. As the wiper moves clockwise, more of the resistance is introduced

Figure 5.2 The resistance between the outer terminals of the variable resistor will not change with the position of the sliding terminal.

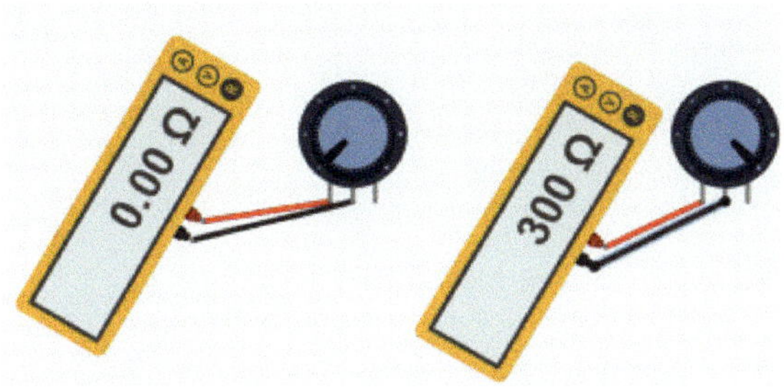

Figure 5.3 Resistance value of the variable resistor for knob fully counterclockwise and fully clockwise.

Figure 5.4 Ohmmeters connected between one outer terminal and the wiper contact, and between the other outer terminal and the wiper contact.

between the terminals, increasing the ohmmeter reading. With the wiper fully clockwise, it contacts the opposite outer terminal from the ohmmeter. Now the full 300-Ω resistance is measured.

By sweeping between the extremes, the ohmmeter can verify the total rated resistance that is accessible by the wiper movement. This provides a simple test to check the operation and range of a potentiometer or rheostat using just an ohmmeter.

Measuring the resistance with an ohmmeter between one outer terminal and the wiper contact, and between the other outer terminal and the wiper contact, will always sum up to the full resistance of the variable resistor (Figure 5.4).

Regardless of the position of the wiper, the sum of the two measurements always equals the full resistance of the variable resistor. This verifies the resistive divider function.

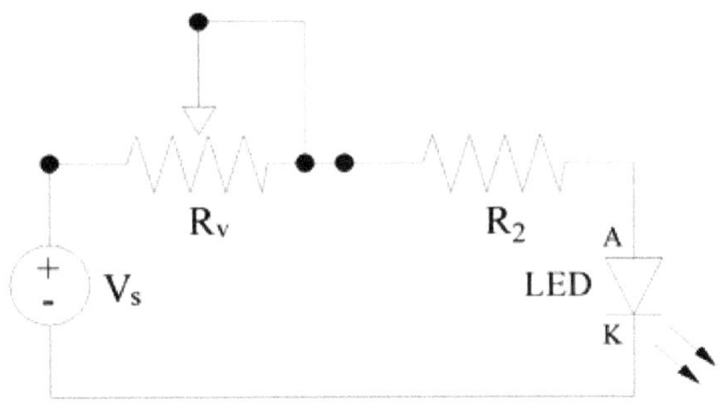

Figure 5.5 Circuit with rheostat, limiting resistor, and LED.

5.2.1 Problem 1

In the circuit depicted in Figure 5.5, the rheostat R_v is used to control the brightness of the LED (light emitting diode). The rheostat has a resistance of 300 Ω, and its purpose is to vary the amount of current flowing through the LED, which in turn controls the brightness of the LED.

To prevent the LED from drawing too much current and potentially getting damaged, a limiting resistor R_2 with a resistance of 150 Ω is included in the circuit. The limiting resistor acts as a current limiter and ensures that the LED operates within its maximum current rating, which is 20 mA. The LED is a polarized component, which means that it has a specific orientation for correct operation. The anode (A) of the LED should be connected to the positive side of the power supply, and the cathode (K) should be connected to the negative side of the power supply. If the LED is connected in reverse polarity, it will not light up. To analyze the circuit, perform the following tasks:

1. In Tinkercad, construct the circuit depicted in Figure 5.5. Use a supply voltage V_s = 5 V, and add an ammeter to measure the current I_{LED} flowing through the LED and voltmeters across R_2 and the LED.
2. Sweep the rheostat resistance to obtain I_{LED} values from 10 mA down to 7 mA in 1-mA increments. For each I_{LED}, measure the voltages V_2 across R_2 and V_{LED} across the LED.
3. For each value of I_{LED}, calculate the expected value of V_2 as $I_{\text{LED}} \cdot R_2$ and compare to the measured values.

5.2.1.1 Solutions to Problem 1

1. The circuit with the rheostat, limiting resistor, and LED is simulated with Tinkercad (Figure 5.6).

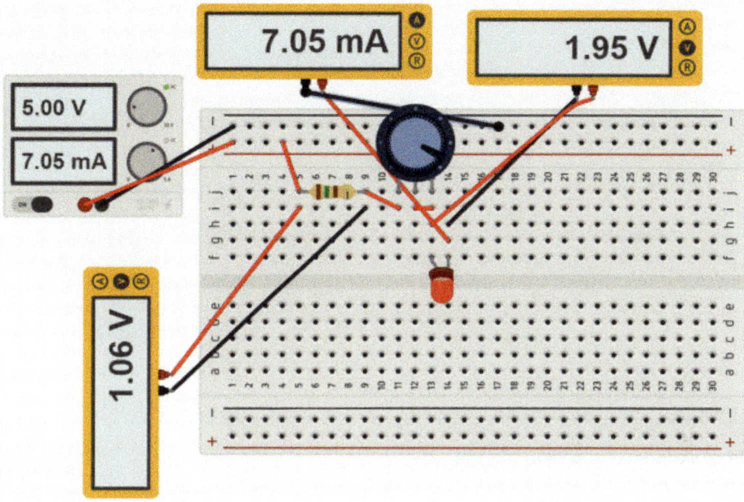

Figure 5.6 Circuit with rheostat, limiting resistor, and LED in Tinkercad.

Table 5.1 Summary of readings and calculations

I_{LED} (mA)	I_{LED} (actual) (mA)	V_{2_meas} (V)	V_{LED_meas} (V)	$V_2 = I_{LED} \cdot R_2$ (V)
10.0	10.0	1.5	1.99	1.5
9.0	8.99	1.35	1.97	1.34
8.0	8.02	1.20	1.96	1.20
7.0	7.05	1.06	1.95	1.06

2. Table 5.1 summarizes the measured currents and voltages in the LED dimming circuit for different rheostat settings. By sweeping the rheostat resistance, the LED current I_{LED} is varied from 10 mA down to 7.05 mA, achieving the desired dimming effect. For each actual value of I_{LED}, the measured voltages V_2 across R_2 and V_{LED} across the LED are also listed in Table 5.1.
3. The theoretical values of V_2 match the measured values (Table 5.1), validating the measurements.

5.2.2 Problem 2

In the circuit of Figure 5.7, a 300-Ω potentiometer R_p is used to dim the LED. A limiting resistor R_2 = 150 Ω prevents the maximum current rating of the LED (20 mA) from being exceeded.

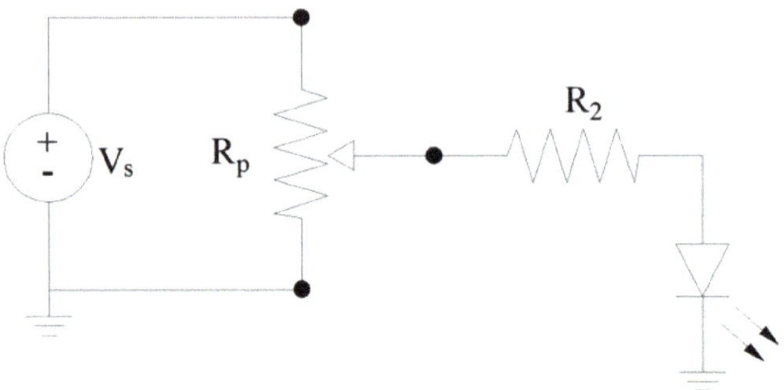

Figure 5.7 Circuit with potentiometer, limiting resistor, and LED.

1. In Tinkercad, construct the circuit depicted in Figure 5.7. Use a supply voltage V_s = 5 V, and add an ammeter to measure the current ILED flowing through the LED and voltmeters across R_2 and the LED.
2. Sweep the rheostat resistance to obtain I_{LED} values from 10 mA down to 7 mA in 1-mA increments. For each I_{LED}, measure the voltages V_2 across R_2 and V_{LED} across the LED.
3. For each value of I_{LED}, calculate the expected value of V_2 as $I_{LED}{\cdot}R_2$, and compare to the measured values.

5.2.2.1 Solutions to Problem 2

1. The circuit with the potentiometer, limiting resistor, and LED is simulated with Tinkercad (Figure 5.8).

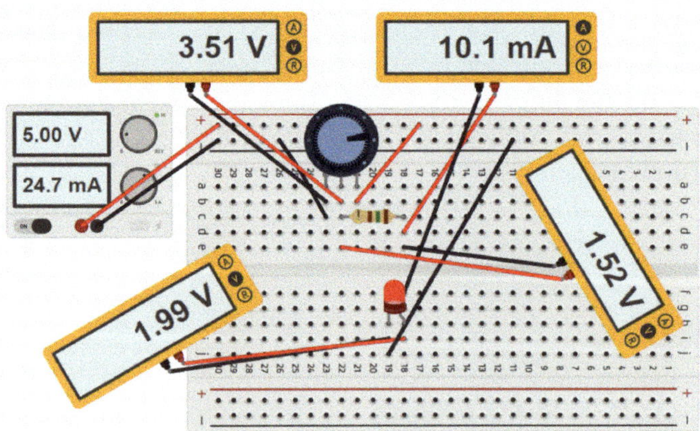

Figure 5.8 Simulated circuit with the potentiometer.

Table 5.2 Summary of readings and calculations

I_{LED} (mA)	I_{LED} (actual) (mA)	V_{2_meas} (V)	V_{LED} (V)	V_P (V)	$V_2 = I_{LED} \cdot R_2$ (V)
10	10.1	1.52	1.99	3.51	1.52
9	8.89	1.33	1.97	3.31	1.33
8	7.77	1.16	1.96	3.12	1.16
7	6.73	1.01	1.95	2.96	1.01

2. Table 5.2 summarizes the measured currents and voltages in the LED dimming circuit for different rheostat settings. By sweeping the potentiometer resistance, the LED current I_{LED} is varied from 10.1 mA down to 6.73 mA, achieving the desired dimming effect. For each actual value of I_{LED}, the measured voltages V_2 across R_2 and V_{LED} across the LED are also listed in Table 5.2.
3. The theoretical values of V_2 match the measured values (Table 5.2), validating the measurements.

5.3 Lab Procedure 1

Construct the circuit in Figure 5.9 in Tinkercad, using a 5-V voltage source and a 500-Ω potentiometer.

1. Before connecting R_1, adjust the potentiometer wiper to set V_{PG} = 2.5 V, as measured by a voltmeter between points P and ground.

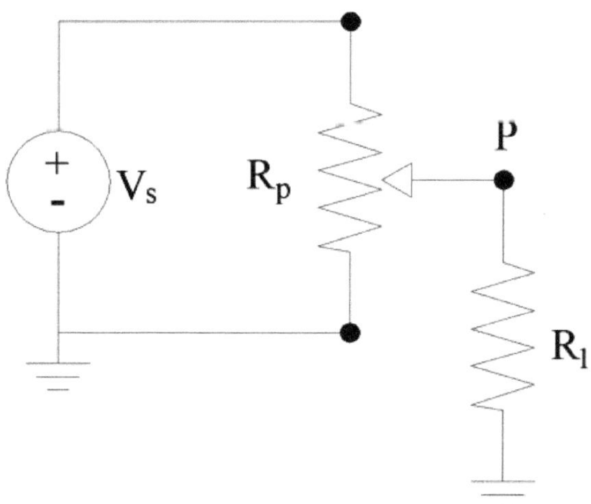

Figure 5.9 Simulated circuit with the potentiometer.

Table 5.3 Summary of readings

	$R_1 = 10\ \Omega$	$R_1 = 10\ \text{k}\Omega$
V_{PG}		

2. Connect $R_1 = 10\ \Omega$ and then $R_1 = 10\ \text{k}\Omega$. Measure the output voltage V_{PG} in each case. Record measured values in Table 5.3.
3. The potentiometer acts as a variable voltage divider, and its output voltage V_{PG} depends on the position of the sliding contact. Provide an analytical justification of the results.

6

Superposition Principle in DC Circuits

6.1 Student Learning Outcomes

- Explain the superposition principle conceptually.
- Construct circuits with multiple sources in Multisim. Analyze each source separately and then combine using superposition to calculate branch voltages and currents.
- Show that the superposition principle does not hold for power.

6.2 Prelab

The superposition principle is a fundamental concept in electrical engineering and circuit analysis. It states that in a linear circuit containing multiple independent sources (i.e., voltage sources or current sources), the total response (voltage or current) at any point in the circuit is equal to the algebraic sum of the individual responses caused by each independent source acting alone, with all other sources turned off. Voltage sources are deenergized by replacing them with *short-circuits* (zero voltage) when considering the response of other sources. For AC current sources, they are replaced with *open-circuits* (zero current).

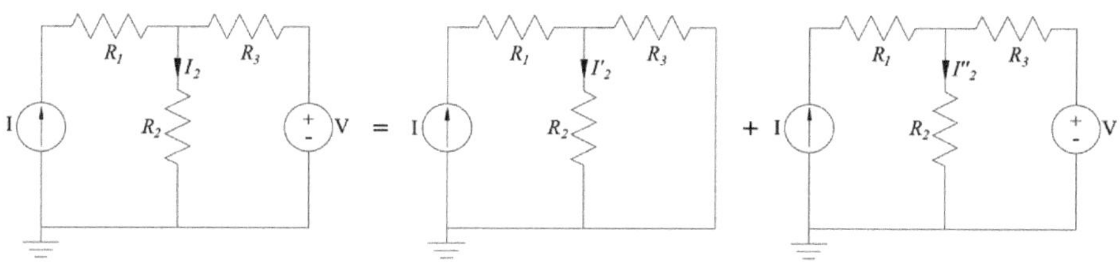

Figure 6.1 Circuit decomposed in two single-source circuits for the application of the superposition principle.

For example, in Figure 6.1, the current I_2 circulating through the resistor R_2 is found by analyzing current source *I* and then voltage source *V* separately to get responses I_2' and I_2'', and then adding:

$$I_2 = I_2' + I_2''. \tag{6.1}$$

A key limitation is that superposition only applies to linear circuits. Nonlinear elements like diodes will invalidate the principle. Importantly, superposition cannot be used to determine power, since power is proportional to the square of voltage or current. The principle only holds for linear quantities like voltage and current.

6.2.1 Problem 1

Construct the circuit in Figure 6.2 in Multisim. For measurements, apply passive and active sign conventions. To analyze the circuit, use the superposition principle and perform the following tasks:

1. Determine the current I_2' circulating through the resistor R_2 when the voltage generator V_{s2} is deenergized. The calculated value of I_2' shall be verified with Multisim simulation.
2. Determine the current I_2'' circulating through the resistor R_2 when the voltage generator V_{s1} is deenergized. The calculated value of I_2'' shall be confirmed with Multisim simulation.
3. Calculate the current I_2 flowing through the resistor R_2 when both sources are active. The calculated value of I_2 shall be validated using Multisim simulation.

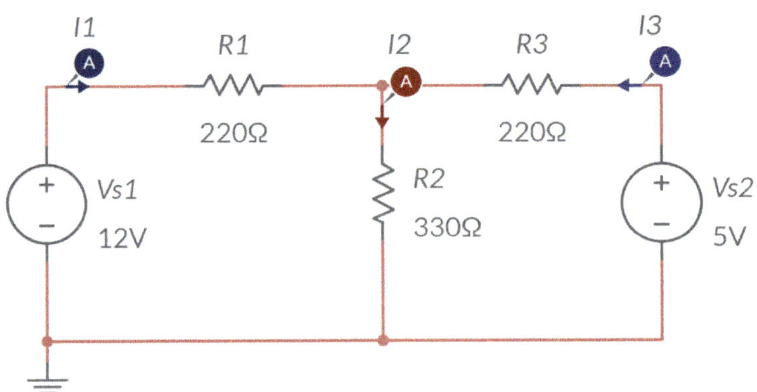

Figure 6.2 Linear resistive circuit with two voltage generators.

4. Calculate the electric powers P', P'', and P dissipated across R_2, associated with the currents I_2', I_2'', and I_2, respectively. Subsequently, verify the inapplicability of the superposition principle to electric power calculations.

6.2.1.1 Solutions to Problem 1

1. When the voltage generator V_{s2} is short-circuited, resistors R_2 and R_3 are connected in parallel, forming an equivalent resistance R_{23}, which is calculated using the following formula:

$$R_{23} = \frac{R_2 \cdot R_3}{R_2 + R_3} = 132\ \Omega. \tag{6.2}$$

Utilizing the voltage divider rule, the voltage drop V_{R2}' across resistor R_2 can be determined as follows:

$$V_{R_2}' = V_{s1} \cdot \frac{R_{23}}{R_1 + R_{23}} = 4.5\ \text{V}. \tag{6.3}$$

To find the current I_2' circulating through resistor R_2 when the voltage source V_{s2} is deenergized, we can apply Ohm's law:

$$I_2' = \frac{V_{R_{23}}'}{R_2} = 13.6\ \text{mA}. \tag{6.4}$$

The calculated value of I_2' is corroborated by the Multisim simulation results, as illustrated in Figure 6.3. This agreement between theory and simulation validates the accuracy of our analysis.

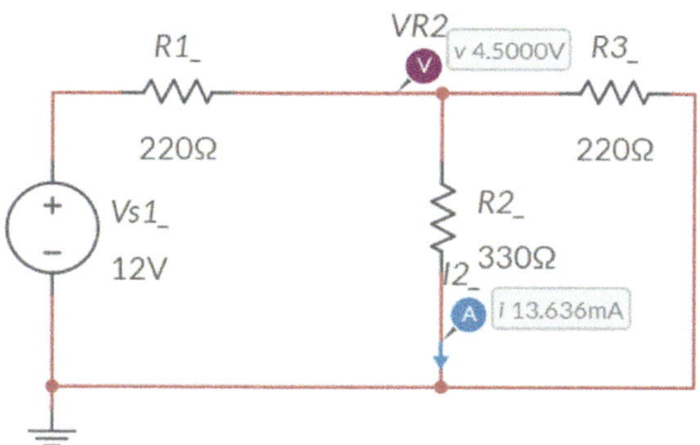

Figure 6.3 Multisim simulation showing the measured value of I_2'.

2. When the voltage generator V_{s1} is short-circuited, resistors R_1 and R_2 are connected in parallel, forming an equivalent resistance R_{12}, which is calculated using the following formula:

$$R_{12} = \frac{R_1 \cdot R_2}{R_1 + R_2} = 132\ \Omega. \tag{6.5}$$

Utilizing the voltage divider rule, the voltage drop V''_{R_2} across resistor R_2 can be determined as follows:

$$''_{R_2} = V_{s2} \cdot \frac{R_{12}}{R_3 + R_{12}} = 1.87\ \text{V}. \tag{6.6}$$

To find the current I''_2 circulating through the resistor R_2 when the voltage source V_{s1} is deenergized, we can apply Ohm's law:

$$I''_2 = \frac{V''_{R_2}}{R_2} = 5.6\ \text{mA}. \tag{6.7}$$

The calculated value of I''_2 is corroborated by the Multisim simulation results, as illustrated in Figure 6.4. This agreement between theory and simulation validates the accuracy of our analysis.

3. When both voltage sources V_{s1} and V_{s2} are active, the total current I_2 flowing through resistor R_2 is the sum of the currents I'_2 and I''_2:

$$I_2 = I'_2 + I''_2 = 19.2\ \text{mA}. \tag{6.8}$$

The value of I_2, calculated using the superposition principle, is in accordance with the Multisim simulation results, as shown in Figure 6.5.

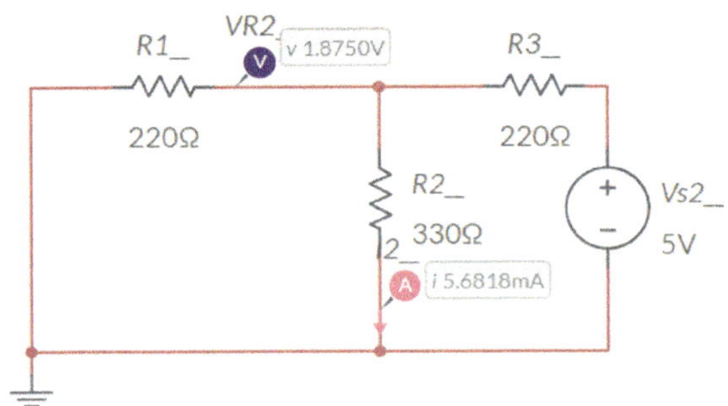

Figure 6.4 Multisim simulation showing the measured value of I''_2.

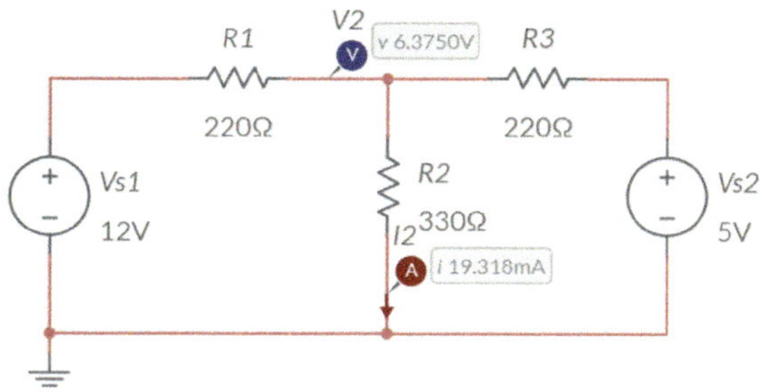

Figure 6.5 Multisim simulation showing the measured value of I_2.

This agreement further validates the correctness of our analysis. The superposition principle allows us to determine the total current in a circuit with multiple sources by considering their individual contributions separately and then summing them up.

4. The electric powers P', P'', and P associated with the currents I_2', I_2'', and I_2 are calculated as follows:

$$P' = I_2'^2 \cdot R_2 = 61 \text{ mW} \tag{6.9}$$

$$P'' = I_2''^2 \cdot R_2 = 10.4 \text{ mW} \tag{6.10}$$

$$P = I_2^2 \cdot R_1 = 121.6 \text{ mW}. \tag{6.11}$$

Upon inspection, we find that P is not equal to the sum of P' and P''. This indicates that the superposition principle does not apply to nonlinear quantities, such as electric power.

6.2.2 Problem 2

Construct the circuit in Figure 6.6 in Multisim. For measurements, apply passive and active sign conventions. To analyze the circuit, use the superposition principle and perform the following tasks:

1. Determine the current I_2' circulating through the resistor R_2 when the current source I_{s1} is deenergized. The calculated value of I_2' shall be verified with Multisim simulation.
2. Determine the current I_2'' circulating through the resistor R_2 when the voltage generator V_{s1} is deenergized. The calculated value of I_2'' shall be confirmed with a Multisim simulation.

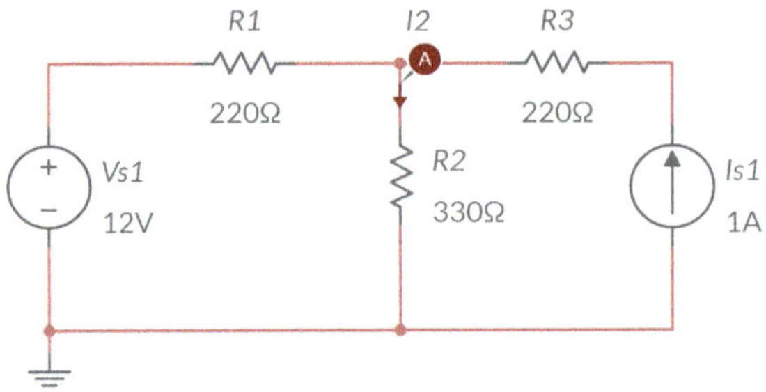

Figure 6.6 Circuit with a DC voltage source and a DC current source.

3. Calculate the current I_2 flowing through the resistor R_2 when both sources are active. The calculated value of I_2 shall be validated using Multisim simulation.

6.2.2.1 Solutions to Problem 2

1. When the current generator I_{s1} is open-circuited, R_1 and R_2 are connected in series forming an equivalent resistance R_{12}, which is calculated using the following formula:

$$R_{12} = R_1 + R_2 = 550\ \Omega. \tag{6.12}$$

To find the current I_2' circulating through resistor R_2 when the current generator I_{s1} is deenergized, we can apply Ohm's law:

$$I_2' = \frac{V_{s1}}{R_{12}} = 21.8\ \text{mA}. \tag{6.13}$$

The calculated value of I_2' is corroborated by the Multisim simulation results, as illustrated in Figure 6.7. This agreement between theory and simulation validates the accuracy of our analysis.

2. When the voltage generator V_{s1} is short-circuited, resistors R_1 and R_2 are connected in parallel, forming an equivalent resistance R_{12}, which is calculated using the following formula:

$$R_{12} = \frac{R_1 \cdot R_2}{R_1 + R_2} = 132\ \Omega. \tag{6.14}$$

To find the current I_2'' circulating through the resistor R_2 when the voltage source V_{s1} is deenergized, we can use the current divider formula:

$$I_2'' = I_{s1}\frac{R_1}{R_1 + R_2} = 400\ \text{mA}. \tag{6.15}$$

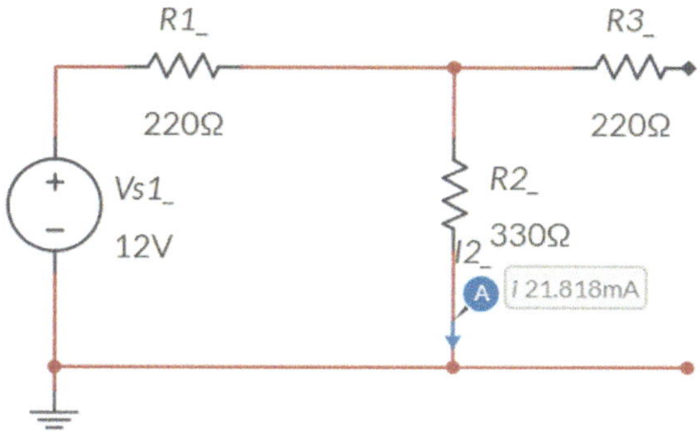

Figure 6.7 Multisim simulation showing the measured value of I_2'.

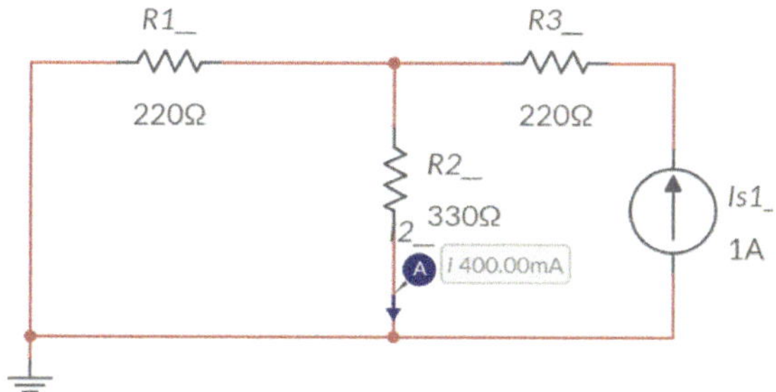

Figure 6.8 Multisim simulation showing the measured value of I_2''.

The calculated value of I_2'' is validated by the Multisim simulation results, as illustrated in Figure 6.8. This agreement between theory and simulation confirms the accuracy of our analysis.

3. When both voltage sources V_{s1} and I_{s1} are active, the total current I_2 flowing through resistor R_2 is the sum of the currents I_2' and I_2'':

$$I_2 = I_2' + I_2'' = 421.8 \text{ mA}. \tag{6.16}$$

The value of I_2, calculated using the superposition principle, is in accordance with the Multisim simulation results, as shown in Figure 6.9.

6.3 Lab Procedure 1

Construct the circuit in Figure 6.10 in Multisim. For measurements, apply passive and active sign conventions. To analyze the circuit, use the superposition principle and perform the following tasks:

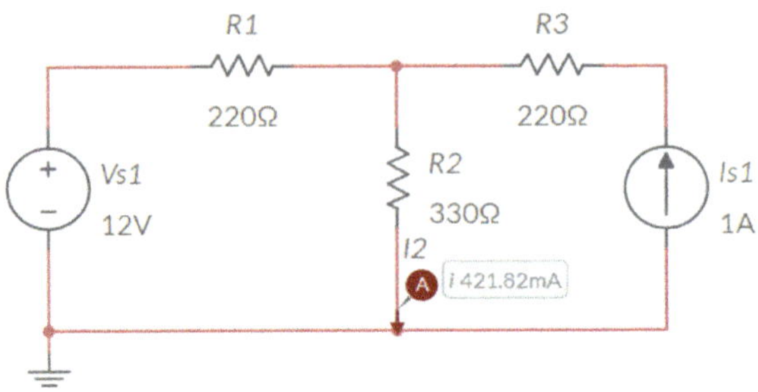

Figure 6.9 Multisim simulation showing the measured value of I_2.

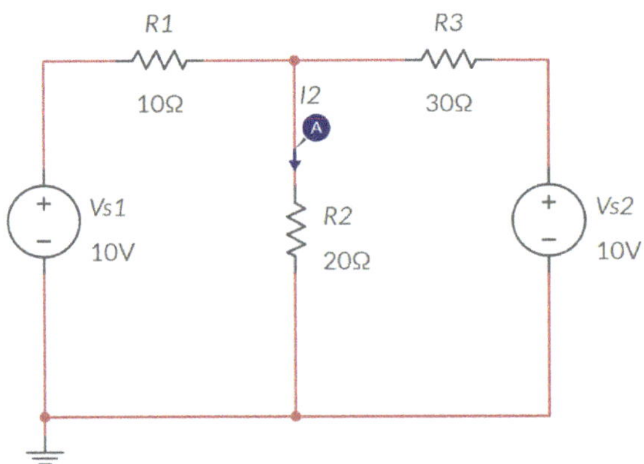

Figure 6.10 Linear resistive circuit with two voltage sources.

1. Determine the current I_2' circulating through the resistor R_2 when the voltage generator V_{s2} is deenergized. The calculated value of I_2' shall be verified with Multisim simulation.
2. Determine the current I_2'' circulating through the resistor R_2 when the voltage generator V_{s1} is deenergized. The calculated value of I_2'' shall be confirmed with Multisim simulation.
3. Calculate the current I_2 flowing through the resistor R_2 when both sources are active. The calculated value of I_2 shall be validated using Multisim simulation.
4. Calculate the electric powers P', P'', and P dissipated across R_2, associated with the currents I_2', I_2'', and I_2, respectively. Subsequently, verify the inapplicability of the superposition principle to electric power calculations.

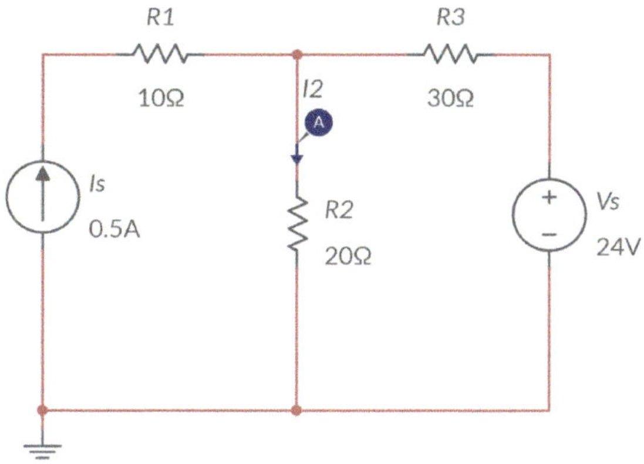

Figure 6.11 Circuit with a DC voltage generator and a DC current generator.

6.4 Lab Procedure 2

Construct the circuit in Figure 6.11 in Multisim. For measurements, apply passive and active sign conventions. To analyze the circuit, use the superposition principle and perform the following tasks:

1. Determine the current I_2' circulating through the resistor R_2 when the current source I_s is deenergized. The calculated value of I_2' shall be verified with Multisim simulation.
2. Determine the current I_2'' circulating through the resistor R_2 when the voltage generator V_s is deenergized. The calculated value of I_2'' shall be confirmed with a Multisim simulation.
3. Calculate the current I_2 flowing through the resistor R_2 when both sources are active. The calculated value of I_2 shall be validated using Multisim simulation.

7

Short-circuits and Open-circuits

7.1 Student Learning Outcomes

- Explain conceptually what a short-circuit and open-circuit represent as two-terminal circuit components.
- Construct *I-V* curves by simulating shorted and open configurations in Multisim. Identify key characteristic differences.
- In Multisim, introduce shorts and opens into circuits. Analyze and measure their effects on networks.

7.2 Prelab

A *short-circuit* and *open-circuit* represent ideal two-terminal circuit components with extreme *I-V* characteristics. A short-circuit has zero resistance and always maintains 0 V across its terminals, regardless of current flow. Its *I-V* curve is a vertical line at $V = 0$ (Figure 7.1(a)) as current is unconstrained.

An *open-circuit* has infinite resistance and zero current flow for any applied voltage. Its horizontal *I-V* line at $I = 0$ indicates no conduction path (Figure 7.1(b)).

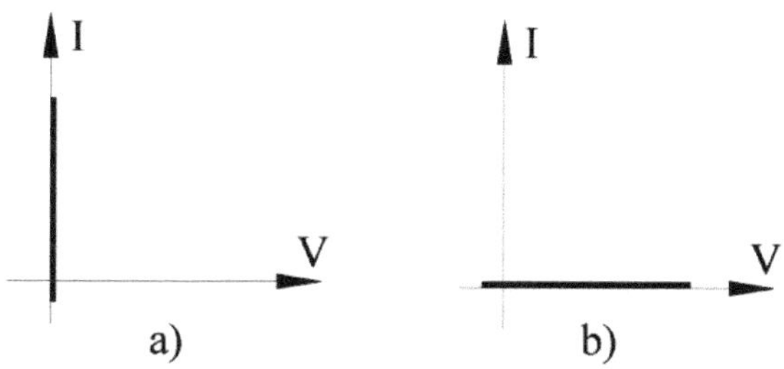

Figure 7.1 *I-V* characteristics for a (a) short-circuit and an (b) open-circuit.

In reality, no physical conductor can have exactly zero resistance (for short-circuits) or infinite resistance (for open-circuits), but under certain conditions and idealizations, they can be approximated as such.

A low-resistance conductor approximates a short if it has large cross-sectional area S, short length l, and low resistivity ρ ($R = \rho \frac{l}{S}$), therefore allowing high currents.

An open-like conductor has small cross-sectional area, long length, and high resistivity.

Neither component dissipates power as the product of I and V is zero. Shorts and opens are not classified as passive or active components

For practical purposes, *short-circuits* and *open-circuits* are vital concepts in circuit analysis, troubleshooting, and understanding the behavior of electrical systems. Their unique I-V curves and null power make them useful modeling abstractions. Practical shortening and opening of circuits facilitates analysis and testing.

7.2.1 Problem 1

Construct the circuit in Figure 7.2 in Multisim. For current measurements, apply passive and active sign conventions. To analyze the circuit, perform the following tasks:

1. For values of the voltage source V_{s1} from 1 to 10 V in increments of 1 V, calculate the current I_{sc} circulating through the short-circuit. Confirm the calculated values of I_{sc} with Multisim simulation.
2. In Multisim, measure the values of the voltage V_{sc} across the short-circuit, for each case.

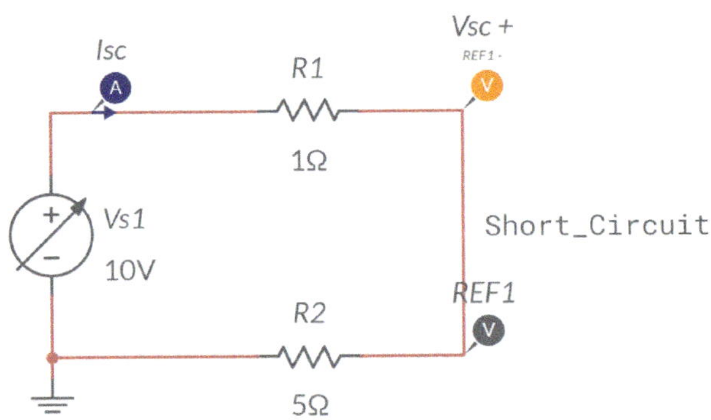

Figure 7.2 Network with a short-circuit.

3. Plot the I-V characteristic curve with I_{sc} on the y-axis and V_{sc} on the x-axis.

7.2.1.1 Solutions to Problem 1

1. For each value of V_{s1}, the current I_{sc} is determined by Ohm's law, using the equivalent resistance of the series R_1–R_2 combination:

$$I_{\text{sc}} = \frac{V_{s1}}{R_1 + R_2}. \tag{7.1}$$

Table 7.1 summarizes the calculated short-circuit currents I_{sc} for swept values of V_{s1} from 1 to 10 V.

The measured I_{sc} values agree closely with the calculated currents, validating the simulation (Figure 7.3).

Table 7.1 Summary of readings and calculations

V_{s1} (V)	I_{sc} (A)	V_{sc} (V)
1	0.17	0
2	0.33	0
3	0.50	0
4	0.67	0
5	0.83	0
6	1.00	0
7	1.17	0
8	1.33	0
9	1.50	0
10	1.67	0

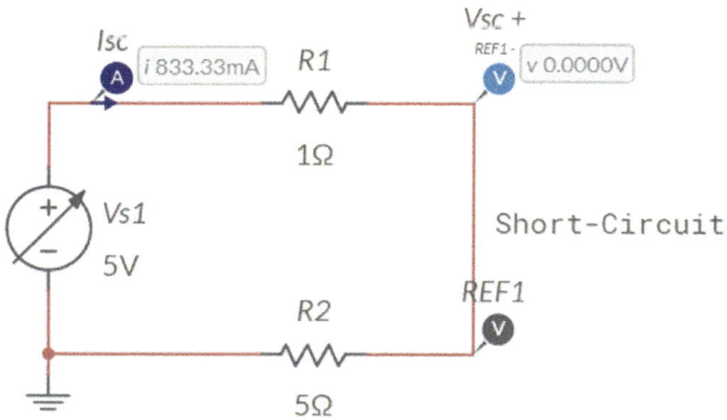

Figure 7.3 Multisim simulation showing the measured value of I_{sc} and V_{sc} for V_{s1} = 5 V.

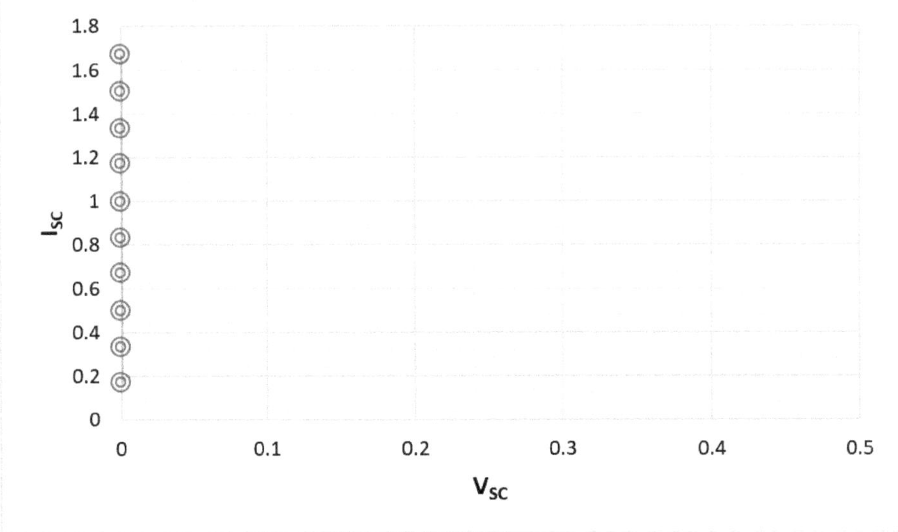

Figure 7.4 *I-V* characteristic curve of the short-circuit.

2. The voltage V_{sc} across the short is measured as 0 V in all cases, as expected for an ideal short-circuit (Table 7.1).
3. Based on Table 7.1, the plot of the $I_{sc}-V_{sc}$ characteristic curve of the short-circuit is given in Figure 7.4.

The increasing I_{sc} with constant null V_{sc} demonstrates the ideal behavior of a short-circuit, which is unimpeded current flow with zero voltage drop.

7.2.2 Problem 2

Construct the circuit in Figure 7.5 in Multisim. For current and voltage measurements, apply passive and active sign conventions. To analyze the circuit, perform the following tasks:

1. For values of the voltage source V_{s2} from 1 to 10 V in increments of 1 V, calculate the voltage V_{oc} across the open-circuit. Confirm the calculated values of V_{oc} with Multisim simulation.
2. In Multisim, measure the current I_{oc} flowing in the circuit, for each case.
3. Plot the *I-V* characteristic curve of the open-circuit with I_{oc} on the *y*-axis and V_{0c} on the *x*-axis.

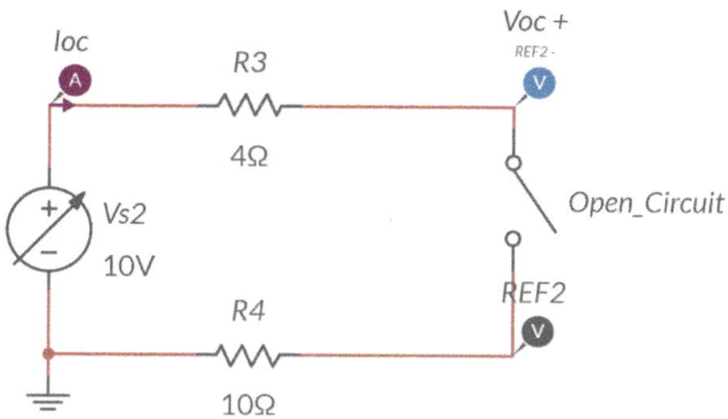

Figure 7.5 Network with an open-circuit.

7.2.2.1 Solutions to Problem 2

1. For each value of V_{s2}, the voltage V_{oc} across the open-circuit is always equal to the source voltage. The values of V_{oc} are confirmed with Multisim simulation (Figure 7.6).
2. Table 7.2 summarizes the measured open-circuit currents I_{oc} for swept values of V_{s2} from 1 to 10 V. Regardless of the value of V_{s2}, the current I_{oc} is always zero, as expected for an ideal open-circuit.

The null values of I_{oc} are confirmed with Multisim simulation (Figure 7.6).

3. Based on Table 7.2, the plot of the I_{oc}−V_{oc} characteristic of the open-circuit is given in Figure 7.7.

The increasing V_{oc} with constant null I_{oc} demonstrates the ideal behavior of an open-circuit, that is, zero current flow with unimpeded voltage drop.

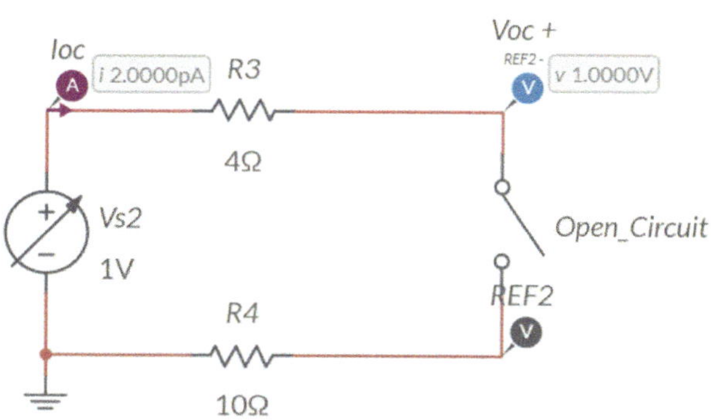

Figure 7.6 Multisim simulation showing the measured values of I_{oc} and V_{oc} for V_{s1} = 1 V.

Table 7.2 Summary of readings

V_{s2} (V)	V_{oc} (V)	I_{sc} (A)
1	1	0
2	2	0
3	3	0
4	4	0
5	5	0
6	6	0
7	7	0
8	8	0
9	9	0
10	10	0

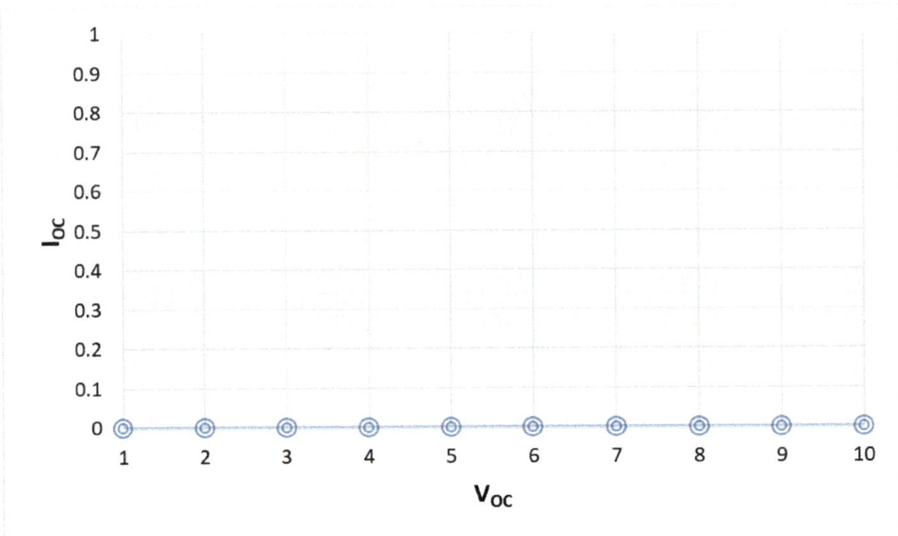

Figure 7.7 *I-V* characteristic curve of the open-circuit.

7.2.3 Problem 3

Construct the circuit in Figure 7.8 in Multisim. For current and voltage measurements, apply passive and active sign conventions. To analyze the circuit, perform the following tasks:

1. With switch open, calculate the voltage V_{SW} across it and the total current I_t circulating in the circuit. In Multisim, simulate the circuit with open switch, measure V_{SW} and I_t, and verify agreement.

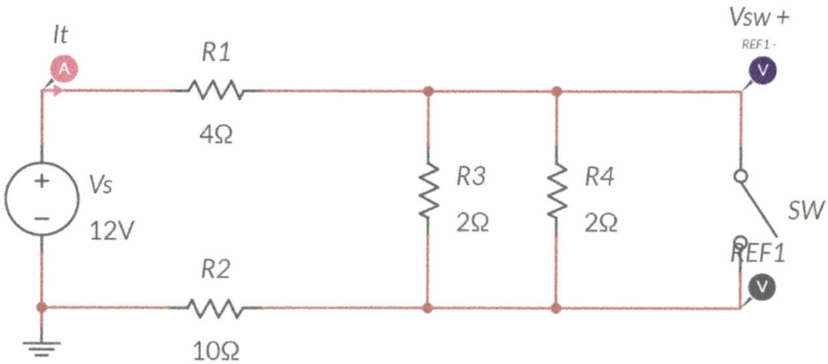

Figure 7.8 Series–parallel combination circuit with switch.

2. With switch closed, calculate the voltage $V_{\rm sw}$ across it and the total current I_t circulating in the circuit. In Multisim, simulate the circuit with open switch, measure $V_{\rm sw}$ and I_t, and verify agreement.
3. For closed switch, calculated the total resistance R_{T2}. Simulate the circuit in Multisim and confirm the calculated value.

7.2.3.1 Solutions to Problem 3

1. With the switch open, the voltage $V_{\rm sw}$ across the open gap is calculated using the voltage divider formula:

$$V_{\rm sw} = V_s \frac{R_{34}}{R_1 + R_2 + R_{3-4}} = 0.8 \text{ V}, \tag{7.2}$$

where R_{34} is the parallel combination of R_3 and R_4.

With the switch open, the total resistance seen by the voltage source V_s can be calculated with the following formula:

$$R_t = R_1 + R_2 + \frac{R_3 R_4}{R_3{+}R_4} = 15 \ \Omega. \tag{7.3}$$

The total current I_t can be calculated using Ohm's law:

$$I_t = \frac{V_{s3}}{R_t} = \frac{12 \text{ V}}{15 \ \Omega} = 0.8 \text{ A}. \tag{7.4}$$

The measured I_t and $V_{\rm sw}$ values agree with the calculated values, validating the calculation (Figure 7.9).

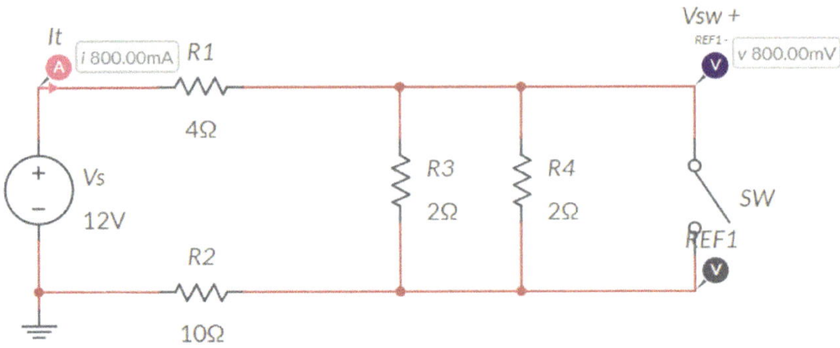

Figure 7.9 Multisim simulation showing the measured value of V_{sw} and I_t with switch off.

1. With switch closed, resistors R_3 and R_4 are short-circuited; thus, the voltage V_{sw} across the switch is zero (i.e., the short-circuit is an ideal two-terminal with zero resistance), and the total current I_t circulating in the circuit with the switch in the closed position is given by: With the switch closed, resistors R_3 and R_4 are shorted together. For an ideal short-circuit, the voltage V_{sw} across the closed switch is zero. This yields a total resistance of $R_1 + R_2$. Then the total current is calculated by Ohm's law:

$$I_t = \frac{V_s}{R_1 + R_2} = \frac{12\ \text{V}}{14\ \Omega} = 0.857\ \text{A}. \tag{7.5}$$

These expected values for V_{sw} and for I_t are confirmed via Multisim simulation, as shown in Figure 7.10.

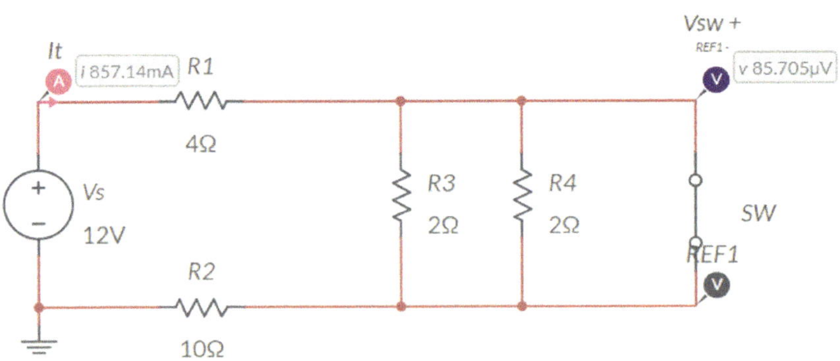

Figure 7.10 Multisim simulation confirming the measured value of V_{sw} and I_t with switch on.

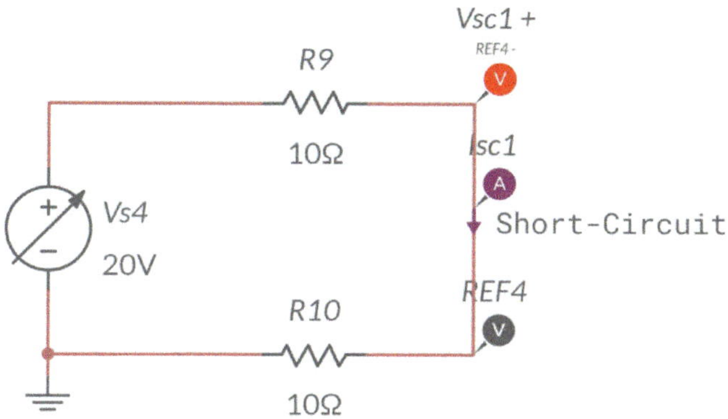

Figure 7.11 Network with a short-circuit.

The nonzero voltage for V_{sw} is due to the on-resistance of the switch in Multisim of 100 μΩ.

7.3 Lab Procedure 1

Construct the circuit in Figure 7.11 in Multisim. For voltage and current measurements, apply passive and active sign conventions. To analyze the circuit, perform the following tasks:

1. For values of the voltage source V_{s4} from 10 to 20 V in increments of 1 V, calculate the current I_{sc1} circulating through the short-circuit. Confirm the calculated values of I_{sc1} with Multisim simulation. Record calculated values in Table 7.3.
2. In Multisim, measure the values of the voltage V_{sc1} across the short-circuit, for each case. Record calculated values in Table 7.3.
3. Plot the I-V characteristic curve with I_{sc4} on the y-axis and V_{sc4} on the x-axis.

7.4 Lab Procedure 2

Construct the circuit in Figure 7.12 in Multisim. For current and voltage measurements, apply passive and active sign conventions. To analyze the circuit, perform the following tasks:

1. For values of the voltage source V_{s5} from 10 to 20 V in increments of 1 V, determine the voltage V_{oc1} across the open-circuit. Confirm the calculated values of V_{oc1} with Multisim simulation. Record readings in Table 7.4.

Table 7.3 Summary of readings and calculations

V_{s4} (V)	I_{sc1} (A)	V_{sc1} (V)
10		
11		
12		
13		
14		
15		
16		
17		
18		
19		
20		

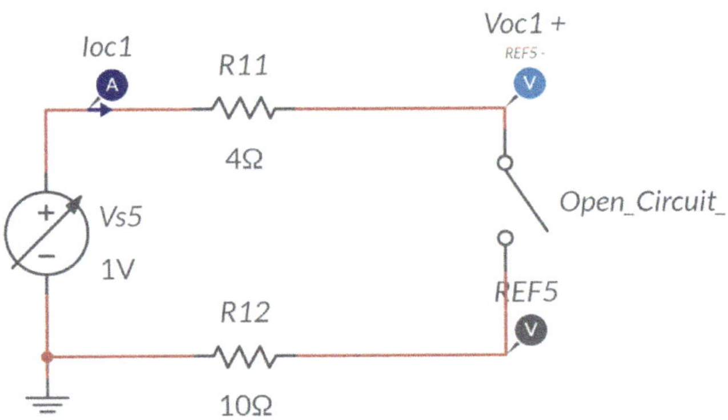

Figure 7.12 Network with an open-circuit.

Table 7.4 Summary of readings and calculations

V_{s5} (V)	V_{oc1} (V)	I_{oc1} (A)
10		
11		
12		
13		
14		
15		
16		
17		
18		
19		
20		

2. In Multisim, measure the current I_{ocs} flowing in the circuit, for each case. Record readings in Table 7.4.
3. Plot the *I*-*V* characteristic curve of the open-circuit with I_{oc1} on the *y*-axis and V_{oc1} on the *x*-axis.

8

Thevenin's Theorem and Maximum Power Transfer in DC Circuits

8.1 Student Learning Outcomes

- Analyze DC linear circuits and determine equivalent Thevenin source voltage and resistance using Thevenin's theorem.
- Construct equivalent Thevenin models in Multisim. Verify circuit behavior is unchanged through simulation.
- Derive the condition for maximum power transfer load resistance based on Thevenin resistance. In Multisim, verify agreement with theoretical expectations.

8.2 Prelab

The Thevenin's theorem allows the replacement of any linear circuit (i.e., which may include resistors, inductors, and capacitors) connected to a load R_L at any two terminals A and B with an equivalent circuit consisting of a single voltage source V_{th} in series with a resistance R_{th} (Figure 8.1).

Thevenin's theorem states that any linear circuit with voltage/current sources and passive elements can be replaced at any two terminals A and B by an equivalent model consisting of a single voltage source V_{th} in series with a single resistance R_{th} (Figure 8.1).

The voltage source, V_{th}, represents the open-circuit voltage between the two terminals A and B of the original circuit, with the load R_L disconnected. The resistor, R_{th}, represents the equivalent resistance seen from these terminals when all the independent sources are deactivated. This involves replacing voltage sources with short-circuits and current sources with open-circuits.

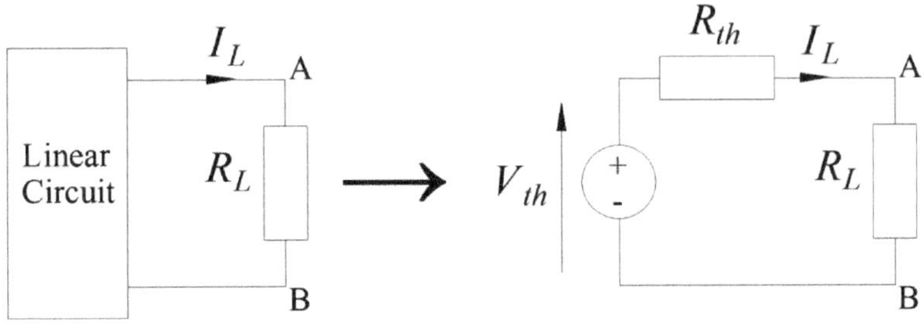

Figure 8.1 A linear circuit and its Thevenin equivalent.

For ideal voltage sources, a short-circuit provides a zero-voltage connection, effectively removing any voltage difference the source may have provided. This temporarily turns off the voltage source. Similarly, an open-circuit draws no current, which deactivates an ideal current source. The open-circuit models an indefinitely large load resistance that inhibits current circulation from the source.

The Thevenin equivalent retains the same behavior as the original circuit for any load R_L attached to the output. It simplifies analysis by reducing a complex network down to a single equivalent source.

The theorem provides a criterion for maximizing the power transferred from the network to a load resistance R_L. Differentiating the load power P_L with respect to R_L and setting the derivative equal to zero, $\frac{\partial P_L}{\partial R_L} = 0$, gives the condition $R_L = R_{\text{th}}$,[1] which means that the maximum power, $P_{L\text{Max}}$, is transferred when the load resistance equals the Thevenin equivalent resistance of the network.

In this condition, the power transferred, $P_{L\text{Max}}$, can be calculated as follows:

$$P_{L\text{Max}} = I_L^2 R_{\text{th}} = \frac{V_{\text{th}}^2}{4R_{\text{th}}}. \tag{8.1}$$

The power transferred to the load, as a function of the load resistance, will have a peak when the load resistance matches the Thevenin equivalent resistance, as depicted in Figure 8.2.

At this point, the *efficiency* of the circuit, defined as the ratio of the power absorbed at the load to the power supplied by the source, is 50%. This means that only half of the power supplied by the source is effectively utilized by the load.

[1]We can confirm that R_{th} is the value corresponding to the maximum of P_L by showing that the second derivative is negative.

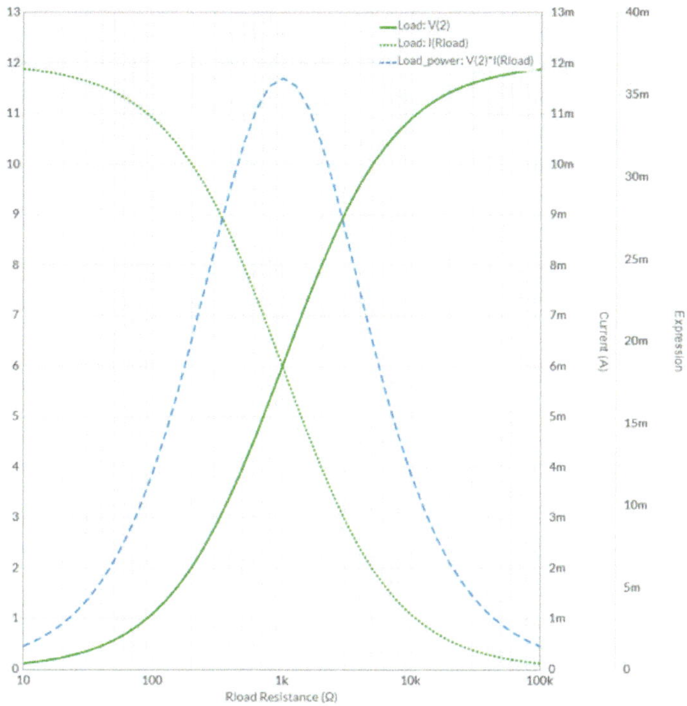

Figure 8.2 Load power as a function of the load resistance.

8.2.1 Problem 1

Construct the two-terminal linear circuit in Figure 8.3 in Multisim. For current and voltage measurements, apply passive and active sign conventions. To analyze the circuit, perform the following tasks:

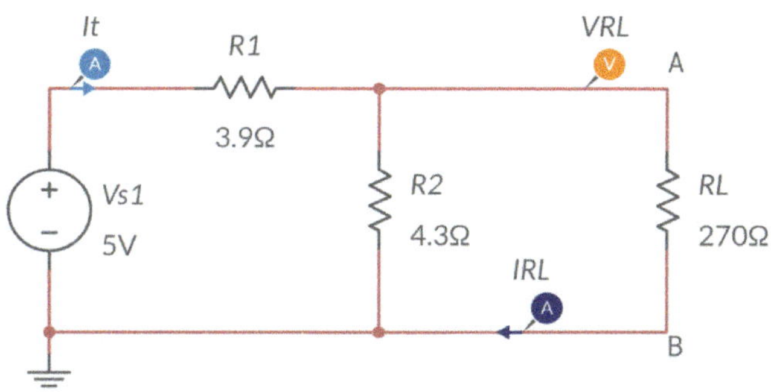

Figure 8.3 A two-terminal linear circuit with a voltage source.

1. In Multisim, with load R_L connected, measure the voltage V_{R_L} across the load R_L and the current I_{R_L} flowing through it. Remove the load R_L and measure the open-circuit voltage V_{th} across the terminals A and B.
2. Calculate the Thevenin equivalent voltage V_{th} at the terminals A and B. Compare calculated and measured values.
3. Calculate the Thevenin equivalent resistance R_{th} at the terminals A and B.
4. In Multisim, simulate the Thevenin equivalent circuit using the calculated values of V_{th} and R_{th}, with the load R_L connected. Measure the values of the current I_{R_L} and the voltage V_{R_L} across the load R_L and compare them with the values measured in the original circuit.

8.2.1.1 Solutions to Problem 1

1. Figure 8.4 shows the Multisim measured values for the voltage V_{R_L} and the current I_{R_L} in the original circuit with load R_L connected.

The measured load voltage and load current represent the reference behavior to validate against the Thevenin model.

Figure 8.5 shows the open-circuit Thevenin voltage V_{th} measured across terminals A and B after disconnecting load R_L.

2. The Thevenin equivalent voltage V_{th} at terminals A and B can be calculated as follows:

$$V_{\text{th}} = V_s \frac{R_2}{R_1 + R_2} = 2.62\ \text{V}. \tag{8.2}$$

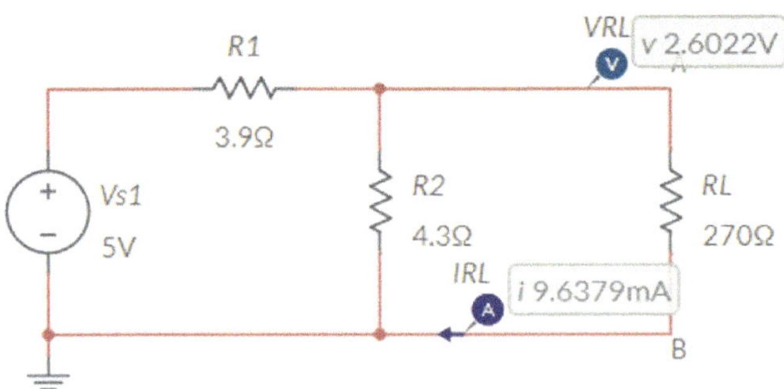

Figure 8.4 Multisim simulation of original circuit providing the measured values of V_{R_L} and I_{R_L}.

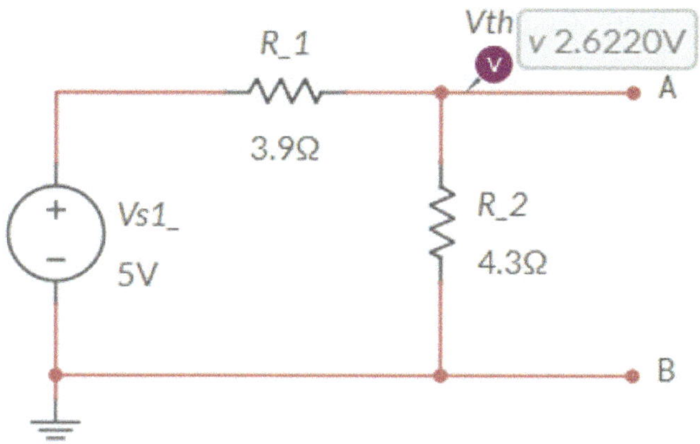

Figure 8.5 Multisim simulation providing the value of V_{th} across the open-circuited terminals A and B.

The calculated V_{th} matches the measurement provided by the simulation in Figure 8.5. This confirms that we have determined the correct Thevenin equivalent voltage for the circuit.

3. To determine the Thevenin resistance R_{th}, the independent voltage source V_{s1} is deactivated by replacing it with a short-circuit. Also, the load R_L is disconnected. This leaves just the parallel resistance combination of R_1 and R_2 connected between terminals A and B. The equivalent resistance is

$$R_{\text{th}} = \frac{R_1 \cdot R_2}{R_1 + R_2} = 2\ \Omega. \tag{8.3}$$

4. Figure 8.6 shows the implementation of the Thevenin equivalent circuit model in Multisim using the calculated values for V_{th} and R_{th}.

In the simulation of Figure 8.6, the values for I_{R_L} and V_{R_L} match those obtained in the simulation of the original circuit. The identical load voltage V_{R_L} and current I_{R_L} seen in the original and Thevenin model circuits experimentally validate the equivalence. The Thevenin equivalent circuit accurately captures the behavior of a complex source network with only the single Thevenin source V_{th} and series resistance R_{th}.

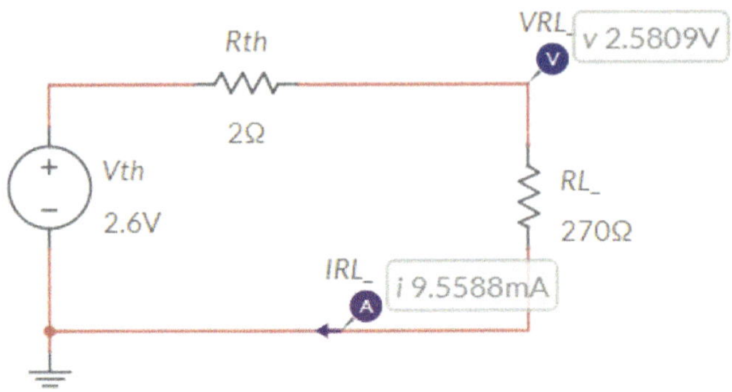

Figure 8.6 Thevenin equivalent circuit in Multisim providing measurements for I_{R_L} and V_{R_L}.

8.2.2 Problem 2

Construct the two-terminal linear circuit in Figure 8.7 in Multisim. For current and voltage measurements, apply passive and active sign conventions. To analyze the circuit, perform the following tasks:

1. In Multisim, with the load R_L connected, measure the voltage V_{R_L} across R_L and the current I_{R_L} circulating through it. Remove the load R_L from the circuit and measure the Thevenin equivalent voltage V_{th} across the open-circuited terminals A and B.
2. Analytically calculate Thevenin voltage V_{th}. Compare calculated value to measured V_{th} from simulation.
3. Analytically calculate Thevenin equivalent resistance R_{th} at the terminals A and B. In Multisim original circuit, determine R_{th} by current injection test. Compare to calculated R_{th}.

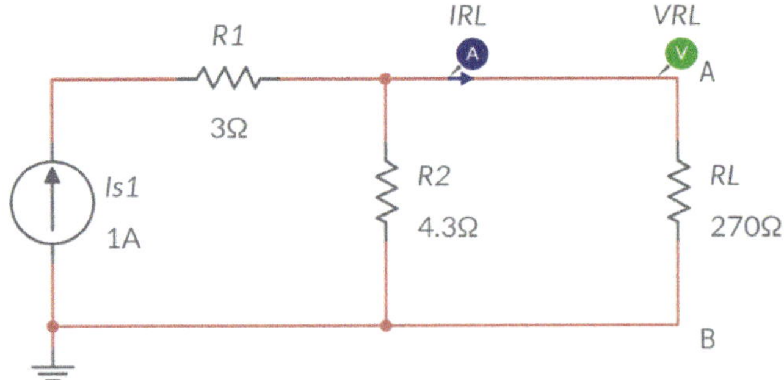

Figure 8.7 Two-terminal linear circuit with a DC current source.

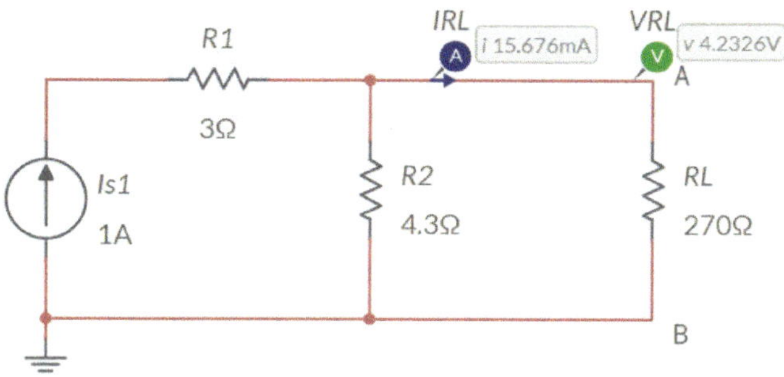

Figure 8.8 Multisim simulation providing the measured values of V_{R_L} and I_{R_L}.

4. In Multisim, construct Thevenin equivalent circuit using calculated V_{th} and R_{th} values. Connect original load R_L. Measure current I_{R_L} and the voltage V_{R_L} across load R_L. Compare to original circuit measurements.

8.2.2.1 Solutions to Problem 2

1. Figure 8.8 shows the Multisim simulation results for the original circuit with load R_L connected. The measured load voltage is V_{R_L} = 4.232 V and the load current is I_{R_L} = 15.676 mA.

Figure 8.9 then shows the open-circuit Thevenin voltage V_{th} measured across terminals A and B after disconnecting the load. V_{th} is measured as 4.3 V.

As expected, V_{th} is higher than V_{R_L} since the load voltage divider effect is removed when R_L is disconnected.

2. The Thevenin voltage V_{th} is calculated by determining the open-circuit voltage across terminals A and B with current source I_{s1} active. V_{th} is calculated by Ohm's law:

$$V_{\text{th}} = I_{s1}R_2 = 1 \text{ A} \cdot 4.3 \ \Omega = 4.3 \text{ V}. \tag{8.4}$$

The above calculated value matches that provided by the simulation (Figure 8.8). This calculated value of 4.3 V matches the V_{th} measurement from the Multisim simulation.

3. To determine the Thevenin resistance R_{th}, the independent current source I_{s1} is deactivated by replacing it with an open-circuit. Also, the load R_L is disconnected. This leaves only resistor R_2 to be considered between terminals A and B. Therefore, the Thevenin resistance is

$$R_{\text{th}} = R_2 = 4.3 \ \Omega. \tag{8.5}$$

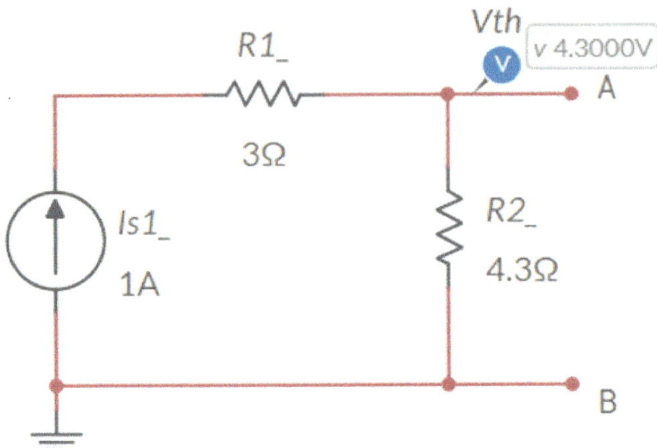

Figure 8.9 Multisim simulation providing the measured value of the open-circuit voltage V_{th} across terminals A and B.

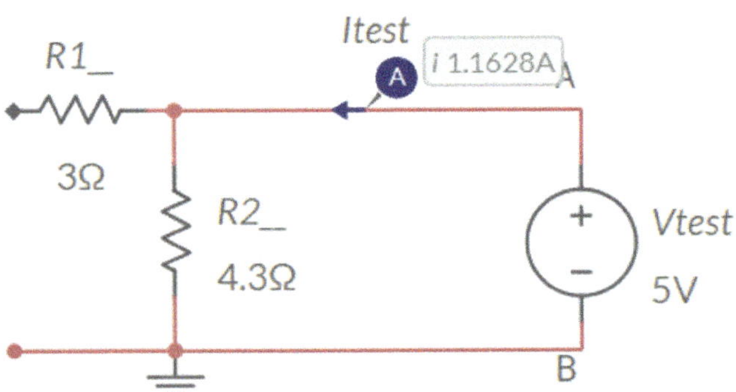

Figure 8.10 Multisim simulation setup for the determination of R_{th}.

This calculated value for R_{th} can be confirmed in Multisim by current injection test using a voltage source V_{test} of arbitrary value and an ammeter, as shown in Figure 8.10.

The network is deenergized by replacing the current source I_{s1} with an open-circuit. With V_{test} = 5 V applied across A and B, R_{th} is given by

$$R_{th} = \frac{V_{test}}{I_{test}} = 4.3\ \Omega. \tag{8.6}$$

This matches the analytically calculated R_{th}, validating the determination of the Thevenin resistance.

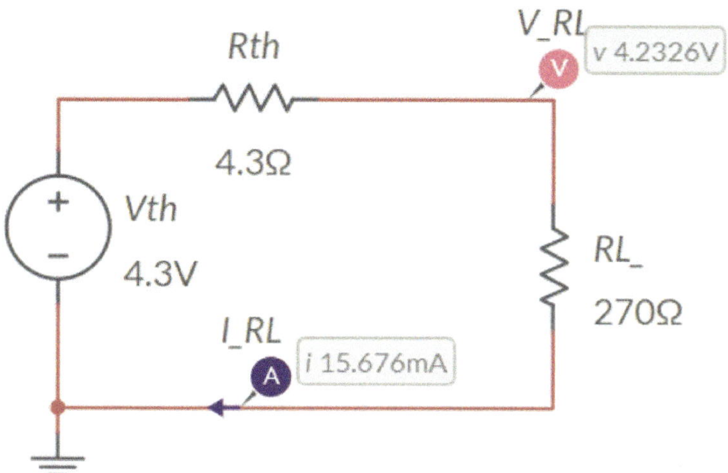

Figure 8.11 Thevenin equivalent circuit in Multisim showing the measured values of V_{R_L} and I_{R_L}.

4. Figure 8.11 shows the implementation of the Thevenin equivalent circuit model in Multisim using the calculated parameters V_{th} = 4.3 V, and R_{th} = 4.3 Ω, with the original load R_L = 270 Ω connected.

The Multisim Thevenin model results match the earlier measurements in the original circuit for the current I_{R_L} and the voltage V_{R_L}. The identical results in the original and Thevenin circuits confirm that the model accurately reproduces the behavior of the complex source network.

8.2.3 Problem 3

In Multisim, construct the Thevenin equivalent circuit in Figure 8.12. For current and voltage measurements, apply passive and active sign conventions. To analyze the circuit, perform the following tasks:

1. Vary load R_{L1} from 1 to 20 Ω in 1-Ω increments. For each R_{L1}, calculate the load voltage $V_{R_{L1}}$ across it. Confirm your calculations with a Multisim simulation.
2. For each R_{L1}, calculate the load power P_{L1} and supply power P_{th}.
3. Generate a plot of the power P_{L1} as a function of R_{L1} based on calculations. Identify maximum P_{L1} and its R_{L1}, and verify consistency with theory.

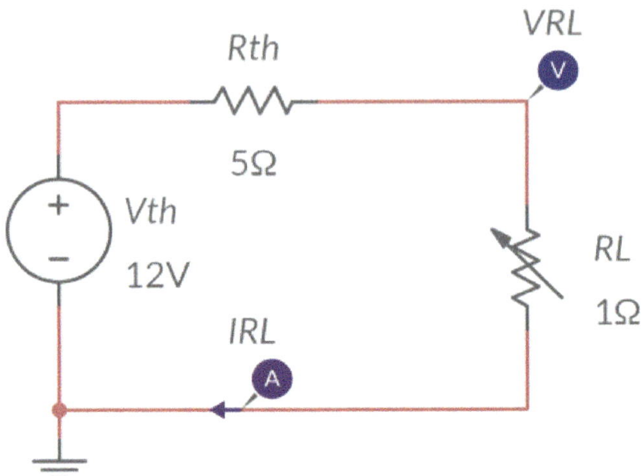

Figure 8.12 Thevenin equivalent circuit with a variable load.

8.2.3.1 Solutions to Problem 3

1. The load voltage V_{R_L} is determined by the voltage divider formula:

$$V_{R_L} = V_{\text{th}} \frac{R_L}{R_{\text{th}} + R_L}. \tag{8.7}$$

Table 8.1 shows the calculated load voltage V_{R_L} for sweeped values of load resistance R_L from 1 to 15 Ω along with the corresponding load power P_L and source power P_{th}, later calculated.

Figure 8.13 provides a Multisim simulation validating the calculated V_{R_L} for R_L = 5 Ω. Agreement between simulated and analytical V_{R_L} helps confirm proper application of the voltage divider formula.

2. The load power P_L can be calculated as follows:

$$P_L = \frac{V_{R_L}^2}{R_L}. \tag{8.8}$$

The source power P_{th} can be calculated as follows:

$$P_{\text{th}} = \frac{V_{\text{th}}^2}{R_{\text{th}} + R_L}. \tag{8.9}$$

From Table 8.1, it is evident that the maximum power transfer occurs for R_L= 5 Ω, for which P_L = 7.2 W, V_{R_L} = 6 V, and P_{th} = 14.2 W. This condition

Table 8.1 Summary of readings and calculations

R_L (Ω)	V_{R_L} (V)	P_L (W)	P_{th} (W)
1	2.0	4.0	24.0
2	3.4	5.9	20.6
3	4.5	6.8	18.0
4	5.3	7.1	16.0
5	6.0	7.2	14.4
6	6.5	7.1	13.1
7	7.0	7.0	12.0
8	7.4	6.8	11.1
9	7.7	6.6	10.3
10	8.0	6.4	9.6
11	8.3	6.2	9.0
12	8.5	6.0	8.5
13	8.7	5.8	8.0
14	8.8	5.6	7.6
15	9.0	5.4	7.2

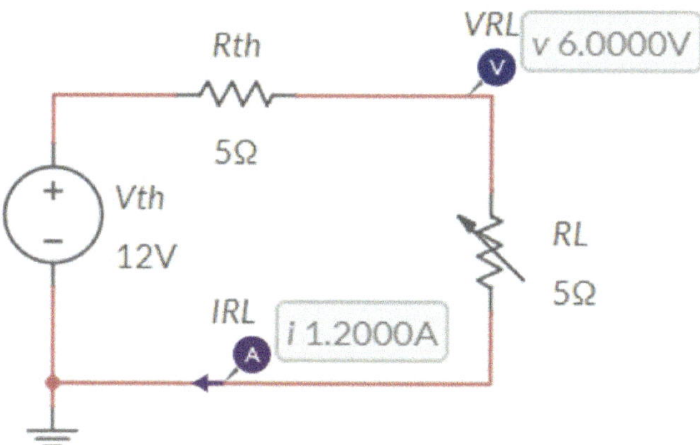

Figure 8.13 Multisim simulation showing the measured value of V_{R_L} for R_L = 5 Ω.

satisfies the criterion for maximum power transfer, where the load voltage equals half the Thevenin voltage, and the absorbed load power equals half the power supplied by the Thevenin generator (i.e., $P_{\text{th}} = 2P_L$).

3. Figure 8.14 shows the plot of load power P_L versus load resistance R_L based on the calculated values from Table 8.1.

A clear peak in the power curve is visible at $R_L = 5$ Ω. This value confirms that when the load resistance matches the Thevenin equivalent resistance, the power transfer is maximized.

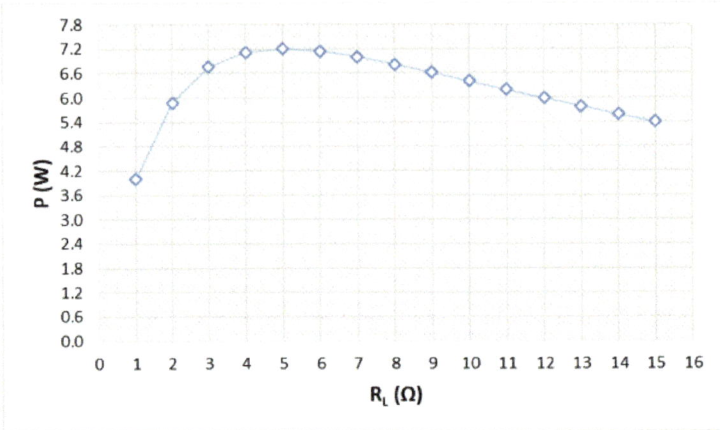

Figure 8.14 Plot of the load power P_L versus load resistance R_L.

8.3 Lab Procedure 1

Construct the two-terminal linear circuit in Figure 8.15 in Multisim. For current and voltage measurements, apply passive and active sign conventions. To analyze the circuit, perform the following tasks:

1. In Multisim, with load R_L connected, measure the voltage V_{R_L} across the load R_L and the current I_{R_L} flowing through it. Remove the load R_L and measure the open-circuit voltage V_{th} across the terminals A and B.
2. Calculate the Thevenin equivalent voltage V_{th} at the terminals A and B. Compare calculated and measured values.

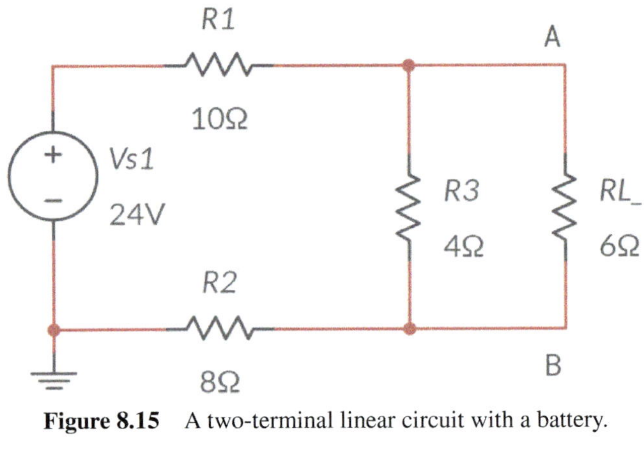

Figure 8.15 A two-terminal linear circuit with a battery.

3. Calculate the Thevenin equivalent resistance R_{th} at the terminals A and B.
4. In Multisim, simulate the Thevenin equivalent circuit using the calculated values of V_{th} and R_{th}, with the load R_L connected. Measure the current I_{R_L} and the voltage V_{R_L} across the load R_L and compare them with the values measured in the original circuit.

8.4 Lab Procedure 2

In Multisim, construct the Thevenin equivalent circuit in Figure 8.16. For current and voltage measurements, apply passive and active sign conventions. To analyze the circuit, perform the following tasks:

1. Vary load R_{L1} from 1 to 20 Ω in 1-Ω increments. For each R_{L1}, calculate the load voltage $V_{R_{L1}}$ across it. Confirm your calculations with a Multisim simulation.
2. For each R_{L1}, calculate the load power P_{L1} and supply power P_{th}.
3. Generate a plot of the power P_{L1} as a function of R_{L1} based on calculations. Identify maximum P_{L1} and its R_{L1}, and verify consistency with theory.

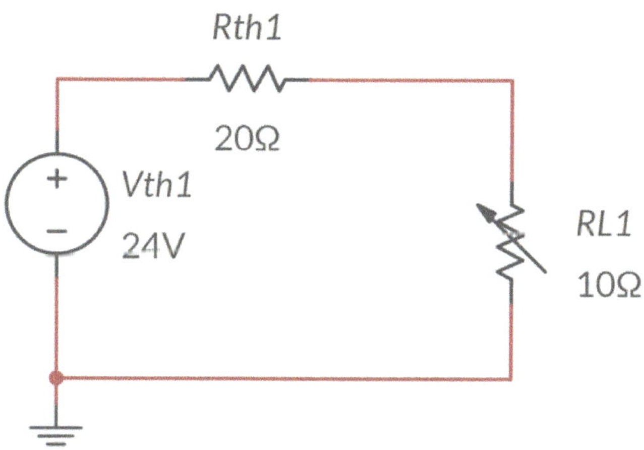

Figure 8.16 Thevenin equivalent circuit with a variable load.

9

Norton's Theorem in DC Circuits

9.1 Student Learning Outcomes

- Analyze DC linear circuits and determine equivalent Norton's current voltage and resistance using Norton's theorem.
- Construct equivalent Norton models in Multisim. Verify circuit behavior is unchanged through simulation.

9.2 Prelab

The Norton's theorem allows the replacement of any linear circuit (i.e., which may include resistors, inductors, and capacitors) connected to a load R_L at any two terminals A and B with an equivalent circuit consisting of a single current source I_N in parallel with a resistance R_N (Figure 9.1).

Similar to Thevenin's theorem, Norton's theorem states that any linear circuit with sources and passive elements can be replaced by an equivalent model consisting of an ideal current source I_N in parallel with a resistance R_N (Figure 9.1).

I_N is the short-circuit current measured through the output terminals A and B with zero load resistance (i.e., an ideal short-circuit). R_N is the equivalent resistance looking back into the terminals with all sources deactivated

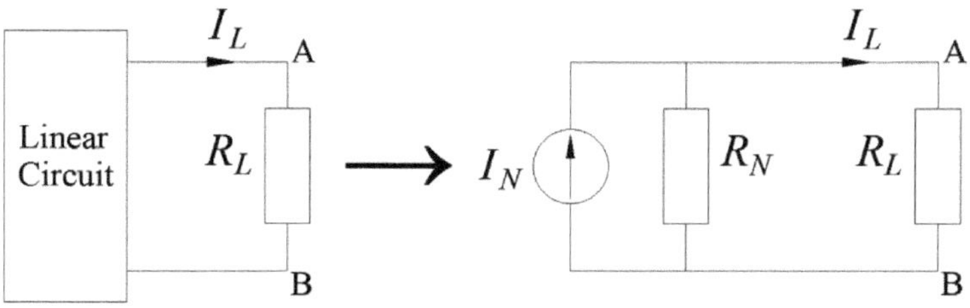

Figure 9.1 A linear circuit and its Norton equivalent.

(i.e., replaced with short-circuits and open-circuits, respectively) same as the Thevenin resistance.

The Norton equivalent model retains the same behavior as the original circuit for any load attached to the output terminals. By applying Norton's theorem, complex linear circuits can be transformed into simplified circuits comprising only a single current source and a parallel resistance. This simplification facilitates circuit analysis and allows engineers to determine the behavior of the circuit with respect to different loads.

The Norton and Thevenin representations are mathematically equivalent. Either model can be used to simplify analysis of complex circuits.

9.2.1 Problem 1

Construct the two-terminal linear circuit in Figure 9.2 in Multisim. For current and voltage measurements, apply passive and active sign conventions. To analyze the circuit, perform the following tasks:

1. In Multisim, with load R_L connected, measure the voltage V_{R_L} across the load R_L and the current I_{R_L} flowing through it. Remove the load R_L and measure the Norton current I_N flowing through the short-circuited terminals A and B.
2. Calculate the Norton equivalent current I_N flowing through the short-circuited terminals A and B. Compare calculated and measured values.
3. Calculate the Norton equivalent resistance R_N at the terminals A and B.

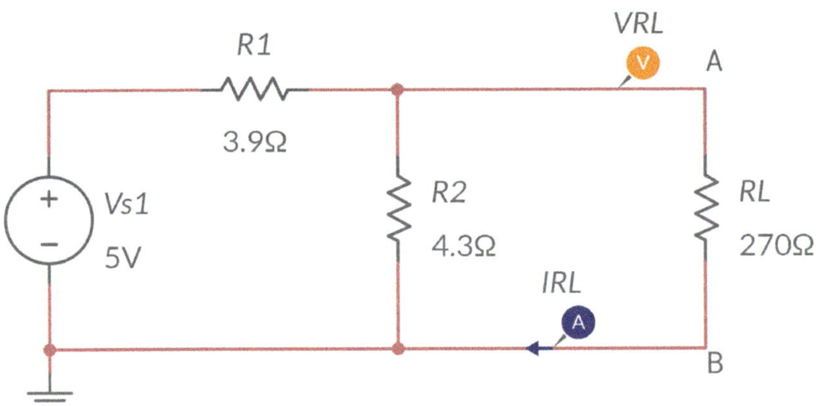

Figure 9.2 A two-terminal linear circuit with a voltage source.

4. In Multisim, simulate the Norton equivalent circuit using the calculated values of I_N and R_N, with the load R_L connected. Measure the values of the current I_{R_L} and the voltage V_{R_L} across the load R_L and compare them with the values measured in the original circuit.

9.2.1.1 Solutions to Problem 1

1. Figure 9.3 shows the Multisim measured values for the voltage V_{R_L} and the current I_{R_L} in the original circuit with load R_L connected.

The measured load voltage and load current represent the reference behavior to validate against the Norton model.

Figure 9.4 shows the Norton equivalent current I_N flowing through the short-circuited terminals A and B.

2. The Norton equivalent current I_N flowing through the short-circuited terminals A and B can be calculated as follows:

$$I_N = \frac{V_{s1}}{R_1} = 1.28 \text{ A}. \tag{9.1}$$

The calculated I_N matches the measurement provided by the simulation in Figure 9.4. This confirms that we have determined the correct Norton equivalent current for the circuit.

3. To determine the Norton equivalent resistance R_N, the independent voltage source V_{s1} is deactivated by replacing it with a short-circuit.

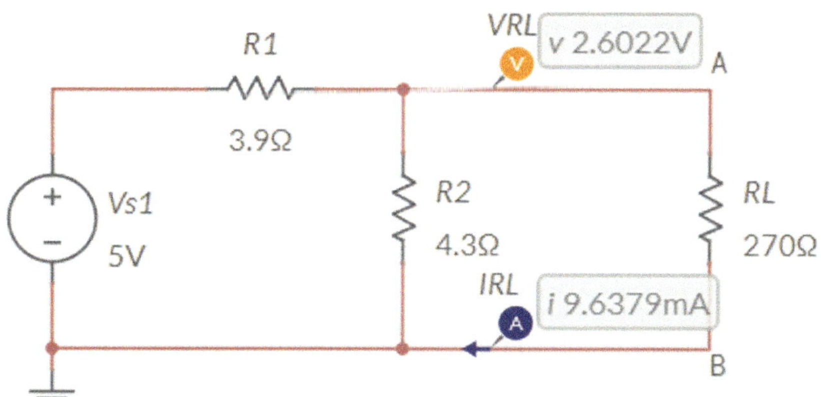

Figure 9.3 Multisim simulation of original circuit providing the measured values of V_{R_L} and I_{R_L}.

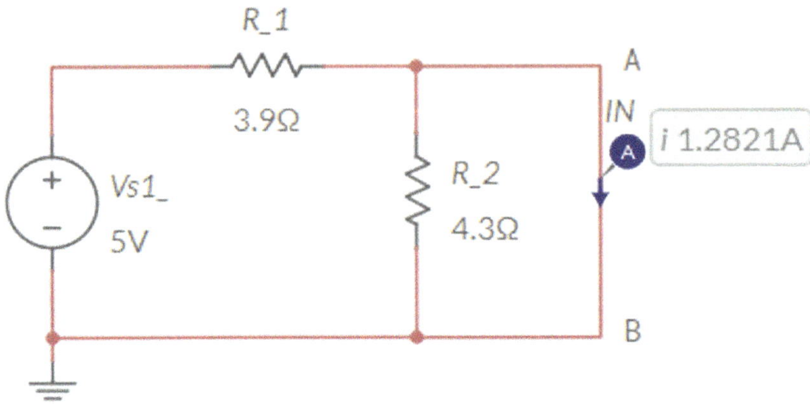

Figure 9.4 Multisim simulation providing the value of I_N flowing through the short-circuited terminals A and B.

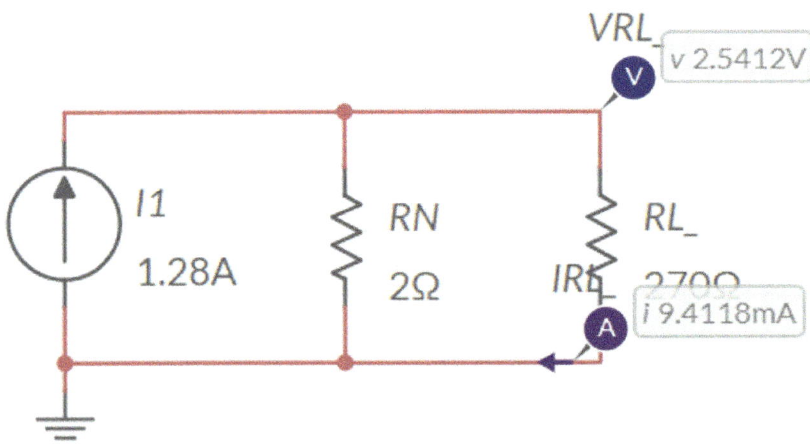

Figure 9.5 Norton equivalent circuit in Multisim providing measurements for I_{R_L} and V_{R_L}.

Also, the load R_L is disconnected. This leaves just the parallel resistance combination of R_1 and R_2 connected between the terminals A and B. The equivalent resistance is

$$R_{\text{th}} = \frac{R_1 \cdot R_2}{R_1 + R_2} = 2\,\Omega. \tag{9.2}$$

4. Figure 9.5 shows the implementation of the Norton equivalent circuit model in Multisim using the calculated values for V_N and R_N.

In the simulation of Figure 9.5, the values for I_{R_L} and V_{R_L} closely match those obtained in the simulation of the original circuit (Figure 9.3). The Norton equivalent circuit accurately captures the behavior of a complex network with only the single Norton source V_{th} and parallel resistance R_N.

9.2.2 Problem 2

Construct the two-terminal linear circuit in Figure 9.6 in Multisim. For current and voltage measurements, apply passive and active sign conventions. To analyze the circuit, perform the following tasks:

1. In Multisim, with the load R_L connected, measure the voltage V_{R_L} across R_L and the current I_{R_L} circulating through it. Measure the Norton equivalent current I_N flowing through the short-circuited terminals A and B.
2. Analytically calculate the Norton equivalent current I_N flowing through the short-circuited terminals A and B. Compare calculated value to measured I_N from simulation.
3. Analytically calculate the Norton equivalent resistance R_N at the terminals A and B.
4. In Multisim, construct the Norton equivalent circuit using calculated I_N and R_N values. Connect original load R_L. Measure current I_{R_L} and the voltage V_{R_L} across load R_L. Compare to original circuit measurements.

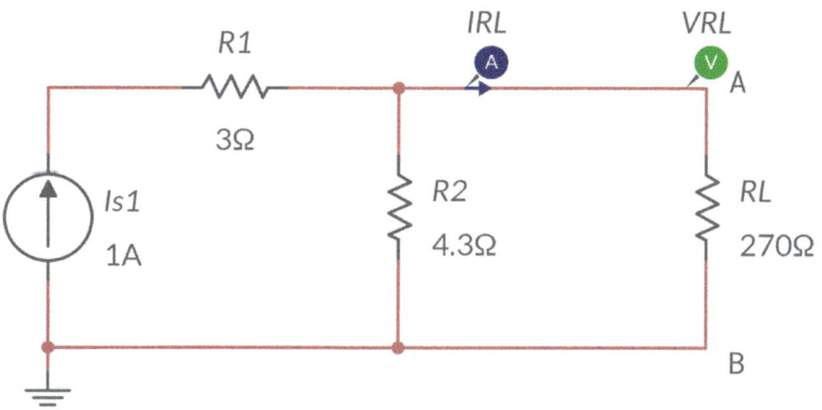

Figure 9.6 A two-terminal linear circuit with a DC current source.

9.2.2.1 Solutions to Problem 2

1. Figure 9.7 shows the Multisim simulation results for the original circuit with load R_L connected. The measured load voltage is V_{R_L} = 4.232 V and the load current is I_{R_L} = 15.676 mA.

The value of the Norton equivalent current I_N flowing through the short-circuited terminals A and B is shown in Figure 9.8. Figure 9.8 then shows the Norton equivalent current I_N measured through the short-circuited terminals A and B. I_N is measured as 1 A.

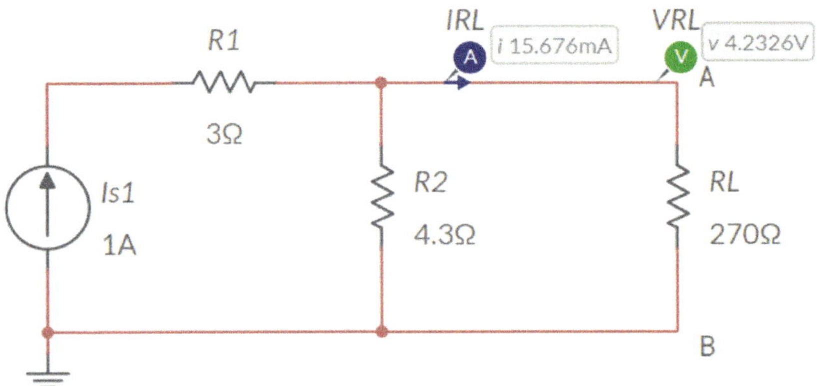

Figure 9.7 Multisim simulation providing the measured values of V_{R_L} and I_{R_L}.

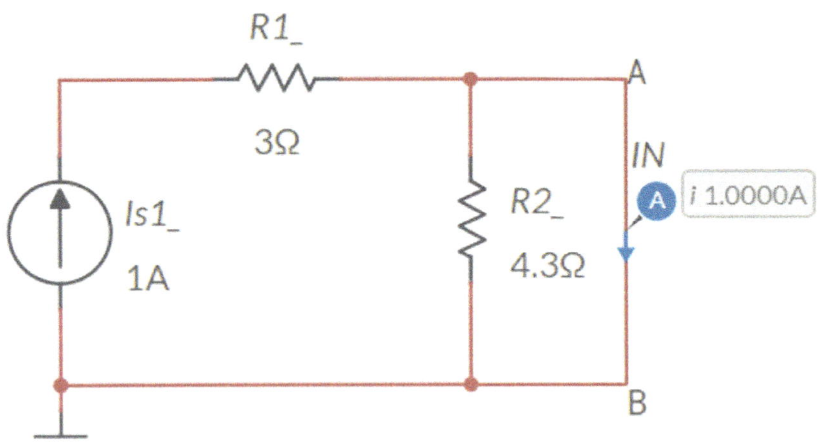

Figure 9.8 Multisim simulation providing the measured value of the Norton equivalent current I_N flowing through the short-circuited terminals A and B.

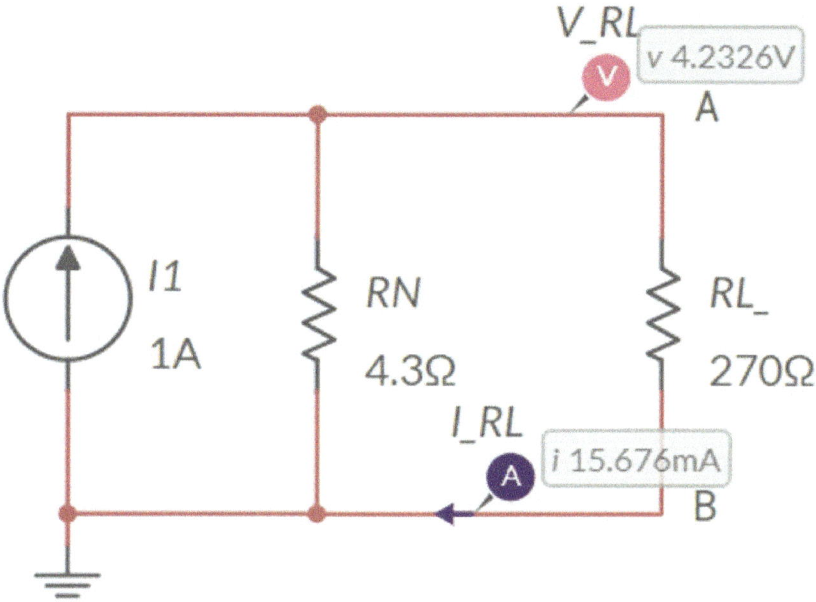

Figure 9.9 Multisim simulation of the Norton equivalent circuit providing the measured values of V_{R_L} and I_{R_L}.

2. The Norton equivalent current I_N flowing through the short-circuited terminals A and B is equal to the current of the generator, I_N= 1 A, regardless of the value of R_1. This calculated value for I_N matches the 1-A measurement from the Multisim simulation shown in Figure 9.8.
3. To determine the Norton equivalent resistance R_N, the independent current source I_{s1} is deactivated by replacing it with an open-circuit. Also, the load R_L is disconnected. This leaves only resistor R_2 to be considered between terminals A and B. Therefore, the Norton resistance is

$$R_N = R_2 = 4.3\ \Omega. \tag{9.3}$$

4. Figure 9.9 shows the implementation of the Norton equivalent circuit model in Multisim using the calculated parameters I_N = 1 A and R_N = 4.3 Ω, with the original load R_L = 270 Ω connected.

The Multisim Norton model results match the earlier measurements in the original circuit for the current $I_{R_L}$and the voltage V_{R_L}. The identical results in the original and Norton circuits confirm that the Norton model accurately reproduces the behavior of the complex source network.

9.2.3 Problem 3

Construct the two-terminal linear circuit in Figure 9.10 in Multisim. For current and voltage measurements, apply passive and active sign conventions. To analyze the circuit, perform the following tasks:

1. Use Multisim to measure the voltage V_{AB} across the terminals A and B and the Norton current I_N flowing through the short-circuited terminals A and B.
2. Calculate the Norton current I_N flowing through the short-circuited terminals A and B. Compare the value of I_N with that obtained from the Multisim simulation to ensure consistency and accuracy.
3. Calculate the Norton equivalent resistance R_N at the terminals A and B.
4. With the calculated values of I_N and R_N, simulate the Norton equivalent circuit and measure the voltage V_{AB}. Compare this measured value with the one previously calculated in the original circuit to verify the accuracy of the Norton equivalent circuit.

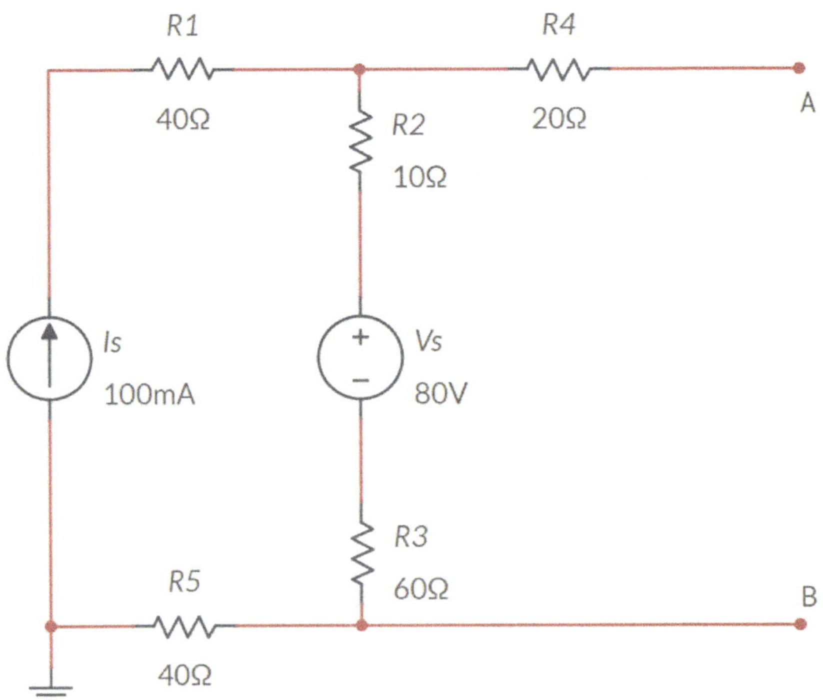

Figure 9.10 Two-terminal linear circuit with multiple sources.

9.2.3.1 Solutions to Problem 3

1. The values of the voltage V_{AB} across the terminals A and B are shown in Figure 9.11.

The Norton equivalent current I_N flowing through the short-circuited terminals A and B is measured in Figure 9.12.

2. The Norton current I_N can be determined by superposition, analyzing the contributions of the independent sources V_s and I_s individually:

When V_s is short-circuited, the Norton current I'_N flowing through the short is

$$I'_N = I_s \frac{R_2 + R_3}{R_2 + R_3 + R_4} = 77.7 \text{ mA}. \tag{9.4}$$

When I_s is open-circuited, the Norton current I''_N flowing through the short is

$$I''_N = \frac{V_s}{R_2 + R_3 + R_4} = 888.9 \text{ mA}. \tag{9.5}$$

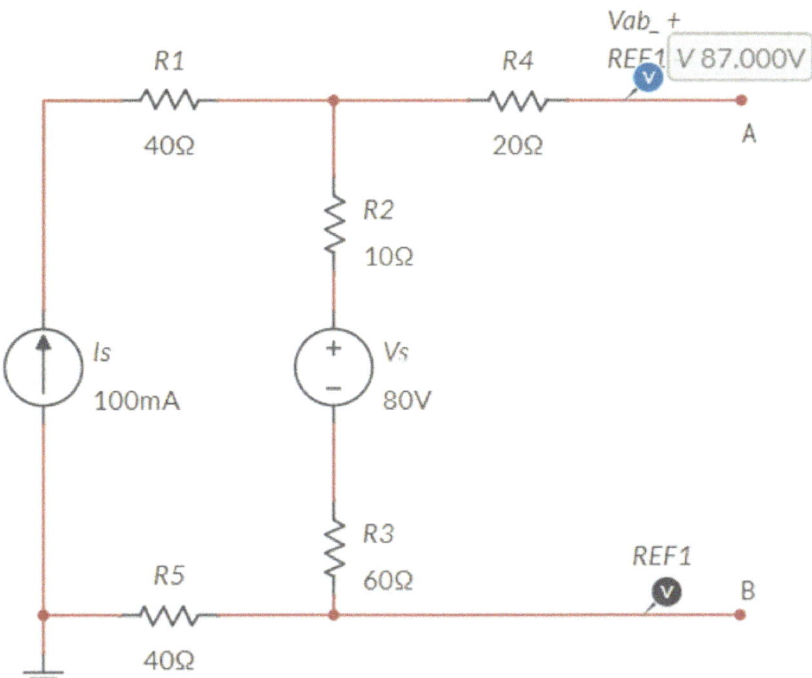

Figure 9.11 Multisim simulation of original circuit providing the measured values of V_{AB}.

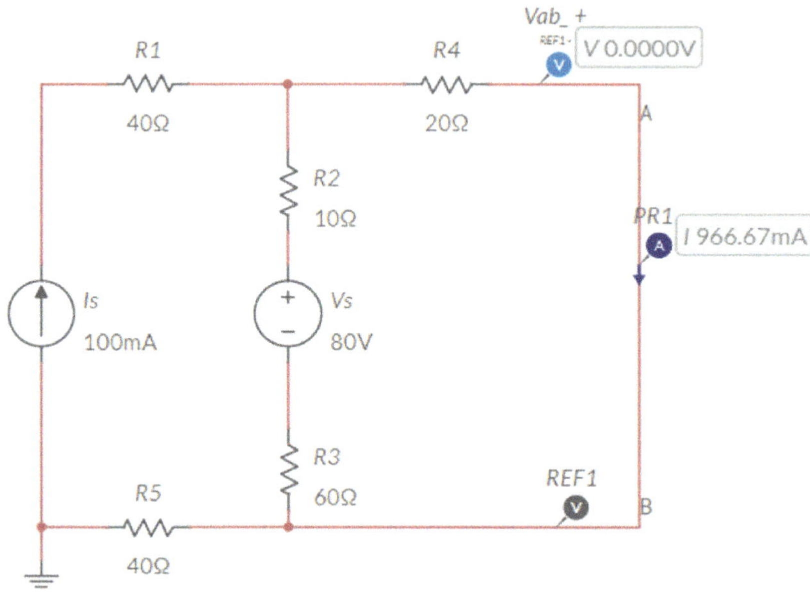

Figure 9.12 Multisim simulation providing the measured value of I_N flowing through the short-circuited terminals A and B.

By superposition, the total Norton current is

$$I_N = I'_N + I''_N = 966.6 \text{ mA}. \tag{9.6}$$

This calculated I_N matches the Multisim simulation measurement shown in Figure 9.12, confirming the analysis.

3. To determine the Norton resistance R_N, the independent sources V_s and I_s are deactivated by replacing them with a short-circuit and with an open-circuit, respectively. This leaves only the series resistor networks R_2, R_3, and R_4 between terminals A and B. The equivalent resistance of resistors in series is

$$R_N = R_2 + R_3 + R_4 = 90\ \Omega. \tag{9.7}$$

It is important to note that the Norton equivalent resistance R_N is the same as the Thevenin equivalent resistance R_{th} for this circuit. This is a result of the duality between Thevenin's and Norton's theorems, where the two equivalent circuits have the same voltage–current relationships and can be interchanged with each other.

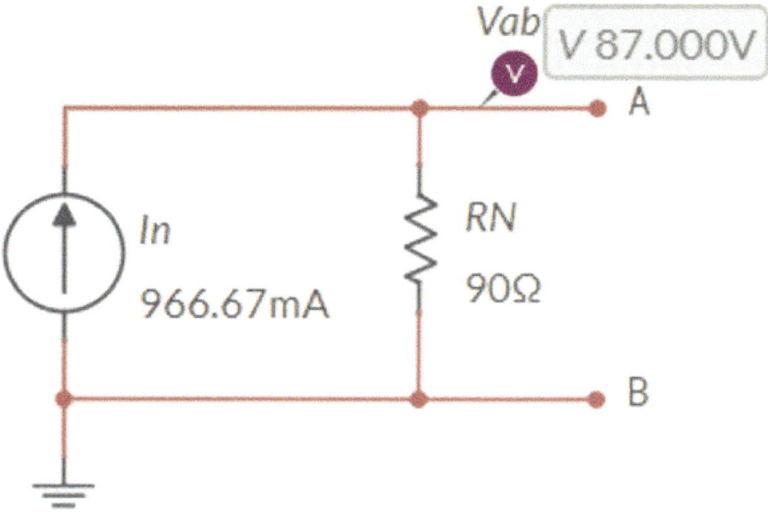

Figure 9.13 Multisim simulation of the Norton equivalent circuit providing the measured voltage V_{ab} across the terminals A and B.

1. The Norton equivalent circuit simulation in Figure 9.13, using the calculated parameters I_N and R_N, confirms that the voltage V_{ab} across the terminals A and B is the same as the one obtained in the simulation of the original circuit (Figure 9.11). This validates the accuracy of the Norton equivalent circuit and demonstrates the effectiveness of Norton's theorem in simplifying and analyzing complex linear circuits.

9.3 Lab Procedure 1

Construct the two-terminal linear circuit in Figure 9.14 in Multisim. For current and voltage measurements, apply passive and active sign conventions. To analyze the circuit, perform the following tasks:

1. In Multisim, with load R_L connected, measure the voltage V_{R_L} across the load R_L and the current I_{R_L} flowing through it. Remove the load R_L and measure the Norton current I_N flowing through the short-circuited terminals A and B.
2. Analytically calculate the Norton current I_N flowing through the short-circuited terminals A and B. Compare calculated and measured values.
3. Analytically calculate the Norton equivalent resistance R_N at the terminals A and B.

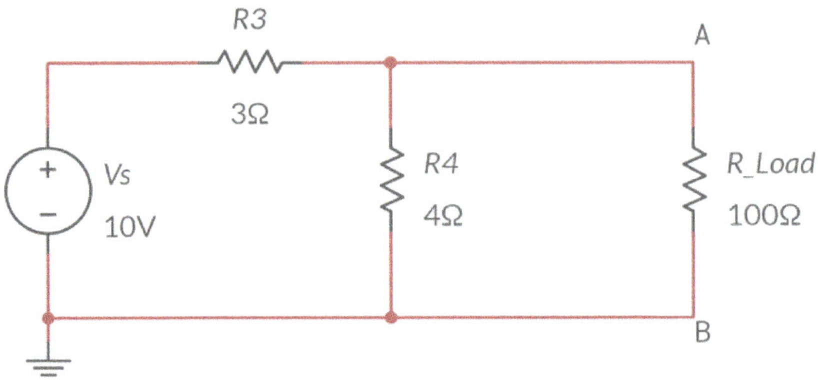

Figure 9.14 Two-terminal linear circuit with a voltage source.

9.4 Lab Procedure 2

Construct the two-terminal linear circuit in Figure 9.15 in Multisim. For current and voltage measurements, apply passive and active sign conventions. To analyze the circuit, perform the following tasks:

1. In Multisim, with the load R_L connected, measure the voltage V_{R_L} across R_L and the current I_{R_L} circulating through it. Remove the load R_L and measure the Norton equivalent current I_N flowing through the short-circuited terminals A and B.

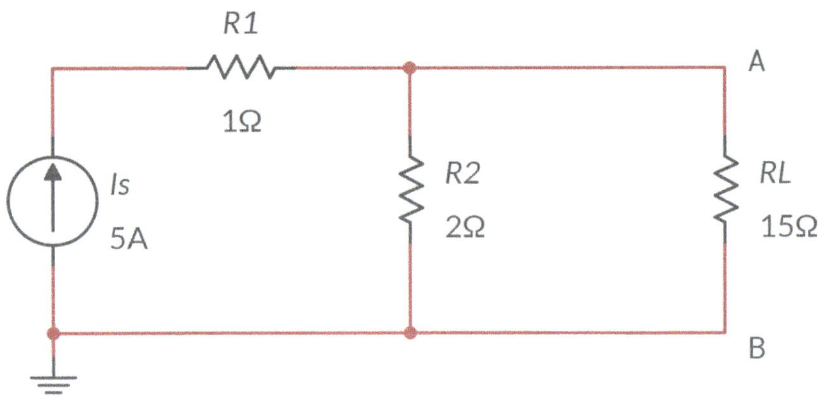

Figure 9.15 A two-terminal linear circuit with a DC current generator.

2. Analytically calculate the Norton equivalent current I_N flowing through the short-circuited terminals A and B. Compare calculated value to measured I_N from simulation.
3. Analytically calculate the Norton equivalent resistance R_N at the terminals A and B.
4. In Multisim, construct the Norton equivalent circuit using calculated I_N and R_N values. Connect original load R_L. Measure current I_{R_L} and the voltage V_{R_L} across load R_L. Compare to original circuit measurements.

10

RC Circuits in DC

10.1 Student Learning Outcomes

- Construct *RC* circuits in Multisim and simulate transient responses to DC excitation using time-domain analyses.
- Calculate values of voltages and currents at specified times during transients in *RC* circuits. In Multisim, use virtual oscilloscope to measure voltage and current values during transients.
- Simulate transient responses in *RC* circuits using a square wave generator as a time-varying voltage source instead of a simple DC voltage source.

10.2 Prelab

Capacitors are linear two-terminal components, where the linear relationship between the current $i_C(t)$ and the voltage $v_C(t)$ is expressed through a derivative, which is a linear operator. The current–voltage relationship for the capacitor is:

$$i_C(t) = C\frac{dv_C(t)}{dt}. \tag{10.1}$$

Voltage and currents will be variable with time during the *transient* state, that is, the brief time frame before voltage and current reach their final values in the *steady* state.

When a switch in the circuit is closed at $t = 0$ (Figure 10.1), the capacitor starts charging, and the voltage $v_C(t)$ across its plates increases exponentially. This charging process continues until the capacitor reaches its full voltage and at this point, the current $i_C(t)$ through the capacitor becomes zero, and the circuit is in steady-state conditions.

At $t = 0$, the moment the switch is closed, the capacitor acts as a short-circuit because it has no initial charge, and its voltage is initially zero.

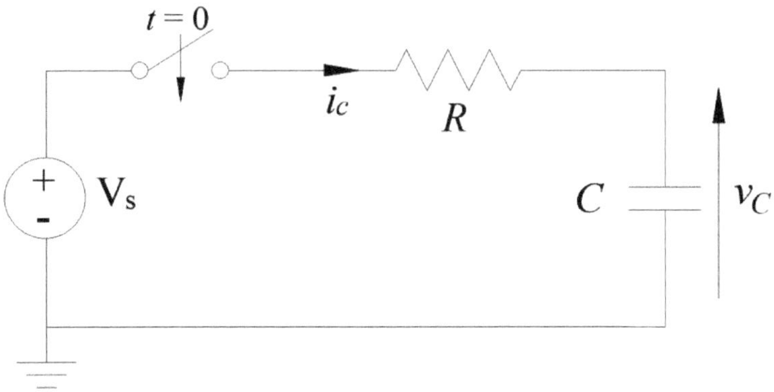

Figure 10.1 *RC* circuit with a DC source.

Therefore, the current $i_C(t)$ is limited only by the resistance *R* in the circuit. As the capacitor starts to charge, the voltage $v_C(t)$ across its plates increases, and the current $i_C(t)$ begins to decay exponentially. When the capacitor is fully charged, the voltage across its plates reaches the same values as the voltage source. At this point, the current $i_C(t)$ becomes zero, and the circuit reaches steady-state conditions. In steady state, the capacitor acts as an open-circuit because no current flows through it.

The charging time of a capacitor is theoretically infinite, but for engineering purposes, it is conventionally considered as fully charged after approximately 5 times the time constant (τ = R·C), at which the capacitor reaches approximately the 99% of its final voltage.

The voltage $v_C(t)$ across the capacitor (initially uncharged) during charging and the corresponding current $i_C(t)$ can be represented by the following formulae:

$$v_C(t) = V_s\,(1 - e^{-\frac{t}{RC}}) \tag{10.2}$$

$$i_C\,(t) = \frac{V_s}{R}\left(e^{-\frac{t}{RC}}\right). \tag{10.3}$$

A capacitor charged at the voltage V_i will also discharge following an exponential curve when connected to a resistor. If the decreasing voltage v_c ends at zero, we will have in formulae:

$$v_c\,(t) = V_i e^{-\frac{t}{RC}}, \tag{10.4}$$

where V_i is the initial voltage across the capacitor at the start of the discharge.

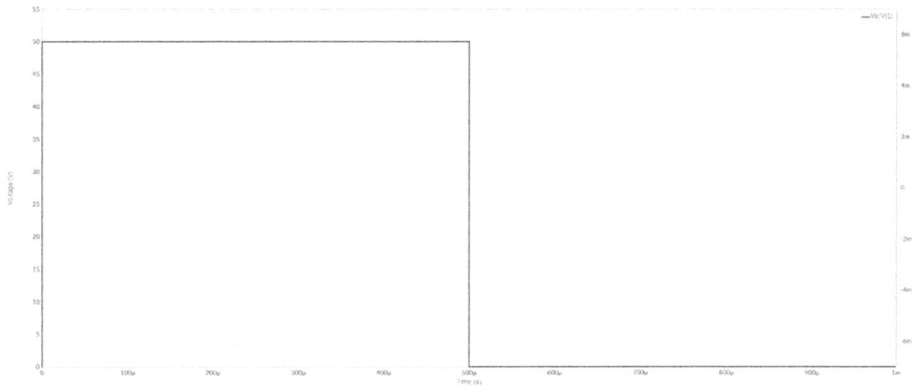

Figure 10.2 Square wave of voltage generator.

In laboratory settings, the study of *RC* circuit transients is often performed using a *square wave* voltage generator $v_s(t)$ (Figure 10.2), instead of a battery and switch.

During the *on* time of the square wave, the voltage across the capacitor increases exponentially toward the *on* value of the square wave. Conversely, when the square wave goes to zero, the voltage across the capacitor decreases exponentially. To allow substantial charge and discharge, the half-period of the square wave ($T/2$) must be greater than or equal to 5τ. Additionally, the internal resistance of the square wave generator should be negligible compared to the resistance (R) in the circuit.

10.2.1 Problem 1

Construct the *RC* circuit in Figure 10.3 in Multisim, where the capacitor is initially uncharged. For current and voltage measurements, apply passive and active sign conventions. To analyze the circuit, perform the following tasks:

1. Verify that the frequency of the square wave voltage generator allows the substantial charge and discharge of the capacitor.
2. In Multisim, plot the voltage $v_c(t)$ across capacitor and current $i_c(t)$ through it as functions of time during the charging and discharging of the capacitor.
3. Determine analytically the value of $v_c(t)$ during the charge of the capacitor for $t = 400$ µs. Compare theoretical value based on exponential equation to Multisim plot.

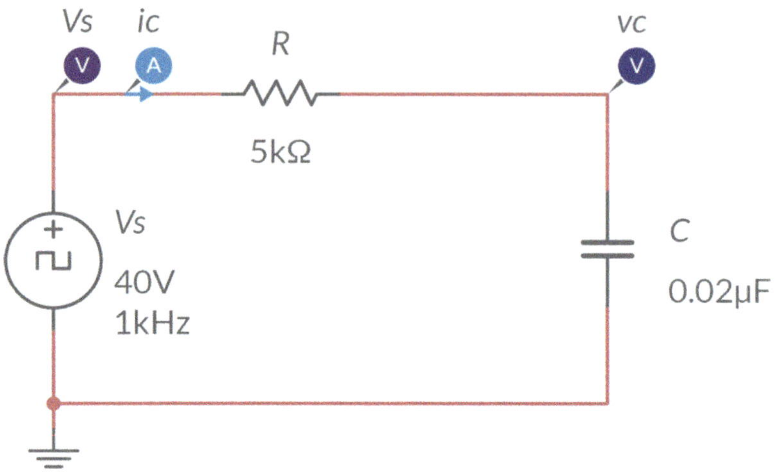

Figure 10.3 *RC* linear circuit with a square wave generator.

4. Determine analytically the value of $i_c(t)$ during the charge of the capacitor for t = 300 μs. Compare theoretical value based on exponential equation to Multisim plot.
5. Determine the value of $v_c(t)$ during the discharge of the capacitor for t = 700 μs. Compare theoretical value based on exponential equation to Multisim plot.

10.2.1.1 Solutions to Problem 1

1. For the circuit of Figure 10.2, the time constant is:

$$\tau = RC = 100\ \mu\text{s}. \tag{10.5}$$

The square wave generator has a half-period:

$$\frac{T}{2} = \frac{1}{2f} = 500\ \mu\text{s}. \tag{10.6}$$

Since the half-period of the square wave voltage generator is equal to 5τ, it allows sufficient time for the capacitor to substantially charge and discharge each half-cycle, enabling observation of the exponential transient behavior.

1. The Multisim plot of the voltage $v_c(t)$ and current $i_c(t)$ as functions of time during the charging and discharging of the capacitor is shown in Figure 10.4.

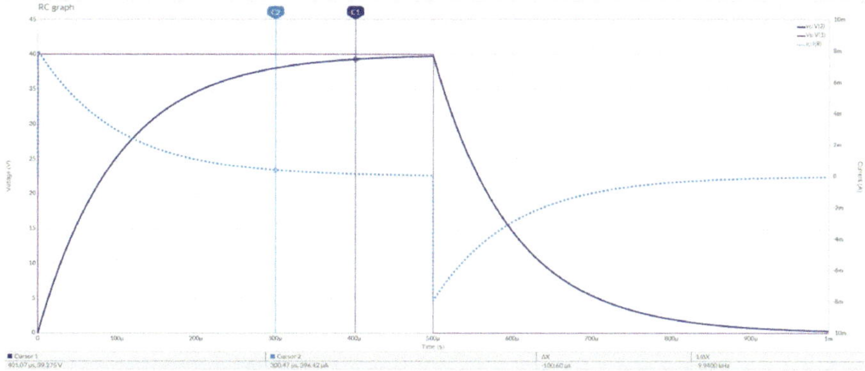

Figure 10.4 Multisim plot of $v_c(t)$, $i_c\,(t)$, and $v_s(t)$.

These transient response plots will be used to take voltage and current measurements at specific times during the charging and discharging intervals as required by this lab.

2. During the charging interval, the capacitor voltage follows the exponential relationship. The value of $v_c(t)$ for t = 400 μs is given by

$$v_c\,(400\ \mu\text{s}) = 40\ \text{V}\ \left(1 - e^{-\frac{400\ \mu\text{s}}{100\ \mu\text{s}}}\right) = 39.27\ \text{V}. \tag{10.7}$$

This calculated voltage matches the value measured at 400 μs on the Multisim plot in Figure 10.4.

3. During the charging interval, the current $i_c(t)$ through the capacitor follows the exponential relationship. The value of $i_c(t)$ for t = 300 μs is given by

$$i_c\,(300\ \mu\text{s}) = \frac{40}{5,000}\left(e^{-\frac{300\ \mu\text{s}}{100\ \mu\text{s}}}\right) = 398\ \mu\text{A}. \tag{10.8}$$

The calculated value closely matches the value measured at 300 μs on the Multisim simulation in Figure 10.4.

1. During the discharging interval, the capacitor voltage $v_c(t)$ follows the exponential decay from its initially charged level V_i. At t = 700 μs, the discharge time is 700 μs − 500 μs = 200 μs, since the capacitor finished charging to V_i = 39.73 V at 500 μs. The value of $v_c(t)$ for t = 700 μs is given by

$$v_c\,(200\ \mu\text{s}) = 39.73\ \text{V}\ e^{-\frac{200\ \mu\text{s}}{100\ \mu\text{s}}} = 5.37\ \text{V}. \tag{10.9}$$

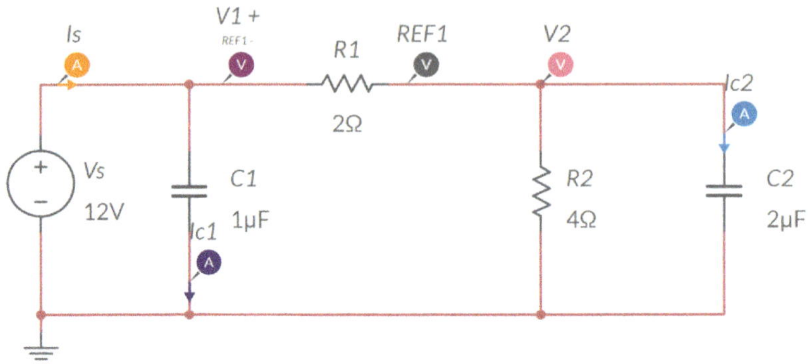

Figure 10.5 *RC* circuit in steady-state conditions.

This result aligns with the voltage measured at 700 µs in the Multisim plot during the discharge cycle (Figure 10.4).

The match verifies the proper application of the exponential discharge relationship to calculate the diminishing capacitor voltage at a given discharge time *t*, based on the initial voltage and *RC* time constant.

10.2.2 Problem 2

In Multisim, construct the resistive–capacitive network in Figure 10.5, where the steady-state conditions have been reached. For current and voltage measurements, apply passive and active sign conventions. To analyze the circuit, perform the following tasks:

1. Calculate the total current I_s supplied by the voltage source V_s and the voltage drop V_1 across the resistor R_1.
2. Confirm your calculations with a Multisim simulation.

10.2.2.1 Solutions to Problem 2

1. In steady-state conditions, both capacitors act as open-circuits, leaving just the series resistance network R_1+R_2. Applying Ohm's law, the total current I_s supplied by the voltage source V_s is:

$$I_s = \frac{V_s}{R_1 + R_2} = 2 \text{ A}. \tag{10.10}$$

Applying Ohm's law, the voltage across R_1 can be calculated as follows:

$$V_1 = I_s R_1 = 4 \text{ V}. \tag{10.11}$$

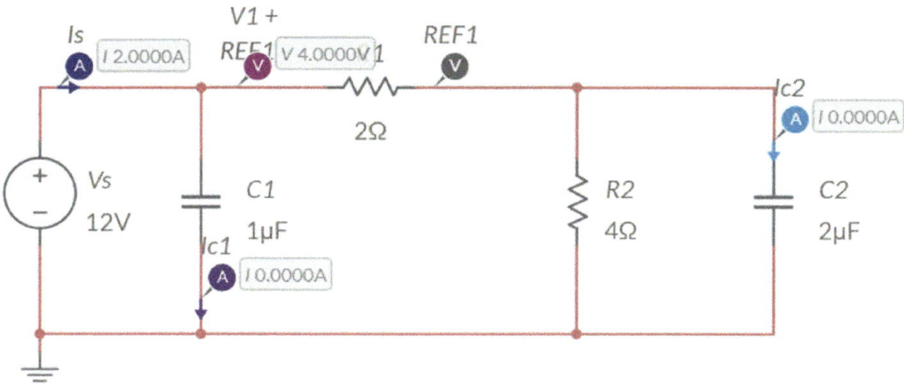

Figure 10.6 Multisim simulation providing measured values for I_s and V_1.

2. The Multisim simulation will allow quantitative verification of these calculated voltages and currents based on modeling the capacitors as open-circuits in steady state. Figure 10.6 shows the Multisim measured results verifying the calculated current I_s and voltage V_1 for the steady-state circuit model.

Additionally, simulated probes on the capacitors reveal zero current, as expected for open-circuited capacitors in steady state. The quantitative agreement between simulated measurements and calculated values validates modeling the capacitors as open-circuits to determine steady-state behavior of circuits.

10.2.3 Problem 3

In Multisim, construct the resistive–capacitive network in Figure 10.7, where the capacitor is initially uncharged. For voltage measurements, apply passive and active sign conventions. To analyze the circuit, perform the following tasks:

1. Calculate the time t_1 to charge the capacitor to $v_c(t_1) = 40$ V, based on RC time constant τ.
2. Calculate the capacitor voltage $v_c(t)$ at t_2 = 150 µs from the inception of the transient.
3. In Multisim, simulate the transient response and plot $v_c(t)$ versus time. Confirm that the measured voltages on the plot agree with calculations.

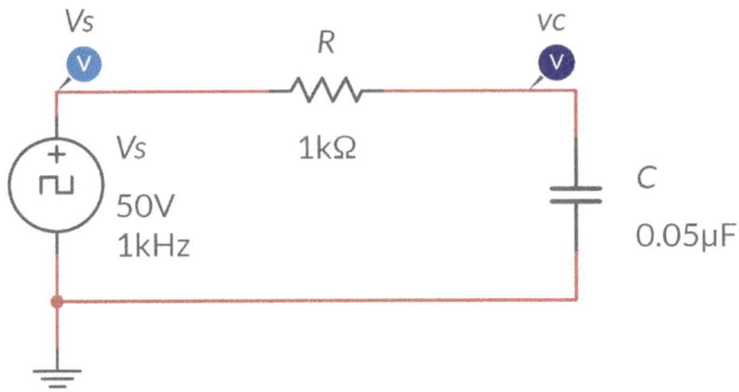

Figure 10.7 *RC* circuit with a square wave generator.

10.2.3.1 Solutions to Problem 3

1. For the *RC* circuit, the time constant τ is:

$$\tau = RC = 50\ \mu\text{s}. \tag{10.12}$$

To obtain the time to charge the capacitor to 40 V, we use the charge equation and set $v_c(t_1) = 40\ \text{V}$. We solve then for t_1:

$$t_1 = -\tau \cdot \ln\left(1 - \frac{40\ \text{V}}{V_s}\right) = 80\ \mu\text{s}. \tag{10.13}$$

2. During the capacitor charging interval, the voltage $v_c(t)$ increases exponentially. At t = 150 µs, the value of $v_c(t)$ is:

$$v_c\,(150\ \mu\text{s}) = 50\ \text{V}\left(1 - e^{-\frac{150\ \mu\text{s}}{50\ \mu\text{s}}}\right) = 47.51\ \text{V}. \tag{10.14}$$

3. The Multisim transient simulation plot of Figure 10.8 allows confirmation of the analytical results.

10.3 Lab Procedure 1

In Multisim, construct the *RC* circuit in Figure 10.9, where the capacitor is initially uncharged. For current and voltage measurements, apply passive and active sign conventions. To analyze the circuit, perform the following tasks:

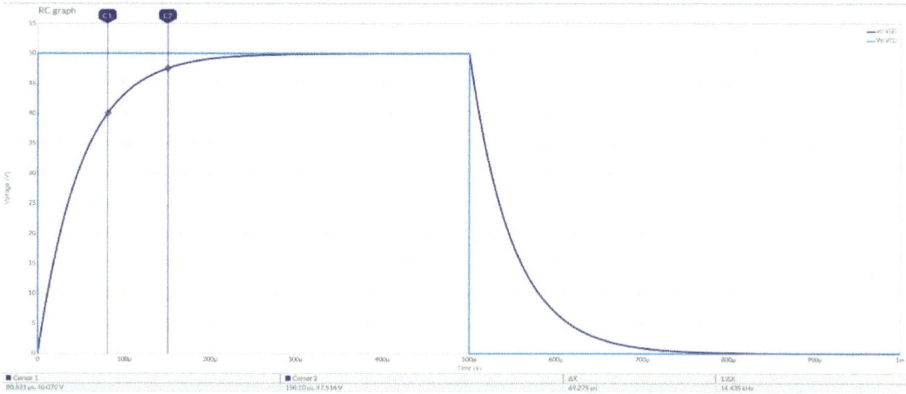

Figure 10.8 Multisim plot of $v_c(t)$ and square wave voltage generator $v_s(t)$.

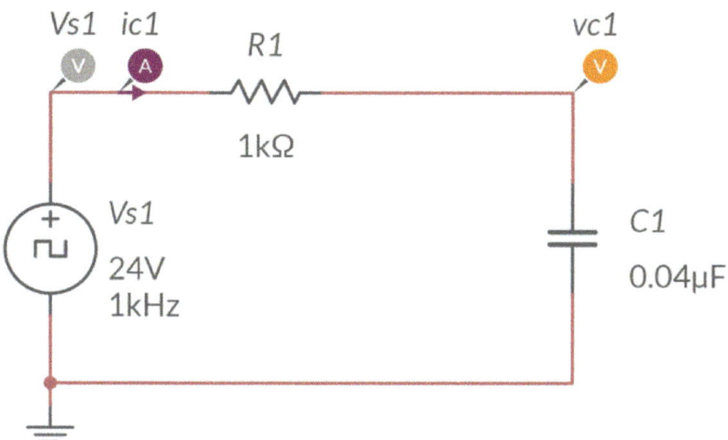

Figure 10.9 *RC* circuit with a square wave generator.

1. Verify that the frequency of the square wave voltage generator allows the substantial charge and discharge of the capacitor.
2. In Multisim, plot the voltage $v_{c1}(t)$ across capacitor and current $i_{c1}(t)$ through it as functions of time during the charging and discharging of the capacitor.
3. Determine analytically the value of $v_{c1}(t)$ during the charge of the capacitor for $t = 100$ µs. Compare theoretical value based on exponential equation to Multisim plot.

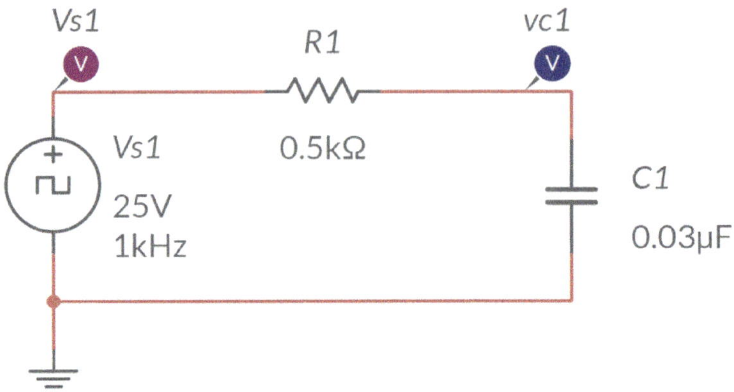

Figure 10.10 *RC* linear circuit with a square wave generator.

4. Determine analytically the value of $i_{c1}(t)$ during the charge of the capacitor for $t = 200$ μs. Compare theoretical value based on exponential equation to Multisim plot.
5. Determine the value of $v_{c1}(t)$ during the discharge of the capacitor for $t = 600$ μs. Compare theoretical value based on exponential equation to Multisim plot.

10.4 Lab Procedure 2

In Multisim, construct the resistive−capacitive network in Figure 10.10, where the capacitor is initially uncharged. For voltage measurements, apply passive and active sign conventions. To analyze the circuit, perform the following tasks:

1. Calculate the time t_1 to charge the capacitor to $v_c(t_1) = 15$ V, based on *RC* time constant τ.
2. Calculate the capacitor voltage $v_{c1}(t)$ at $t_2 = 50$ μs from the inception of the transient.
3. In Multisim, simulate the transient response and plot $v_{c1}(t)$ versus time. Confirm that the measured voltages on the plot agree with calculations.

11

RL Circuits in DC

11.1 Student Learning Outcomes

- Construct *RL* circuits in Multisim and simulate transient responses to DC excitation using time-domain analyses.
- Calculate values of voltages and currents at specified times during transients in *RL* circuits. In Multisim, use virtual oscilloscope to measure voltage and current values during transients.
- Simulate transient responses in *RL* circuits using a square wave generator as a time-varying voltage source instead of a simple DC voltage source.

11.2 Prelab

Inductors are linear two-terminal components, where the linear relationship between the voltage $v_L(t)$ and the current $i_L(t)$ is expressed through a derivative, signifying a linear operator. The current–voltage relationship for the inductor is mathematically represented as follows:

$$v_L\left(t\right) = \mathrm{L}\frac{di_L\left(t\right)}{\mathrm{dt}}. \tag{11.1}$$

Here, the constant of proportionality L is termed *inductance* and measured in Henries (H).

Even in resistive–inductive circuits supplied with a constant source (e.g., a battery), voltage and currents can vary with time during the *transient* state, which is the brief interval before reaching their final constant values in the *steady* state.

In Figure 11.1, when the switch is closed at $t = 0$, the current $i_L(t)$ through the inductor, initially zero, cannot instantaneously (i.e., in a time $dt = 0$) reach its steady-state value of $i_L = \frac{V_s}{R}$. If the current assumed a finite value in a time

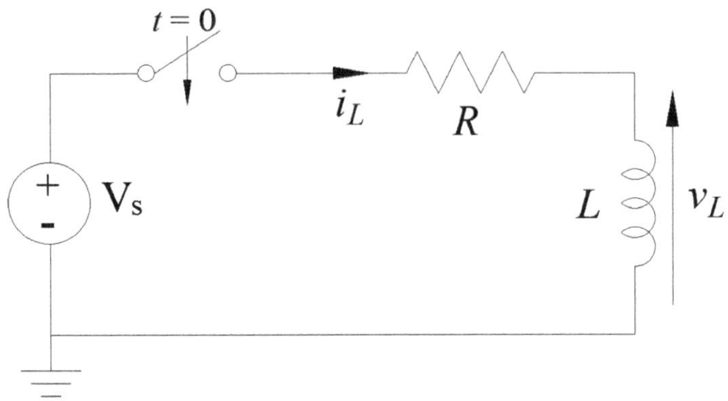

Figure 11.1 *RL* circuit with a DC source.

$dt = 0$, the inductor would store a finite value of energy $W = \frac{1}{2}Li_L^2$. However, this would imply an infinite power $P = \frac{dW}{dt=0} = \infty$, which is physically impossible. Consequently, both $i_L(t)$ and W will grow exponentially during the transient phase until they reach their steady-state values.

At $t = 0$, the inductor acts as an open-circuit, and the voltage v_L across the inductor is equal to the source voltage V_s. As the current $i_L(t)$ through the inductor increases exponentially, the voltage $v_L(t)$ across it decreases exponentially. Once the circuit reaches *steady-state* conditions, the inductor acts as a short-circuit. The charging time of an inductor is theoretically infinite. However, after approximately 5 times the time constant $\tau = L/R$, the current through the inductor reaches 99% of its final value, which is conventionally considered as the full current in practical engineering applications.

Mathematically, if the increasing exponential current in the inductor begins at zero, we express it as follows:

$$i_L(t) = \frac{V_s}{R}(1-\mathrm{e}^{-\frac{t}{\mathrm{L/R}}}). \tag{11.2}$$

If the final value of the decreasing voltage $v_L(t)$ across the inductor is zero, its representation is

$$v_L(t) = V_s \cdot e^{-\frac{t}{L/R}}. \tag{11.3}$$

Similarly, if the final value of the decreasing exponential current $i_L(t)$ through the inductor is zero, it is denoted as follows:

$$i_L\,(t) = \frac{V_s}{R}\mathrm{e}^{-\frac{t}{\mathrm{L/R}}}. \tag{11.4}$$

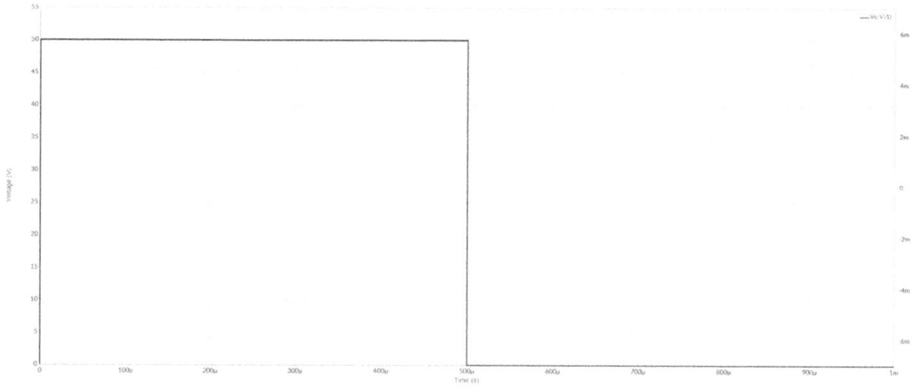

Figure 11.2 Square wave of voltage generator.

To investigate the increasing and decreasing currents and voltages in *RL* circuits, laboratories employ a square wave voltage generator $v_s(t)$, instead of a battery and switch (Figure 11.2).

During the square wave's *on*-time, the current $i_L(t)$ through the inductor increases exponentially toward its final value, which is approximately achieved in 5τ. When the square wave goes to zero, $i_L(t)$ decreases exponentially, reaching zero again in approximately 5τ.

For substantial charge and discharge of the inductor, the square wave must have a suitable half-period $\frac{T}{2}$, ensuring $\frac{T}{2} \geq 5\tau$. Additionally, the internal resistance of the square wave generator should be negligible compared to the resistance *R*.

11.2.1 Problem 1

In Multisim, construct the *RL* circuit in Figure 11.3, where the inductor's initial current is zero. For current and voltage measurements, apply passive and active sign conventions. To analyze the circuit, perform the following tasks:

1. Verify that the frequency of the square wave voltage generator allows the substantial charge and discharge of the inductor.
2. In Multisim, plot the voltage $v_L(t)$ across the inductor and current $i_L(t)$ through it as functions of time during the charging and discharging of the inductor.

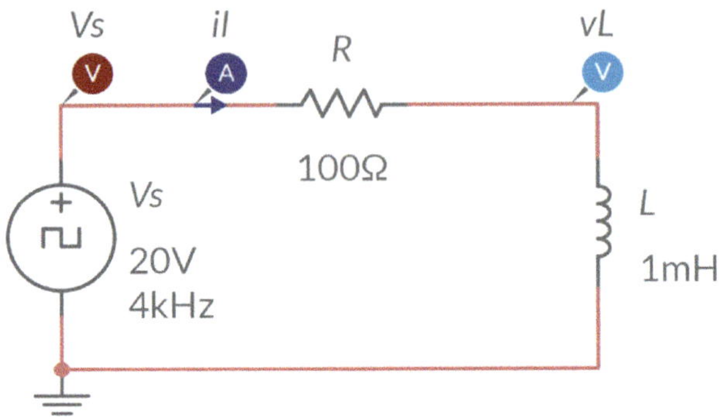

Figure 11.3 *RL* circuit with a square wave voltage generator.

3. Determine analytically the value of $i_L(t)$ during the charge of the inductor for $t = 20\mu s$. Compare the theoretical value based on the exponential equation to Multisim plot.
4. Determine analytically the value of $v_L(t)$ during the charge of the inductor for $t = 40\mu s$. Compare the theoretical value based on the exponential equation to Multisim plot.
5. Determine the value of $i_L(t)$ during the discharge of the inductor for $t = 145\mu s$. Compare the theoretical value based on the exponential equation to Multisim plot.

11.2.1.1 Solutions to Problem 1

1. For the circuit of Figure 11.3, the time constant is

$$\tau = L/R = 10\ \mu s. \tag{11.5}$$

The square wave generator has a half-period:

$$\frac{T}{2} = \frac{1}{2f} = 125\ \mu s. \tag{11.6}$$

Since the half-period of the square wave voltage generator is greater than 5τ, it allows sufficient time for the inductor to substantially charge and discharge each half-cycle, enabling observation of the exponential transient behavior.

2. The Multisim plot of the voltage $v_L(t)$ and current $i_L(t)$ as functions of time during the charging and discharging of the inductor is shown in Figure 11.4.

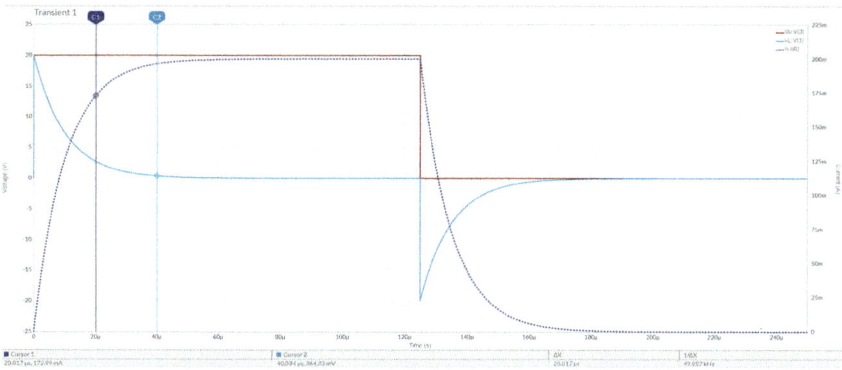

Figure 11.4 Multisim plot of $v_L(t)$, $i_L(t)$, and $v_s(t)$.

From the plot, at t = 0, when the square wave first turns on, the initial inductor current is $i_L(0)$ = 0 since the inductor starts uncharged. Per the governing equation, this causes the initial voltage $v_L(t)$ to equal the full applied voltage $v_s(t)$, as the inductor acts momentarily like an open-circuit.

As current starts flowing, $i_L(t)$ rises exponentially toward its final value, while $v_L(t)$ exponentially decays down to zero. After around 5 time constants, when steady state is reached, $i_L(t)$ settles at v_s/R and the inductor is modeled as a short-circuit.

When the square wave input switches to 0 V, the steady-state current $i_L(t)$ cannot instantly go to 0 A. Instead, $v_L(t)$ reverses polarity to counter the collapsing magnetic field, exhibiting a negative exponential that forces $i_L(t)$ down to zero over multiple time constants.

The transient response plots of Figure 11.4 will be used to take volt age and current measurements at specific times during the charging and discharging intervals, as required by this lab.

3. During the charging interval, the inductor current $i_L(t)$ follows the exponential relationship. The value for $i_L(t)$ for t = 20µs is given by

$$i_L\,(20\ \mu\text{s}) = \frac{20\ V}{100\ \Omega}\left(1\text{–}\mathrm{e}^{-\frac{20\ \mu\text{s}}{10\ \mu\text{s}}}\right) = 172.9\ \text{mA}. \qquad (11.7)$$

This calculated current matches the value measured at 200 µs on the Multisim plot in Figure 11.4.

4. Similarly, the decreasing voltage $v_L(t)$ for t = 40μs is calculated as follows:

$$v_L\left(40\ \mu\text{s}\right) = 20\ \text{V}\left(e^{-\frac{40\ \mu\text{s}}{10\ \mu\text{s}}}\right) = 366\ \text{mV}. \tag{11.8}$$

This calculated value aligns with the corresponding measurements from the Multisim simulation in Figure 11.4, further validating the accuracy of the analysis.

5. During the inductor discharge interval, the current $i_L(t)$ follows an exponential decay from its initial value of 200 mA at t = 125 μs. The discharge time is 145 − 125 μs = 20 μs (Figure 11.5.)

The decreasing current $i_L(t)$ for t = 20μs is calculated as follows:

$$i_L\left(20\ \mu\text{s}\right) = 200\ \text{mA e}^{-\frac{20\ \mu\text{s}}{10\ \mu\text{s}}} = 27\ \text{mA}. \tag{11.9}$$

The above calculated value closely matches the Multisim simulation of Figure 11.4. This result matches the current measured at 145 μs during the discharge cycle in the Multisim plot of Figure 11.5. The agreement verifies the proper application of the exponential discharge relationship to calculate the diminishing inductor current based on the initial value and time constant.

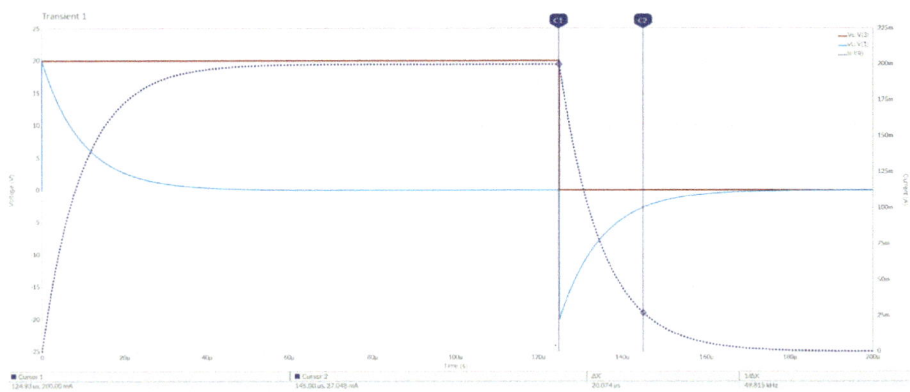

Figure 11.5 Multisim plot of $v_L(t)$, $i_L(t)$, and $v_s(t)$.

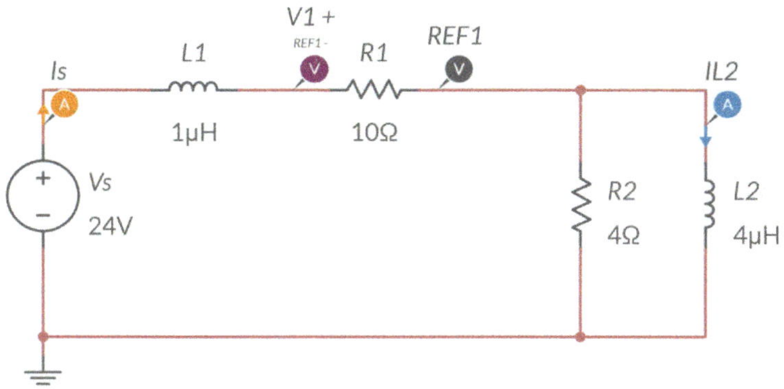

Figure 11.6 *RL* circuit in steady-state conditions.

11.2.2 Problem 2

In Multisim, construct the resistive–inductive network in Figure 11.6, where the steady-state conditions have been reached. For current and voltage measurements, apply passive and active sign conventions. To analyze the circuit, perform the following tasks:

1. Calculate the total current I_s supplied by the voltage source V_s and the voltage drop V_1 across the resistor R_1.
2. Confirm your calculations with a Multisim simulation.

11.2.2.1 Solutions to Problem 2

1. In steady-state conditions, the inductors act as short-circuits. In steady-state conditions, both inductors act as short-circuits. R_2 is therefore shortcircuited, leaving just the resistance R_1. Applying Ohm's law, the total current I_s supplied by the voltage source V_s is

$$I_s = \frac{V_s}{R_1} = 2.4 \text{ A}. \tag{11.10}$$

Applying Ohm's law, the voltage across R_1 can be calculated as follows:

$$V_1 = I_s R_1 = 24 \text{ V}. \tag{11.11}$$

2. The Multisim simulation will allow quantitative verification of these calculated voltages and currents based on modeling the inductors as short-circuits in steady state. Figure 11.7 shows the Multisim measured results verifying the calculated current I_s and voltage V_1 for the steady-state circuit model.

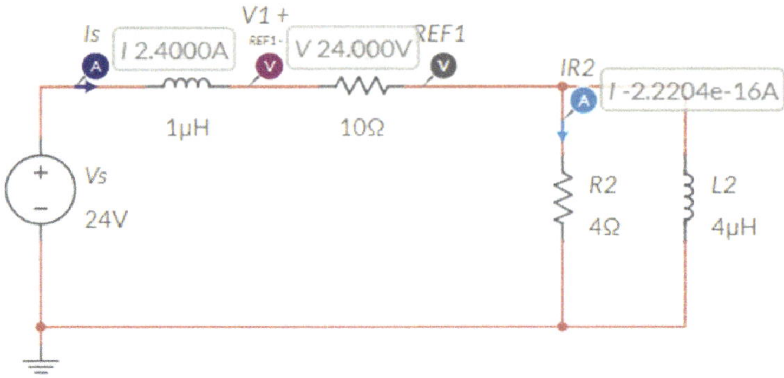

Figure 11.7 Multisim simulation confirming calculated values for I_s and V_1.

Additionally, simulated probes in the inductor's branch reveal zero current, as expected for short-circuited inductor in steady state. The quantitative agreement between simulated measurements and calculated values validates modeling the inductors as short-circuits to determine the steady-state behavior of circuits.

11.2.3 Problem 3

In Multisim, construct the resistive–inductive network in Figure 11.8, where the inductor is initially uncharged. For voltage measurements, apply passive

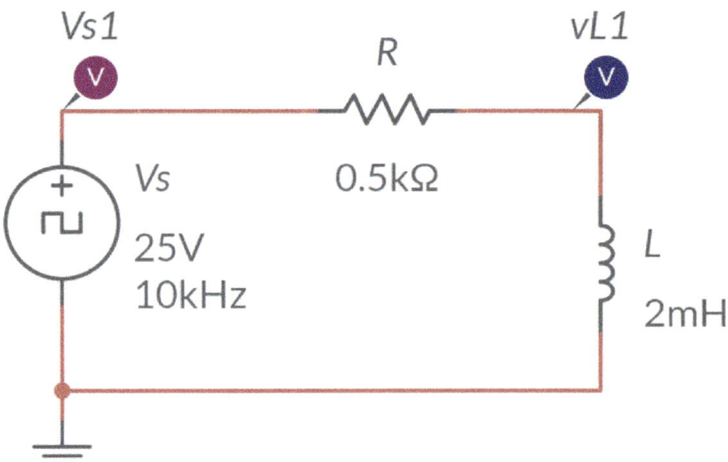

Figure 11.8 *RL* linear circuit with a square wave generator.

and active sign conventions. To analyze the circuit, perform the following tasks:

1. Calculate the time t_1 to charge the inductor to $i_L(t_1) = 31$ mA, based on the time constant τ.
2. Calculate the inductor voltage $v_L(t_2)$ at $t_2 = 8\mu s$ from the inception of the transient.
3. In Multisim, simulate the transient response and plot $i_L(t)$ and $v_L(t)$ versus time. Confirm that the measured voltage and current on the plot agree with calculations.

11.2.3.1 Solutions to Problem 3

1. For the *RL* circuit, the time constant is

$$\tau = L/R = 4\ \mu\text{s}. \tag{11.12}$$

To obtain the time to charge the inductor to 31 mA, we use the charge equation and set $i_L(t_1) = 31$ mA. We then solve for t_1:

$$t_1 = -\tau \cdot \ln\left(1 - 0.031\ \frac{V_s}{R}\right) = 3.8\ \mu\text{s}. \tag{11.13}$$

2. During the inductor charging interval, the voltage $v_L(t)$ decreases exponentially. At $t = 8\mu s$, the value of $v_L(t)$ after the inception of the transient is given by

$$v_L\ (8\ \mu\text{s}) = 25\ \text{V}\ e^{-\frac{8\ \mu\text{s}}{4\ \mu\text{s}}} = 3.38\ \text{V}. \tag{11.14}$$

3. The Multisim transient simulation plot of Figure 11.9 allows the confirmation of the analytical results.

11.3 Lab Procedure 1

In Multisim, construct the *RL* circuit in Figure 11.10, where the inductor's initial current is zero. For current and voltage measurements, apply passive and active sign conventions. To analyze the circuit, perform the following tasks:

1. Verify that the frequency of the square wave voltage generator allows the substantial charge and discharge of the inductor.

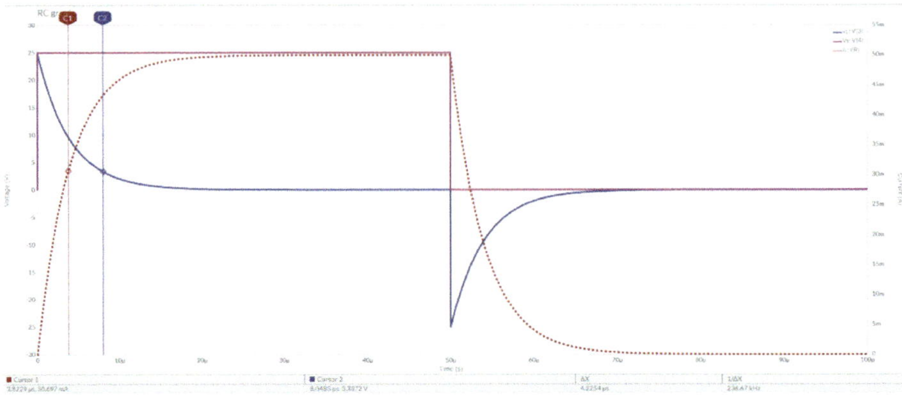

Figure 11.9 Plot of $i_L\ (t)$, $v_L(t)$, and square wave voltage generator $v_s(t)$.

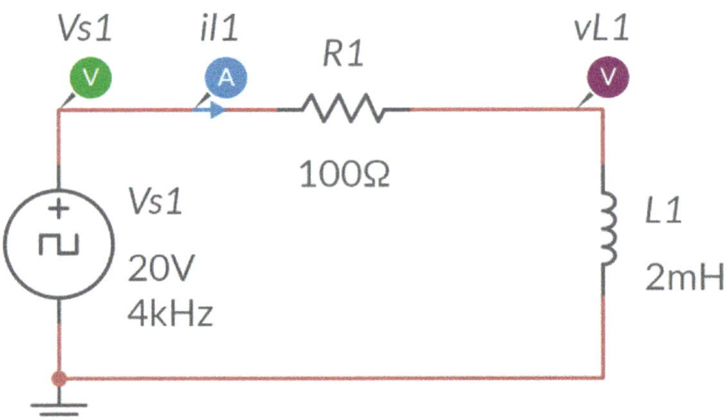

Figure 11.10 *RL* circuit with a square wave voltage generator.

2. In Multisim, plot the voltage $v_L(t)$ across the inductor and current $i_L(t)$ through it as functions of time during the charging and discharging of the inductor.
3. Determine analytically the value of $i_L(t)$ during the charge of the inductor for t = 60μs. Compare the theoretical value based on the exponential equation to Multisim plot.
4. Determine analytically the value of $v_L(t)$ during the charge of the inductor for t = 20μs. Compare the theoretical value based on the exponential equation to Multisim plot.

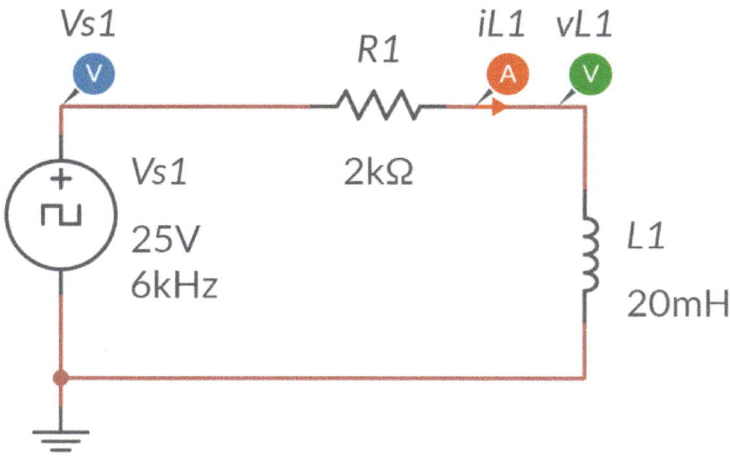

Figure 11.11 *RL* circuit with a square wave voltage generator.

5. Determine the value of $i_L(t)$ during the discharge of the inductor for t = 140μs. Compare the theoretical value based on the exponential equation to Multisim plot.

11.4 Lab Procedure 2

In Multisim, construct the resistive–inductive network in Figure 11.11, where the inductor is initially uncharged. For voltage measurements, apply passive and active sign conventions. To analyze the circuit, perform the following tasks:

1. Calculate the time t_1 to charge the inductor to $i_L(t_1) = 7.9$ mA, based on the time constant τ.
2. Calculate the inductor voltage $v_L(t_2)$ at t_2 = 15μs from the inception of the transient.
3. In Multisim, simulate the transient response and plot $i_L(t)$ and $v_L(t)$ versus time. Confirm that the measured voltage and current on the plot agree with calculations.

12

Wheatstone Bridge

12.1 Student Learning Outcomes

- Construct Wheatstone bridge circuits in Multisim using combinations of known and unknown resistors and use virtual voltmeters to detect balanced and unbalanced conditions.
- Calculate unknown resistance based on ratios of known resistances that achieve voltage balance.

12.2 Prelab

The Wheatstone bridge is a widely used circuit for measuring the value of an unknown resistance, R_x, as shown in Figure 12.1. The bridge, excited by a voltage source V_s, is considered "balanced" when the potential difference across nodes A and B is zero, meaning $V_A = V_B$. This *null* condition allows

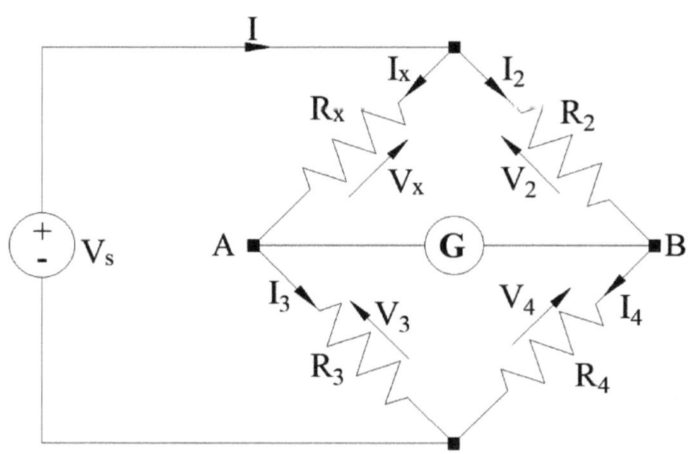

Figure 12.1 Wheatstone bridge.

for precise resistance measurements, with an error of less than 0.1%. The galvanometer, denoted as G, serves as a very sensitive ammeter with a zero-midscale point, capable of detecting currents in either direction. When the circulating current in the galvanometer is zero, it signifies the condition $V_A = V_B$, leading to $I_x = I_3$ and $I_2 = I_4$.

To analyze the balanced Wheatstone bridge, we apply the passive sign convention and apply Kirchhoff's voltage law to the top and bottom portions of the circuit, yielding the following equations:

$$R_x I_x = R_2 I_2 \tag{12.1}$$

$$R_4 I_4 = R_3 I_3. \tag{12.2}$$

At balance, $I_x = I_3$ and $I_2 = I_4$, and this gives the unknown resistance R_x in terms of the other resistors in the bridge:

$$R_x = \frac{R_2 I_2}{I_x} = \frac{R_2 I_4}{I_3} = R_3 \frac{R_2}{R_4}. \tag{12.3}$$

Typically, R_2 and R_4 are fixed resistors, and their ratio remains constant. As a result, R_3 serves as a variable resistor that can be adjusted until the galvanometer current becomes zero, providing an accurate measurement of the unknown resistance R_x.

The Wheatstone bridge allows determining unknown resistors precisely by balancing ratios in a null-detection test circuit.

12.2.1 Problem 1

In Multisim, construct the balanced Wheatstone bridge in Figure 12.2. For current and voltage measurements, apply passive and active sign conventions. To analyze the circuit, perform the following tasks:

1. Analyze the balanced Wheatstone bridge in Figure 12.2 to calculate the unknown resistance R_x using the balance ratio formula. Construct the circuit in Multisim and verify that the bridge is in *null* voltage condition.
2. With the bridge balanced, calculate the total current I_t supplied by the voltage source V_s. Confirm results with Multisim measurement.
3. With the bridge balanced, calculate analytically currents I_x, I_2, I_3, and I_4. In Multisim, measure each branch current and verify agreement with calculated values.

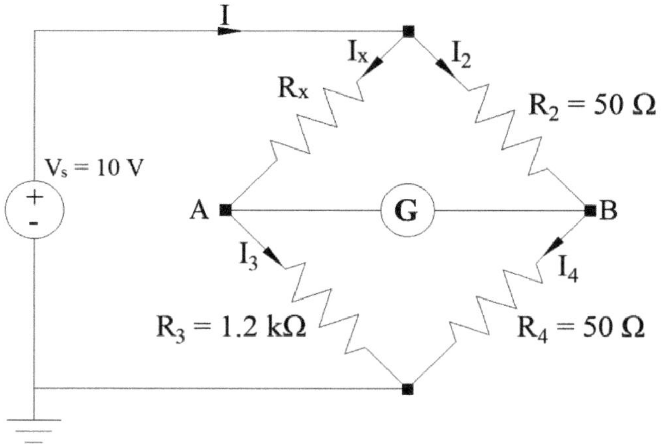

Figure 12.2 Wheatstone bridge circuit.

12.2.1.1 Solutions to Problem 1

1. For the circuit in Figure 12.2, the value of R_x can be calculated by applying the balance ratio formula:

$$R_x = R_3 \frac{R_2}{R_4} = 1.2\ \text{k}\Omega. \tag{12.4}$$

This calculated value is confirmed by the Multisim simulation in Figure 12.3, where V_A and V_B are both measured to be 5 V, indicating that the bridge is in a *null* condition, validating the calculated value for R_x.

2. With the bridge balanced, the total resistance R_P seen by the source can be determined using the following formula:

$$R_P = \frac{(R_x + R_3)(R_2 + R_4)}{R_2 + R_3 + R_4 + R_x} = 96\ \Omega. \tag{12.5}$$

The total current I_t can then be calculated using Ohm's law:

$$I_t = \frac{V_s}{R_P} = 104.17\ \text{mA}. \tag{12.6}$$

This value is confirmed by the Multisim simulation in Figure 12.3.

3. With the bridge balanced, the currents I_x, I_2, I_3, and I_4 are given by

$$I_x = I_3 = \frac{V_s}{R_x + R_3} = 4.16\ \text{mA} \tag{12.7}$$

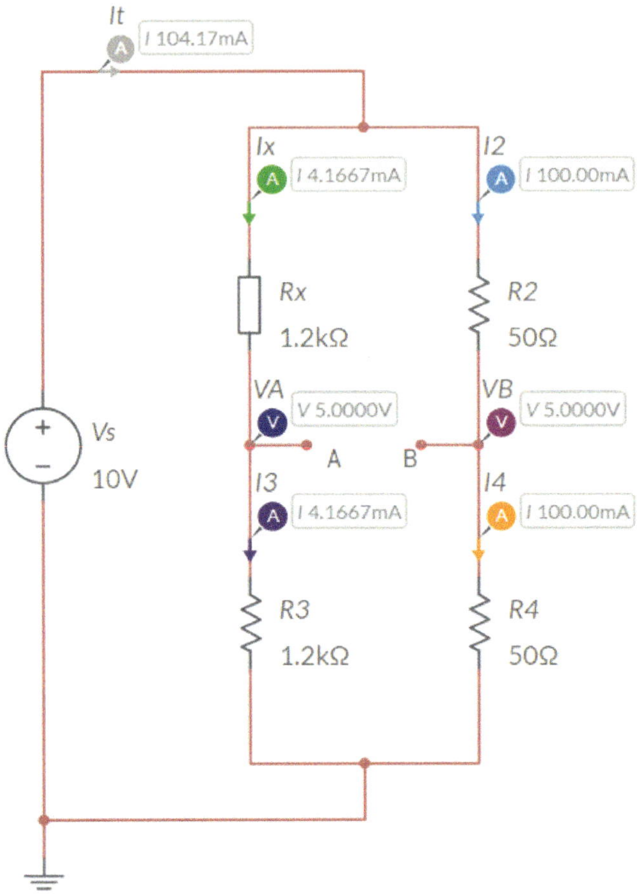

Figure 12.3 Multisim simulation providing measured values in balanced conditions.

$$I_2 = I_4 = \frac{V_s}{R_2 + R_4} = 100 \text{ mA}. \tag{12.8}$$

These calculated values are in agreement with the results of the Multisim simulation in Figure 12.3. The simulation further validates that in *null* condition, $I_x = I_3$ and $I_2 = I_4$.

12.2.2 Problem 2

In Multisim, construct the balanced Wheatstone bridge in Figure 12.4. For current and voltage measurements, apply passive and active sign conventions. To analyze the circuit, perform the following tasks:

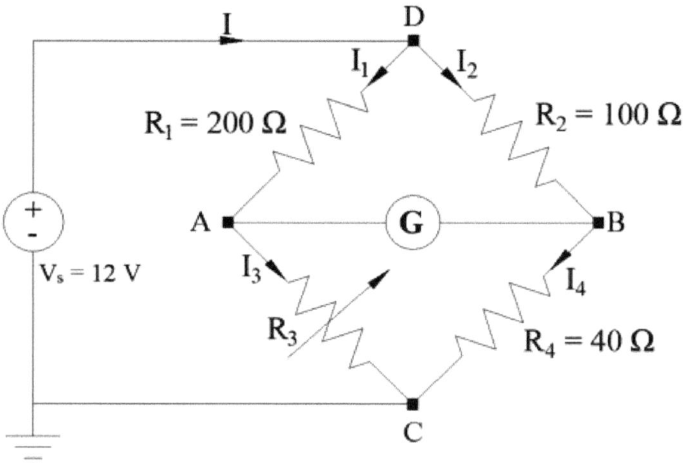

Figure 12.4 Wheatstone bridge circuit.

1. Calculate the R_3 value that gives a balanced *null* voltage V_{AB}, using the resistor ratio formula. In Multisim, verify that the calculated R_3 makes V_{AB} = 0 V, indicating balance.
2. With the bridge balanced, calculate the voltages V_{DA}, V_{DB}, V_{AC}, and V_{BC}. Measure these voltages in Multisim using voltage probes to confirm calculations.
3. With the bridge balanced, calculate the currents I_1, I_2, I_3, and I_4. Measure these currents in Multisim using current probes to confirm calculations.

12.2.2.1 Solutions to Problem 2

1. Applying the balance ratio formula gives the value of the resistor R_3 that balances the bridge:

$$R_3 = \frac{R_1 R_4}{R_2} = 80\ \Omega. \tag{12.9}$$

Figure 12.5 shows the Multisim simulation confirming that with R_3 = 80 Ω, the voltage V_{AB} = 0 V indicating balance. The calculation and Multisim confirm that the variable resistor R_3 should have a value of 80 Ω to balance the bridge.

2. With the bridge balanced, the voltages across the resistors are calculated using the voltage divider formula:

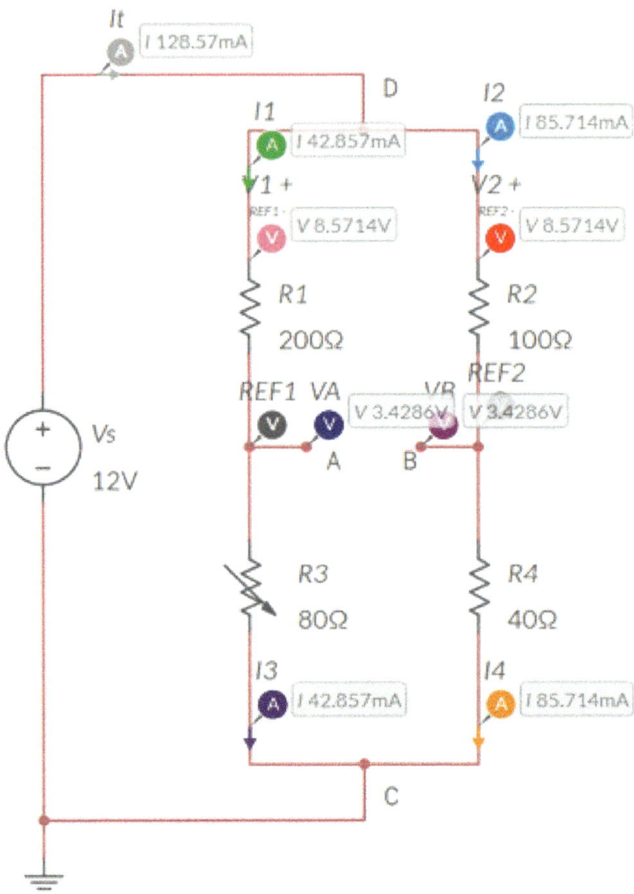

Figure 12.5 Multisim simulation providing measured values of voltages and currents.

$$V_{\mathrm{DA}} = V_s \frac{R_1}{R_1 + R_3} = 8.57\ \mathrm{V} \tag{12.10}$$

$$V_{\mathrm{DB}} = V_s \frac{R_2}{R_2 + R_4} = 8.57\ \mathrm{V} \tag{12.11}$$

$$V_{\mathrm{AC}} = V_s \frac{R_3}{R_1 + R_3} = 3.42\ \mathrm{V} \tag{12.12}$$

$$V_{\mathrm{BC}} = V_s \frac{R_4}{R_2 + R_4} = 3.42\ \mathrm{V}. \tag{12.13}$$

The voltages are equal in pairs because of the *null* condition of the bridge (i.e., V_{AB} = 0). For further confirmation, we apply Kirchhoff's voltage law to the top and bottom portion of the bridge:

$$V_{\text{AB}} + V_{\text{DA}} - V_{\text{DB}} = 0 \Rightarrow V_{\text{DA}} = V_{\text{DB}} \tag{12.14}$$

$$V_{\text{AB}} - V_{\text{AC}} + V_{\text{BC}} = 0 \Rightarrow V_{\text{BC}} = V_{\text{AC}}. \tag{12.15}$$

Multisim measurements in Figure 12.5 validate the calculated voltages, verifying the analysis.

3. With the Wheatstone bridge balanced, the currents I_1, I_2, I_3, and I_4 through the resistors are calculated using Ohm's law:

$$I_1 = \frac{V_{\text{DA}}}{R_1} = 42.85 \text{ mA} \tag{12.16}$$

$$I_2 = \frac{V_{\text{DB}}}{R_2} = 85.70 \text{ mA} \tag{12.17}$$

$$I_3 = \frac{V_{\text{AC}}}{R_3} = 42.75 \text{ mA} \tag{12.18}$$

$$I_4 = \frac{V_{\text{BC}}}{R_4} = 85.50 \text{ mA}. \tag{12.19}$$

Multisim measurements in Figure 12.5 validate these calculated currents.

12.3 Lab Procedure 1

In Multisim, construct the balanced Wheatstone bridge in Figure 12.6. For current and voltage measurements, apply passive and active sign conventions. To analyze the circuit, perform the following tasks:

1. Analyze the balanced Wheatstone bridge in Figure 12.2 to calculate the unknown resistance R_x using the balance ratio formula. Construct the circuit in Multisim and verify that the bridge is in *null* voltage condition.
2. With the Wheatstone bridge balanced, calculate the total current I_t supplied by the voltage source V_s. Confirm results with Multisim measurement.
3. With the Wheatstone bridge balanced, calculate analytically currents I_x, I_2, I_3, and I_4. In Multisim, measure each branch current and verify agreement with calculated values.

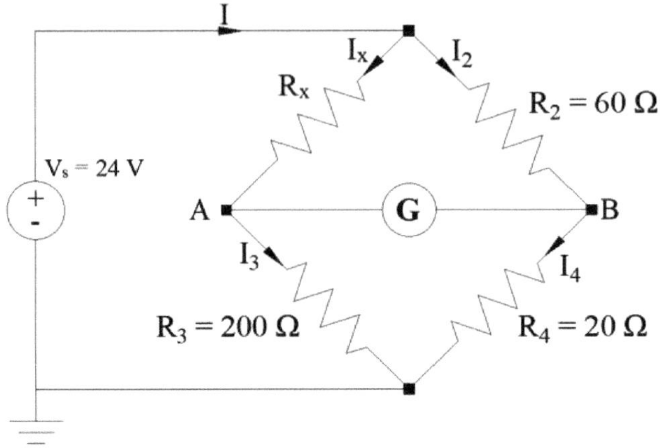

Figure 12.6 Wheatstone bridge circuit.

Part II

Operational Amplifier Circuits

13

Amplifier Fundamentals

13.1 Student Learning Outcomes

- Identify the components and characteristics of voltage and current amplifiers, including input and output resistances, gain factors, and loading effects.
- Apply voltage and current divider formulae to calculate outputs and gains for given voltage and current amplifier configurations.
- Utilize Multisim simulations to confirm and validate calculations of voltage and current amplifiers, as well as input and output loading effects.

13.2 Prelab

13.2.1 Voltage amplifier

The voltage amplifier shown in Figure 13.1 is a linear four-terminal device that exhibits the property of amplifying an input voltage and producing an output voltage equal to the product of the input voltage and a constant *gain* factor, denoted as A_v, expressed in units of V/V (i.e., volt per volt).

The amplifier input can be modeled as a Thevenin equivalent driving an input resistance R_i (Figure 13.1). On the other hand, the output port is represented by a *voltage-controlled voltage source* (VCVS) with its value,

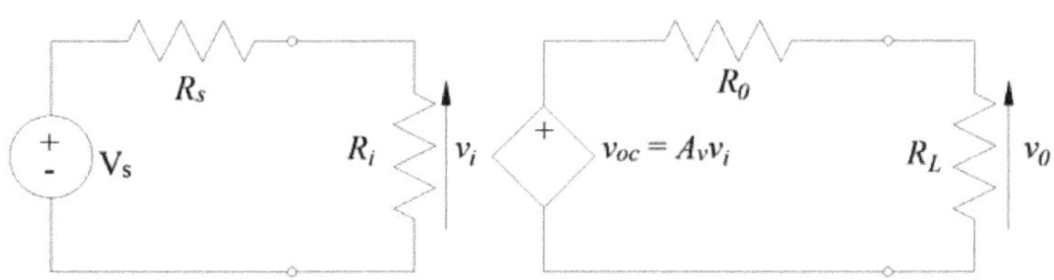

Figure 13.1 Voltage amplifier circuit.

v_{oc}, dependent on the input voltage v_i through the gain factor A_v. This voltage source is connected in series with an output resistance R_0.

The output voltage v_o can be derived using the voltage divider formula as follows:

$$v_o = A_v v_i \frac{R_L}{R_o + R_L}. \tag{13.1}$$

This expression demonstrates that the voltage amplifier imposes a reduction on the output voltage, and the ideal case would feature $R_o = 0$, resulting in an unattenuated output voltage of $v_o = A_v v_i$. Additionally, in the absence of a load (i.e., $R_L = \infty$, or open-circuited output circuit), the output voltage remains unattenuated.

Similarly, the input voltage v_i as a function of the source voltage v_s can be obtained using the voltage divider formula:

$$v_i = V_s \frac{R_i}{R_s + R_i}. \tag{13.2}$$

Here, the voltage division reduces the source voltage V_s applicable to the input port of the voltage amplifier. In the ideal scenario, R_i would be infinite, resulting in an unattenuated input voltage $V_s = v_i$.

The above voltage attenuations, both in input and output, are referred to as *loading*. The loading can be expressed as a percentage:

$$\text{Input loading } \% = \frac{v_i}{V_s} 100 \tag{13.3}$$

$$\text{Output loading } \% = \frac{v_0}{v_{oc}} 100. \tag{13.4}$$

In practical applications, some loading is inevitable but can be minimized by ensuring that $R_i >> R_s$ and $R_o << R_L$.

The overall *source-to-load* voltage gain $A_v T$ is given by

$$A_v T = \frac{v_o}{V_s} = A_v \frac{R_i}{R_s + R_i} \frac{R_L}{R_o + R_L}. \tag{13.5}$$

13.2.2 Current amplifier

The current amplifier depicted in Figure 13.2 is another four-terminal device that receives an input current and amplifies it by a constant gain factor, A_I (in units of A/A), returning an output current equal to the product of the input current and A_I. In this case, both the input source and the amplifier can be modeled as Norton equivalents.

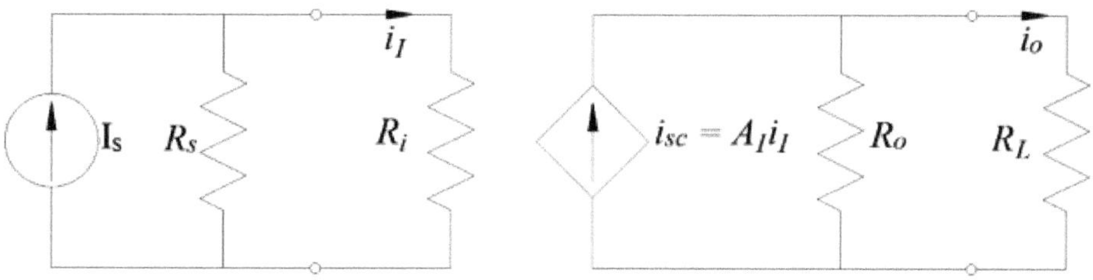

Figure 13.2 Current amplifier circuit.

The input port of the current amplifier is represented by an input resistance R_i (Figure 13.2). On the other hand, the output port is characterized by a *current-controlled current source* (CCCS) whose value depends on the input current i_I multiplied by the constant A_I. This CCCS is connected in parallel with an output resistance R_0.

The expression for i_o can be derived using the current divider formula as follows:

$$i_o = A_I i_I \frac{R_0}{R_o + R_L}. \tag{13.6}$$

This expression illustrates that the current amplifier reduces the output current i_o. In the ideal case, R_o would be infinite, resulting in an unattenuated output current, $i_o = A_I i_I$. Similarly, with a short-circuited output circuit, $R_L = 0$, the output current remains unattenuated.

The input current i_I can be calculated using the current divider formula as follows:

$$i_I = I_s \frac{R_s}{R_s + R_i}. \tag{13.7}$$

This formula demonstrates that the current division reduces the source current I_s flowing through the input port of the current amplifier. In the ideal scenario, $R_i = 0$ would be zero, resulting in an unattenuated source current, $I_s = i_I$.

Similarly to the voltage amplifier, the above reductions are referred to as *loading*. In practical applications, some loading is unavoidable, but it can be minimized by ensuring that $R_i << R_s$ and $R_o >> R_L$.

The overall *source-to-load* voltage gain A_{IT} is given by

$$A_{IT} = \frac{i_o}{I_s} = A_I \frac{R_s}{R_s + R_i} \frac{R_o}{R_o + R_L}. \tag{13.8}$$

13.2.3 Problem 1

Given a voltage source of 12 V with a source resistance R_s = 10 Ω and a voltage amplifier (Figure 13.1) with input resistance R_i = 2 MΩ, voltage gain A_v = 10^5 V/V, output resistance R_o = 75 Ω, and load resistor R_L = 100 Ω:

1. Calculate the overall voltage gain A_vT from source to load. Verify with Multisim simulation.
2. Quantify input and output loading as percentages. Confirm with Multisim measurements.

13.2.3.1 Solutions to Problem 1

1. To calculate the overall voltage gain A_vT, we first determine the value of the input voltage v_i by applying the voltage divider formula:

$$v_i = V_s \frac{R_i}{R_s + R_i} = 11.99 \text{ V}. \tag{13.9}$$

The value of the output voltage v_o can be obtained as follows:

$$v_o = A_v v_i \frac{R_L}{R_o + R_L} = 685.14 \text{ kV}. \tag{13.10}$$

Figure 13.3 provides Multisim simulation results verifying the calculated v_i and v_o values.

The overall gain of the above voltage amplifier is

$$A_vT = \frac{v_o}{V_s} = 57.14 \text{ kV/V}. \tag{13.11}$$

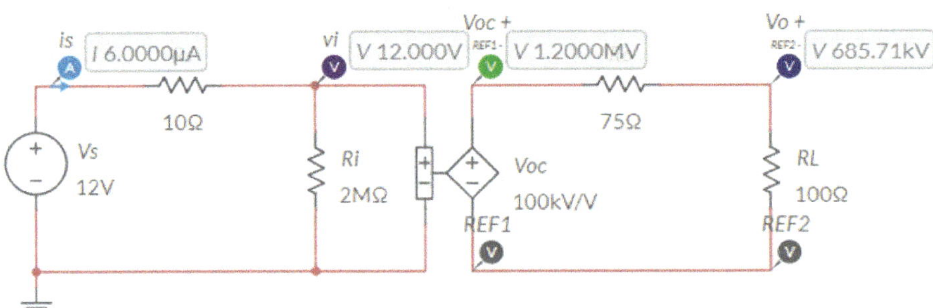

Figure 13.3 Multisim simulation of voltage amplifier providing measured values of voltages and currents.

1. The input loading is quantified as follows:

$$\text{Input loading } \% = \frac{v_i}{V_s}100 = 99.91\%. \tag{13.12}$$

This indicates the source voltage is attenuated to 99.91% of its unloaded value due to loading by the amplifier input resistance.

Similarly, the output loading of the voltage amplifier is

$$\text{Output loading } \% = \frac{v_0}{v_{oc}}100 = \frac{v_0}{A_v v_i}100 = 57.6\%. \tag{13.13}$$

The loaded output voltage is 57.6% of the ideal unattenuated voltage v_{oc}.

13.2.4 Problem 2

Given a current source of 10 A with a source resistance R_s = 50 kΩ and a current amplifier (Figure 13.2) with input resistance R_i = 20 kΩ, current gain A_I = 100 A/A, output resistance R_o = 21 kΩ, and load resistor R_L = 300Ω:

1. Calculate the overall current gain A_{IT} from source to load. Verify with Multisim simulation.
2. Quantify input and output loading as percentages. Confirm with Multisim measurements.

13.2.4.1 Solutions to Problem 2

1. To calculate the overall current gain A_{IT}, we first determine the value of the input current i_I applying the current divider formula:

$$i_I = I_s \frac{R_s}{R_s + R_i} = 7.14 \text{ A}. \tag{13.14}$$

The value of the output current i_o can be calculated as follows:

$$i_o = A_I i_I \frac{R_0}{R_o + R_L} = 620.8 \text{ A}. \tag{13.15}$$

Figure 13.4 provides Multisim simulation results verifying the calculated i_I and i_o values.

The overall gain of the current amplifier is

$$A_{IT} = \frac{i_o}{I_s} = \frac{620.8 \text{ A}}{10 \text{ A}} = 62.08 \frac{\text{A}}{\text{A}}. \tag{13.16}$$

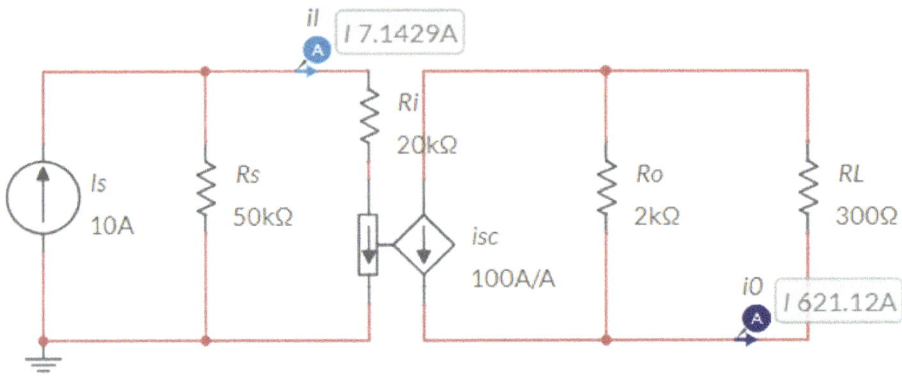

Figure 13.4 Multisim simulation providing measured values for currents.

2. The input loading of the current amplifier is quantified as follows:

$$\text{Input loading }\% = \frac{i_I}{I_s}100 = 71.4\%. \tag{13.17}$$

This indicates that the source current is attenuated to 71.4% of its unloaded value due to loading by the amplifier input resistance. Similarly, the output loading of the voltage amplifier is

$$\text{Output loading }\% = \frac{i_O}{i_{sc}}100 = \frac{i_O}{A_I i_I}100 = 86.9\%. \tag{13.18}$$

The loaded output current is 86.9% of the ideal unattenuated current i_{sc}.

13.3 Lab Procedure 1

Given a voltage source of 24 V with a source resistance R_s = 20 Ω and a voltage amplifier (Figure 13.1) with input resistance R_i = 1 MΩ, voltage gain A_v = $2{\cdot}10^5$ V/V, output resistance R_o = 50 Ω, and load resistor R_L = 200 Ω:

1. Calculate the overall voltage gain A_{VT} from source to load. Verify with Multisim simulation.
2. Quantify input and output loading as percentages. Confirm with Multisim measurements.

14

Saturation of Operational Amplifiers

14.1 Student Learning Outcomes

- Apply knowledge of operational amplifier to calculate input and output voltage ranges that cause saturation.
- Construct operational amplifier circuits in Multisim and use test signals to demonstrate and measure saturation and other nonideal effects.
- Analyze how operational amplifier open-loop gain, power supply levels, and input signals influence distortion and clipping based on simulations. Relate trends to theory.

14.2 Prelab

Operational amplifiers, commonly referred to as *op amps*, are often modeled as *voltage-controlled voltage sources* (VCVS) in electrical circuits (Figure 14.1). The output voltage v_{oc} of an op amp is represented as the product of the open-loop gain A_v and the differential input voltage $v_+ - v_-$:

$$v_{oc} = A_v(v_+ - v_-), \tag{14.1}$$

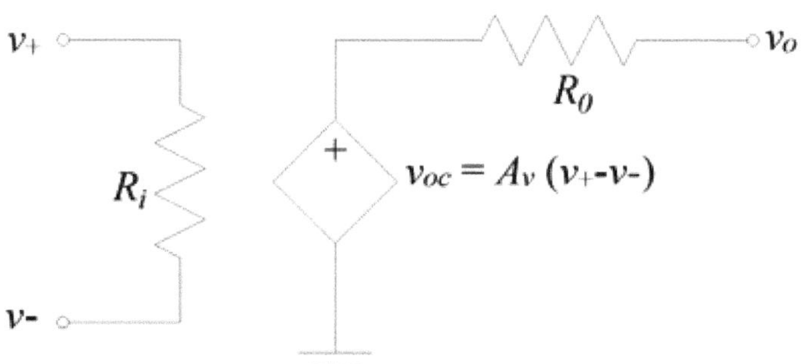

Figure 14.1 Operational amplifiers as *voltage-controlled voltage sources*.

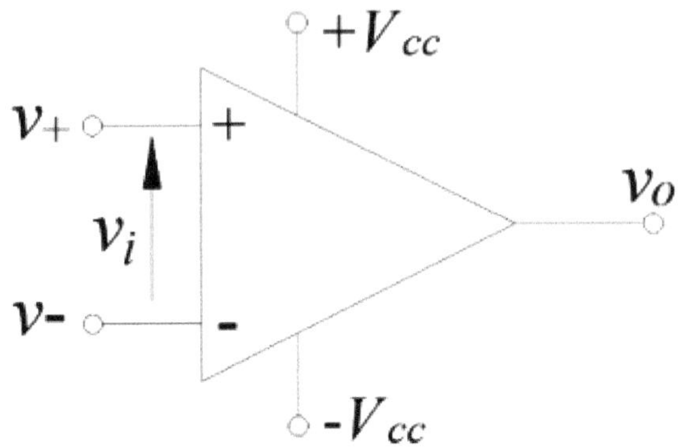

Figure 14.2 Operational amplifier with positive and negative power supply rails $\pm V_{cc}$.

where v_+ is referred to as the *noninverting signal*, whereas v_- is referred to as the *inverting signal*, and A_v is the open-loop voltage gain.

Here, v_+ and v_- are denoted as the *noninverting* and *inverting* signals, respectively.

Op amps typically utilize dual power supplies, one positive (V_{cc}) and one negative ($-V_{cc}$), known as *supply rails* (Figure 14.2). These supply voltages are typically limited to ± 15 V.

In an ideal scenario, the output voltage v_o can be as large as the gain A_v multiplied by the input voltage v_i. However, practical op amps have limitations and v_o cannot exceed the supply voltage and typically 1 or 2 V less. This limitation is not considered in the dependent voltage source model. Consequently, the *transfer function* of an op amp, which depicts the output voltage against the input voltage (Figure 14.3), becomes a *piecewise* function. This means it consists of different functions over various intervals.

The transfer function, as depicted in Figure 14.3, can be expressed in three parts:

$$v_o = +v_{cc}, \text{ if } v_i > \frac{V_{cc}}{A_v} \tag{14.2}$$

$$v_o = A_v v_i, \text{ if } \frac{-V_{cc}}{A_v} < v_i < \frac{+V_{cc}}{A_v} \tag{14.3}$$

$$v_o = -v_{cc}, \text{ if } v_i < \frac{-V_{cc}}{A_v}. \tag{14.4}$$

For input voltages beyond $\pm V_{cc}/A_V$, the output *saturates* at the supply rails. This distortion from the ideal VCVS model can be significant given the high open-loop gains of op amps (e.g., $2{\cdot}10^5$ V/V).

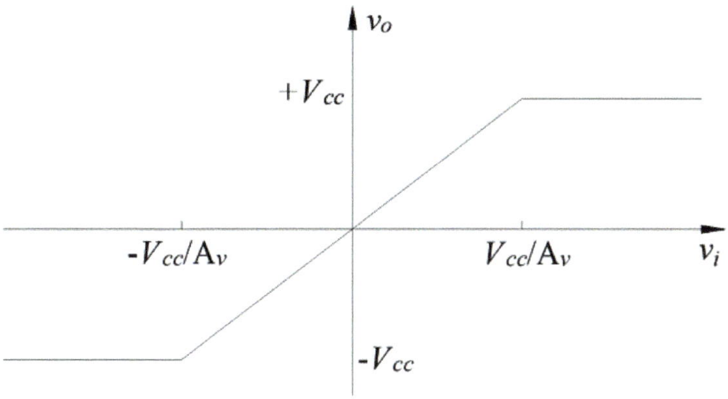

Figure 14.3 Transfer function of the operational amplifier.

14.2.1 Problem 1

In Multisim, construct the circuit in Figure 14.4. The operational amplifier has the following characteristics: A_v = 200 kV/V, R_i = 10 M Ω, R_o = 10 Ω, and voltage swing $v_{o\ \max} = \pm 13.5$ V (the power supply rails are not shown). Perform the following tasks:

1. Calculate the magnitude of the input voltage $v_{iM} = (v_+ - v_-)$ that causes the saturation of the output voltage v_o.
2. Calculate the values of the output voltage v_o for a range of v_i values from $-v_{iM}$ and $+v_{iM}$. Tabulate the results and verify with Multisim simulation.
3. Plot the transfer function of the op amp. Plot $v_o$versus v_i transfer function based on calculations.

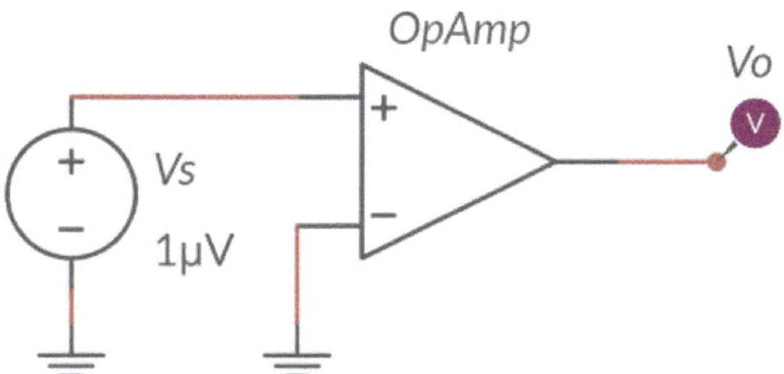

Figure 14.4 Operational amplifier with noninverting signal.

Table 14.1 Summary of readings

v_i [ţV]	v_o [V]		v_i [ţV]	v_o [V]
−77.5	−13.5		10	2.0
−67.5	−13.5		20	4.0
−57.5	−11.5		30	6.0
−47.5	−9.5		40	8.0
−37.5	−7.5		50	10.0
−27.5	−5.5		60	12.0
−17.5	−3.5		67.5	13.5
−7.5	−1.5		77.5	13.5
0	0.0		87.5	13.5

14.2.1.1 Solutions to Problem 1

1. To calculate the magnitude of the input voltage v_{iM} that causes the saturation of the output voltage, we use the following formula:

$$|v_{iM}| = \frac{|V_{cc}|}{A_v} = 67.5\ \mu\text{V}, \tag{14.5}$$

where A_v is the open-loop voltage gain of the operational amplifier and $|V_{cc}|$ is the magnitude of the power supply voltage (± 13.5 V).

1. To calculate the corresponding values of v_o by varying v_i from -77.5 to +87.5 μV, we use the piecewise transfer function defined earlier (Table 14.1).
2. To plot the transfer function of the op amp, we use the calculated values of v_o and the value of v_i of Table 1 (Figure 14.5).

The plot shows the output voltage v_o as a function of the input voltage v_i.

14.2.2 Problem 2

The operational amplifier of Figure 14.6 has the following characteristics: A_v = 100 kV/V, R_i = 10 M Ω, R_o = 10 Ω, and voltage swings $v_{o\,\max} = \pm 15$ V. A sinusoidal voltage generator $v_i\,(t) = 10\sin(628t)$ is connected to the noninverting input. Perform the following tasks:

1. Calculate the input voltage magnitude $v_{iM} = (v_+ - v_-)$ that causes the output voltage v_o saturation.
2. In Multisim, plot the output voltage $v_o(t)$ and the input voltage $v_i(t)$. Examine the output waveform $v_o(t)$ for distortion effects predicted by theory.

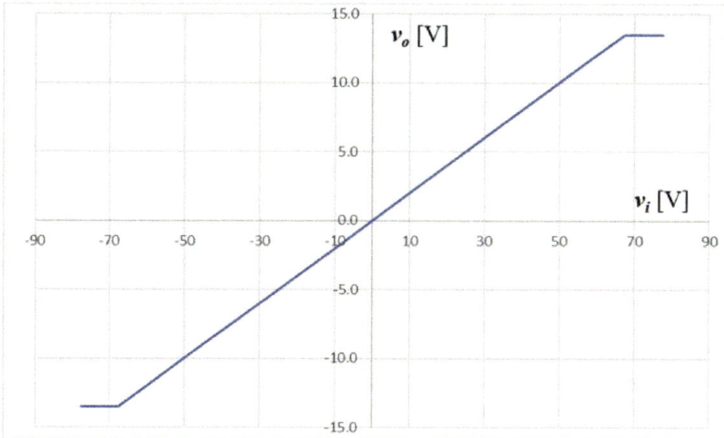

Figure 14.5 Transfer function of the op amp.

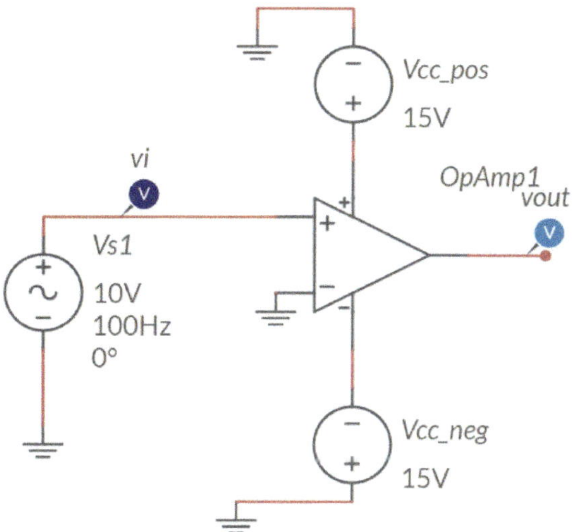

Figure 14.6 Operational amplifier with noninverting input signal.

14.2.2.1 Solutions to Problem 2

1. The magnitude of the input voltage v_{iM} that causes the operational amplifier to reach saturation is given by

$$|v_{iM}| = \frac{|V_{cc}|}{A_v} = 150\ \mu\text{V}. \tag{14.6}$$

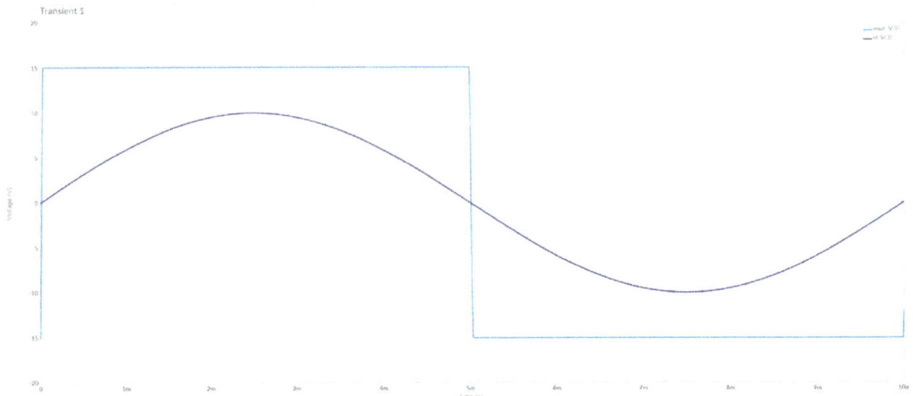

Figure 14.7 Output behavior of the op amp during saturation, producing a square wave output.

2. The inverting input voltage v_{i-} equals zero; consequently, when the input sinewave is positive and its amplitude exceeds +150 μV, the output voltage v_o will reach the positive voltage swing limit (positive saturation). On the other hand, when the input voltage becomes more negative than −150 μV, the output voltage v_o will reach the negative voltage swing limit (negative saturation). As a result, the output voltage v_o exhibits a square wave behavior, where it remains at its maximum positive or negative values depending on the input voltage's polarity (Figure 14.7).

v_o shows a clipped square wave distortion rather than faithfully following the sinusoidal input v_i.

14.2.3 Problem 3

In Multisim, construct the circuit in Figure 14.8. The *comparator* configuration outputs the positive saturation voltage whenever the noninverting signal is greater than the inverting signal, which is the reference input with a 3-V level, and vice versa. The op amp has the following characteristics: A_v = 180 kV/V, R_i = 10 MΩ, R_o = 5Ω, and voltage swing $v_{o\ \max} = \pm 13.5$ V (the power supply rails are not shown).

The operational amplifier *comparator* shown in Figure 14.8 functions in such a way that it produces a positive output saturation voltage when the noninverting signal (v_+) exceeds the inverting signal (v_-), which serves as the reference input with a voltage level of 3 V. Conversely, when the

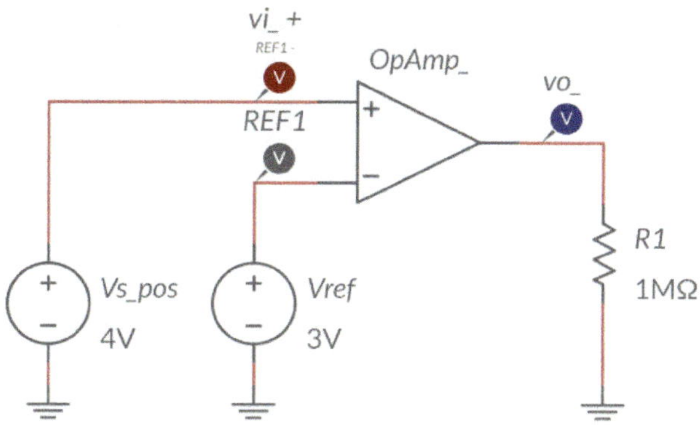

Figure 14.8 Operational amplifier in the comparator configuration.

inverting signal is greater than the noninverting signal, the output voltage v_o reaches negative saturation. The operational amplifier in this comparator has the following characteristics: A_v = 180 kV/V, R_i = 10 M Ω, R_o = 5 Ω, and voltage swing $v_{o\ \mathrm{max}} = \pm 13.5$ V (the power supply rails are not shown in Figure 14.8). Perform the following tasks:

1. Calculate the input voltage magnitude $v_{iM} = (v_+ - v_-)$ that causes output voltage v_o saturation.
2. Use Multisim simulation to demonstrate that for $v_+ > v_{\mathrm{ref}}$, output voltage v_o = +13.5 V.
3. Use Multisim simulation to demonstrate that for $v_+ < v_{\mathrm{ref}}$, output voltage v_o = −13.5 V.
4. Use Multisim simulation to demonstrate that for $v_+ = v_{\mathrm{ref}}$, output voltage v_o = 0 V.

14.2.3.1 Solutions to Problem 3

1. The magnitude of the input voltage v_{iM} that causes the output voltage to saturate is

$$|v_{iM}| = |v_+ - v_-| = \frac{|V_{cc}|}{A_v} = 75\ \mu\text{V}. \tag{14.7}$$

To validate the behavior of the output voltage v_o under different conditions, we conduct Multisim simulations. For $v_+ > v_{\mathrm{ref}}$, the noninverting input voltage is greater than 3 V, and the output voltage v_o of the op amp reaches the positive saturation voltage (+13.5 V). The Multisim simulation of (Figure 14.9) confirms this result.

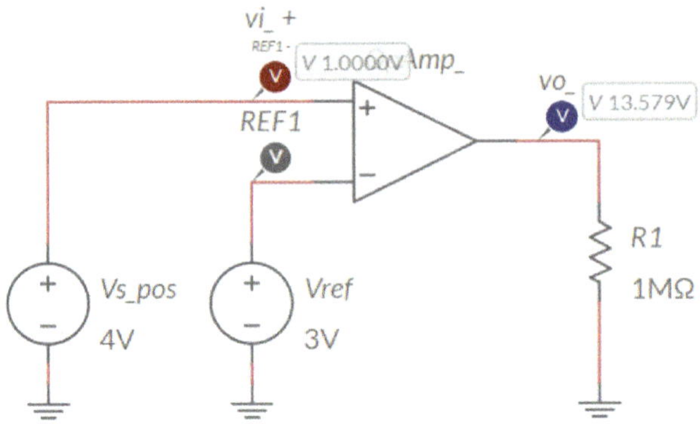

Figure 14.9 For $v_+ > v_{\text{ref}}$ op amp reaches positive saturation output.

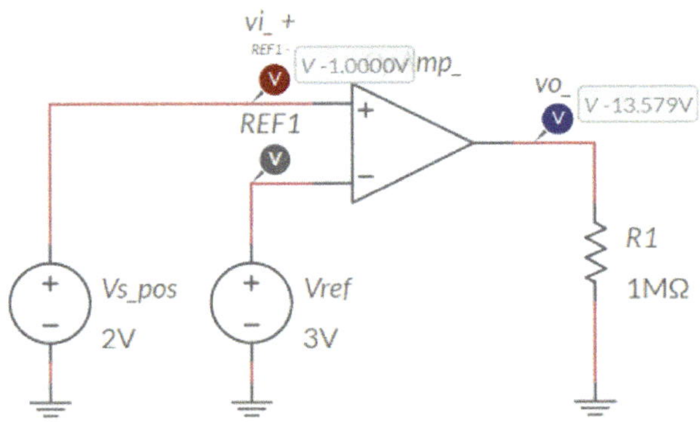

Figure 14.10 For $v_+ < v_{\text{ref}}$ op amp reaches positive saturation output.

1. For $v_+ < v_{\text{ref}}$, the noninverting input voltage is greater than 3 V, and the output voltage v_o of the op amp reaches the negative saturation voltage (-13.5 V). The Multisim simulation of (Figure 14.10) confirms this result.
2. For $v_+ = v_{\text{ref}}$, the noninverting input voltage equals 3 V, and the output voltage of the op amp is zero. The Multisim simulation of (Figure 14.11) confirms this result.

In practice, nonideal input offset voltages present in the differential input stage of the operational amplifier may preclude achieving precisely zero output voltage when the inputs are balanced. Manufacturing imperfections, component variations, and inherent electrical noise can introduce small differences between the noninverting and inverting inputs, even when they are intended to be equal. These unavoidable mismatches can skew the output

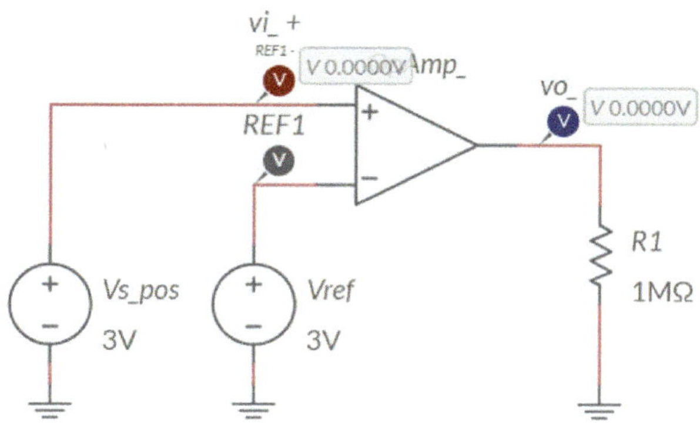

Figure 14.11 For $v_+ = v_{\text{ref}}$, op amp output voltage is zero.

unpredictably to either positive or negative saturation with zero differential input.

14.3 Lab Procedure 1

An operational amplifier has the following characteristics: A_v = 100 kV/V, R_i = 20 M Ω, R_o = 15 Ω, and voltage swing $v_{o\ \text{max}} = \pm 5$ V. Perform the following tasks:

1. Calculate the magnitude of the input voltage $v_{iM} = (v_+ - v_-)$ that causes the saturation of the output voltage v_o.
2. Calculate the values of the output voltage v_o for a range of v_i values from $-v_{iM}$ and $+v_{iM}$. Tabulate the results and verify with Multisim simulation.
3. Plot the transfer function of the op amp. Plot $v_o$versus v_i transfer function based on calculations.

15

Noninverting Operational Amplifier Configuration

15.1 Student Learning Outcomes

- Apply knowledge of negative feedback models to calculate closed-loop gain, input resistance, and output resistance for noninverting operational amplifier configurations.
- Construct noninverting amplifier and voltage follower circuits in Multisim. Take measurements to characterize gain, input/output signals, and resistances.
- Analyze how gain is affected by open- and closed-loop parameters based on Multisim simulations. Relate trends to theory.

15.2 Prelab

The noninverting operational amplifier configuration utilizes *negative feedback*, a process that captures a portion of the output signal and subtracts it from the input signal (Figure 15.1). This feedback mechanism mitigates the adverse effects of device nonlinearities and aging, while simultaneously enhancing the bandwidth of the circuit. Idealizations, valid within the middle frequency range, are commonly employed to determine key parameters of the noninverting configuration, including:

1. The input resistance r_i of the op amp is considered infinite, resulting in zero input current i_I.
2. The output resistance r_o of the op amp is treated as zero, ensuring that the output voltage v_o remains unaffected by any loading effect.
3. The open-loop gain A_{vo} of the basic op amp is presumed infinite, leading to a potential difference $v_+ - v_-$ of zero between the inverting and noninverting inputs.

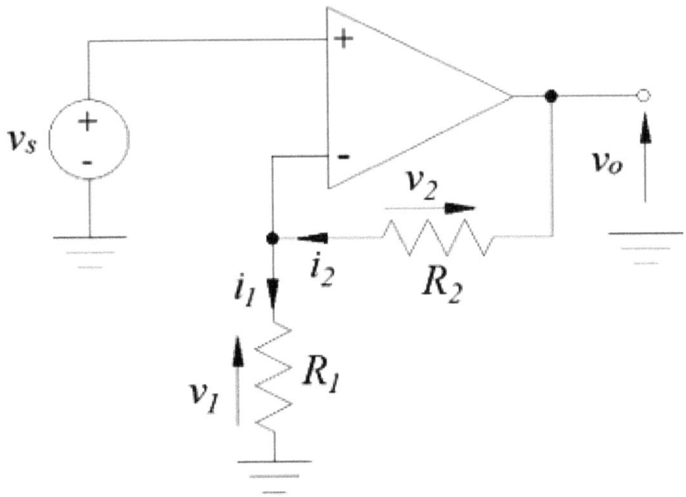

Figure 15.1 Noninverting operational amplifier configuration.

Based on these idealizations and with reference to Figure 15.1, $i_1 = i_2$, consequently establishing the closed-loop gain A_v of the noninverting configuration as follows:

$$A_v = \frac{v_O}{v_S} = \frac{i_1 R_1 + i_1 R_2}{i_1 R_1} = 1 + \frac{R_2}{R_1}. \tag{15.1}$$

The closed-loop gain is always greater than one and is determined by the ratio of two resistors. These resistors inherently offer more stable parameters over time than integrated circuits, ensuring a positive gain and in-phase relationship between the output and input voltages. The closed-loop gain A_v of the noninverting op amp is much less than A_{vo}; however, A_v is much more stable in time because it is independent of transistor parameters.

The application of negative feedback also benefits the closed-loop input and output resistances R_i and R_o, respectively. The closed-loop input resistance is determined as follows:

$$R_i = \frac{v_S}{i_I} = r_i \left(1 + A_{vo}\frac{R_1}{R_1 + R_2}\right). \tag{15.2}$$

Given that A_{vo} is significantly larger than A_v, the closed-loop input resistance R_i is substantially greater than r_i; thus, the negative feedback raises the already high input resistance of the op amp.

On the other hand, the closed-loop output resistance R_o is calculated as follows:

$$R_o = \frac{v_o}{i_o} = \frac{r_o}{1 + A_{vo}\frac{R_1}{R_1+R_2}}. \tag{15.3}$$

Since A_{vo} is much larger than A_v, R_o becomes much smaller than r_o, further decreasing the already low output resistance of the op amp.

15.2.1 The voltage follower

With reference to Figure 15.1, in the special case where R_1 is an open-circuit (i.e., $R_1 = \infty$) and R_2 is a short-circuit (i.e., $R_2 = 0$), resulting in zero divided by infinity, the closed-loop gain A_v becomes one, $R_i = \infty$, and $R_o = 0$. This specific noninverting op amp configuration is known as the *unity-gain amplifier* or the *voltage follower*. Although no amplification occurs (i.e., the output voltage equals the input voltage), this configuration serves as a *resistance transformer*. A voltage source v_s connected to the input terminals experiences an open-circuit, while a load connected to the output terminals perceives an ideal voltage generator with $v_o = v_s$. This buffer feature effectively prevents loading effects and facilitates the transfer of the source voltage to the load without any loss.

15.2.2 Problem 1

Given the noninverting op amp configuration (Figure 15.1) with: A_{v0} = 200 kV/V, R_i = 10 MΩ, R_o = 10 Ω, voltage swing $v_{o\ \max} = \pm 12$ V, R_1 = 1 kΩ, and R_2 = 2 kΩ, perform the following tasks:

1. Calculate input voltage magnitude v_{sM} that causes output voltage saturation. Verify results with Multisim simulation.
2. Determine closed-loop output resistance R_o analytically. Confirm via Multisim measurement.

15.2.2.1 Solutions to Problem 1

1. Applying the noninverting op amp closed-loop gain equation yields

$$A_v = 1 + \frac{R_2}{R_1} = 3\ \text{V/V}. \tag{15.4}$$

The input voltage magnitude v_{sM} that saturates the output is

$$|v_{sM}| = \frac{|V_{cc}|}{A_v} = 4\ \text{V}. \tag{15.5}$$

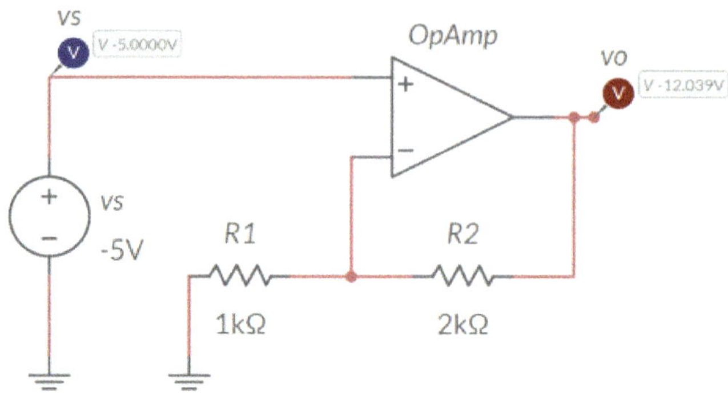

Figure 15.2 Noninverting voltage amplifier in saturation.

Multisim simulation confirms saturation for $v_s \geq |4|$ V (Figure 15.2).

2. The closed-loop output resistance R_o is

$$R_o = \frac{v_o}{i_o} = \frac{r_o}{1 + A_{vo}\frac{R_1}{R_1+R_2}} = 0.15\ \text{m}\Omega. \tag{15.6}$$

This calculated value is confirmed through a Multisim simulation, as shown in Figure 15.3.

The source generator v_s has been short-circuited and a voltage generator v_o of arbitrary magnitude has been applied to the output terminals. By measuring the circulating current i_o and applying the Ohm's law, we obtain the closed-loop output resistance R_o as

$$R_o = \frac{v_o}{i_o} = 0.15\ \text{m}\Omega. \tag{15.7}$$

Therefore, the Multisim simulation validates the previously calculated value of R_o.

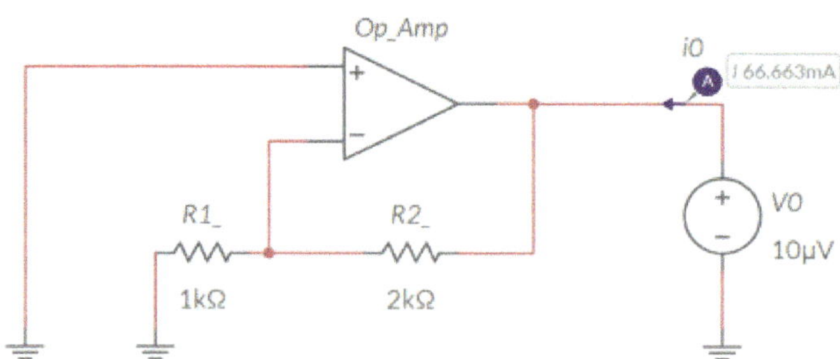

Figure 15.3 Multisim simulation to measure R_o.

15.2.3 Problem 2

Given the following characteristics of the operational amplifier in the noninverting configuration (Figure 15.1): A_{v0} = 200 kV/V, R_i = 10 MΩ, R_o = 10 Ω, voltage swing $v_{o\ \max} = \pm 13.5$ V, R_1 = 2 kΩ, and R_2 = 10 kΩ, perform the following tasks:

1. Calculate the input voltage v_{sM} that causes output voltage saturation. Verify with Multisim.
2. For $v_s = 1$ V, calculate the voltage v_1 across the resistor R_1 and the current i_1 circulating through it. Confirm with Multisim simulation.

15.2.3.1 Solutions to Problem 2

1. The closed-loop gain A_v is

$$A_v = 1 + \frac{R_2}{R_1} = 6 \text{ V/V}. \tag{15.8}$$

The input voltage v_{sM} that causes the saturation of the output voltage is

$$|v_{sM}| = \frac{|V_{cc}|}{A_v} = 2.25 \text{ V}. \tag{15.9}$$

Multisim simulation confirms saturation for $v_s \geq |2.25|$ V (Figure 15.4).

2. For v_s = 1 V, the output voltage v_o is

$$v_o = A_v v_s = 6 \text{ V}. \tag{15.10}$$

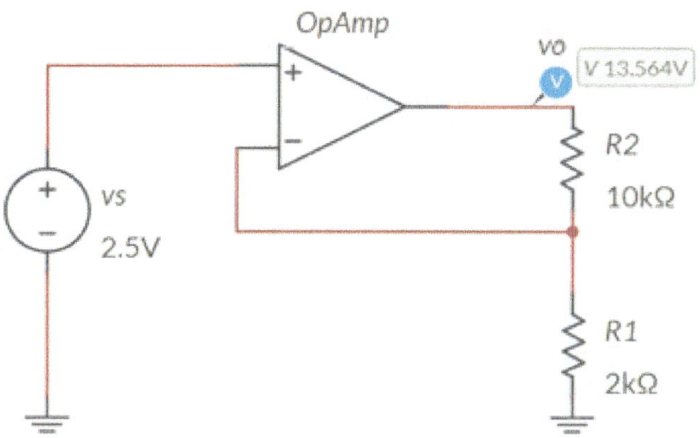

Figure 15.4 Noninverting voltage amplifier in saturation.

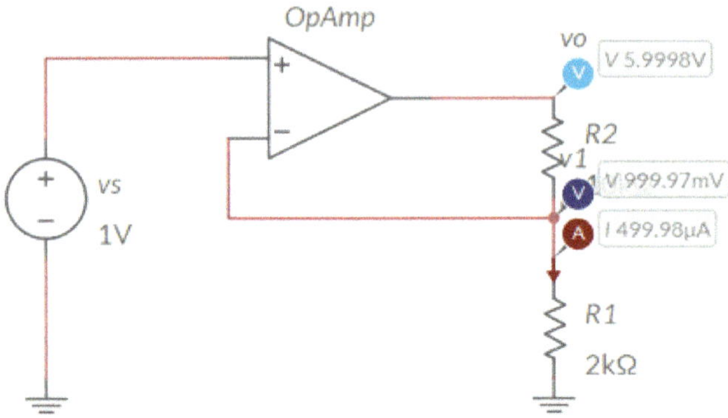

Figure 15.5 Multisim simulation showing the measured values of v_1 and i_1.

The voltage v_1 and the current i_1 are, respectively,

$$v_1 = v_o \frac{R_1}{R_1 + R_2} = 1\ \text{V} \tag{15.11}$$

$$i_1 = \frac{v_1}{R_1} = 0.5\ \text{mA}. \tag{15.12}$$

Multisim measurements of v_1 and i_1 match calculations (Figure 15.5), validating the analysis.

15.3 Lab Procedure 1

An operational amplifier in the noninverting configuration of Figure 15.1 has the following characteristics: A_{v0} = 100 kV/V, R_i = 10 MΩ, R_o = 7 Ω, voltage swing $v_{o\ \max} = \pm 15$ V, R_1 = 10 kΩ, and R_2 = 15 kΩ.

1. Calculate the input voltage magnitude v_{sM} that causes output voltage saturation. Verify results with Multisim simulation.
2. Determine closed-loop output resistance R_o analytically. Confirm via Multisim measurement.

15.4 Lab Procedure 2

The operational amplifier in the noninverting configuration of Figure 15.1 has the following characteristics: A_{v0} = 120 kV/V, R_i = 10 MΩ, R_o = 75 Ω, voltage swing $v_{o\ \max} = \pm 10$ V, R_1 = 1 kΩ, and R_2 = 5 kΩ.

1. Calculate input voltage v_{sM} that causes output voltage saturation. Verify with Multisim.
2. For $v_s = 0.5$ V, calculate the voltage v_1 across the resistor R_1 and the current i_1 circulating through it. Confirm with Multisim simulation.

16

Inverting Operational Amplifier Configuration

16.1 Student Learning Outcomes

- Apply knowledge of negative feedback models to calculate closed-loop gain, input resistance, and output resistance for inverting operational amplifier configurations.
- Construct inverting amplifier circuits in Multisim. Take measurements to characterize gain, input/output signals, and resistances.
- Analyze how gain is affected by open- and closed-loop parameters based on Multisim simulations. Relate trends to theory.

16.2 Prelab

The *inverting* operational amplifier configuration is designed based on the negative feedback model (Figure 16.1). Idealizations, which are typically

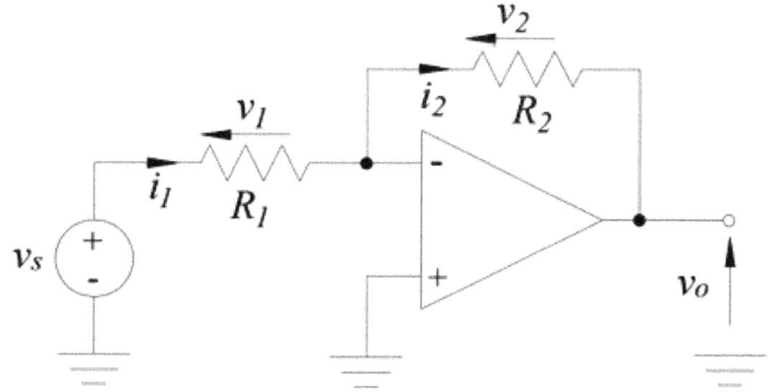

Figure 16.1 Inverting operational amplifier configuration.

valid in the middle frequency range, are utilized to determine the parameters of the inverting configuration, which include:

- The *open-loop* input resistance r_i of the op amp is assumed to be infinite, resulting in zero input current i_I and equal currents i_1 and i_2.
- The *open-loop* output resistance r_o of the op amp is considered zero, leading to no loading effect on the output voltage v_o.
- The *open-loop* gain A_{vo} of the basic op amp is assumed to be infinite, resulting in zero potential difference $(v_+ - v_-)$ between the inverting and noninverting inputs. This implies that the inverting input potential equals the noninverting input potential.

Based on these idealizations and with reference to Figure 16.1, the following observations can be made:

The input voltage $(v_s = I_1R_1)$ causes the entire voltage drop across R_1, as the right terminal of R_1 is at ground potential.

The output voltage $(v_o = -v_2)$ is of the same magnitude and opposite polarity as the voltage drop across R_2, leading to a negative output.

Hence, the *closed-loop* voltage gain A_v of the inverting configuration is given by

$$A_v = \frac{v_o}{v_s} = \frac{-i_1R_2}{i_1R_1} = -\frac{R_2}{R_1}. \tag{16.1}$$

The closed-loop gain is determined by the ratio of two resistors, which inherently exhibit greater stability over time compared to integrated circuits. A_v is negative, indicating a phase difference of 180° between output and input voltages. The closed-loop voltage gain A_v of the inverting op amp is much less than A_{vo}, sacrificing gain for stability.

In the ideal conditions mentioned above, the input voltage v_s drops entirely across R_1, leading to the input resistance R_i being equal to R_1:

$$R_i = \frac{v_s}{i_1} = R_1. \tag{16.2}$$

In contrast to the noninverting configuration, the relationship between A_v and R_1 prevents the input resistance from being raised without affecting the closed-loop voltage gain, and vice versa.

However, the negative feedback is beneficial for the closed-loop output resistance R_o. To determine R_o, the input voltage v_s of the inverting configuration is short-circuited, a small test current i_o is applied to the output terminals, and the output voltage v_o is measured. Assuming that i_o entirely flows into the output resistance r_o, which is a valid assumption in reality, R_o can be calculated as follows:

$$R_o = \frac{v_o}{i_o} = \frac{r_o}{1 + A_{vo}\frac{R_1}{R_1+R_2}}. \quad (16.3)$$

Since $A_{vo} \gg A_v$, then $R_o \ll r_o$, indicating that the negative feedback further decreases the already low open-loop output resistance r_o of the op amp.

16.2.1 Problem 1

An operational amplifier in the inverting configuration of Figure 16.1 has the following characteristics: A_{v0} = 200 kV/V, R_i = 10 MΩ, R_o = 10Ω, voltage swing $v_{o\,\max} = \pm 12$ V, R_1 = 1 kΩ, and R_2 = 6 kΩ. Perform the following tasks:

1. Calculate the magnitude of the input voltage v_{sM} that causes the saturation of the output voltage v_o. Verify with Multisim.
2. Determine analytically the closed-loop input resistance R_i. Verify with Multisim.
3. Calculate the closed-loop output resistance R_o. Verify with Multisim.

16.2.1.1 Solutions to Problem 1

1. The closed-loop gain A_v of the inverting configuration is given by

$$A_v = -\frac{R_2}{R_1} = -6 \text{ V/V}. \quad (16.4)$$

The maximum output voltage swing is ± 12 V. To cause saturation, the input voltage v_s must be such that the output voltage v_o reaches one of these extremes. Since A_v is negative, the saturation occurs for both positive and negative input voltages. The magnitude of v_s that causes the saturation is

$$|v_{sM}| = \frac{|V_{cc}|}{A_v} = 2 \text{ V}. \quad (16.5)$$

Values of the input voltage $v_s \geq |2|$ V will cause the output voltage v_o to saturate to the value ± 12 V.

The Multisim simulation in Figure 16.2 confirms this result, showing the output voltage saturation to -12 V and the inversion of the input.

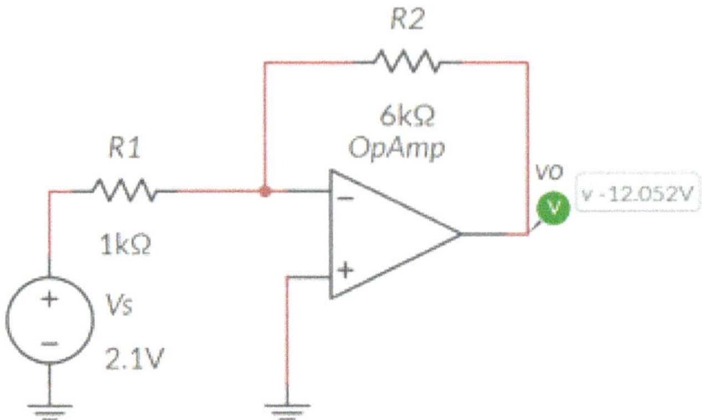

Figure 16.2 Inverting operational amplifier configuration in saturation.

2. The closed-loop input resistance R_i is set by R_1, and thus $R_i = R_1 = 1\ \text{k}\Omega$. To confirm this calculation, a Multisim simulation can be performed as shown in Figure 16.3.

A small test current i_i is applied to the input terminals of the inverting configuration and the input voltage v_i is measured. R_i is determined by applying Ohm's law:

$$R_i = \frac{v_i}{i_i} = 1\ \text{k}\Omega. \tag{16.6}$$

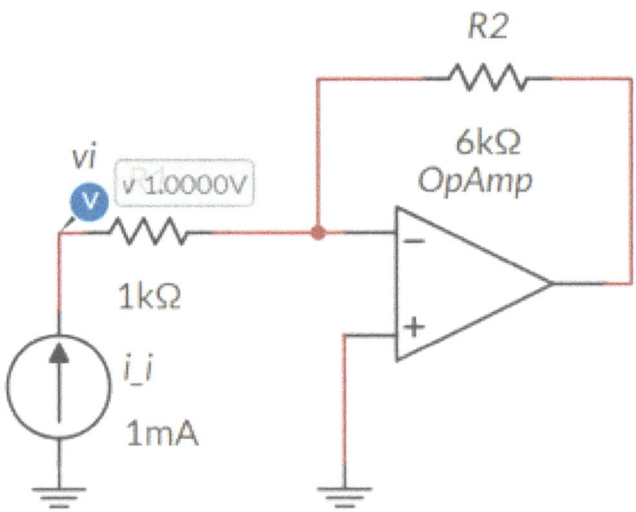

Figure 16.3 Circuit to measure closed-loop R_i of the inverting configuration.

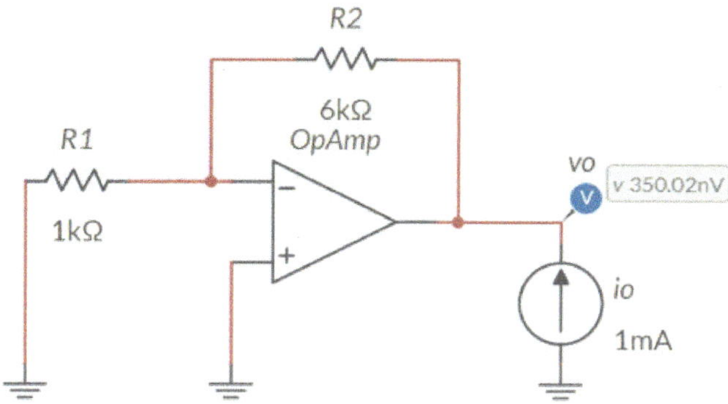

Figure 16.4 Multisim simulation to measure R_o.

This matches the analytical value, confirming R_i.

3. The closed-loop output resistance R_o is given by

$$R_o = \frac{v_o}{i_o} = \frac{r_o}{1 + A_{vo}\frac{R_1}{R_1+R_2}} = 350\ \mu\Omega. \tag{16.7}$$

To confirm this result, the circuit is simulated in Multisim (Figure 16.4). In this simulation, the source generator v_s has been short-circuited, and a current generator i_o has been applied to the output terminals. By reading the output voltage v_o and applying Ohm's law, we calculate the value of R_o:

$$R_o = \frac{v_o}{i_o} = 350\ \mu\Omega. \tag{16.8}$$

The simulation result matches the calculated value. This indicates that the measured output resistance is consistent with the calculated value, validating the accuracy of the calculation.

16.2.2 Problem 2

The operational amplifier in the inverting configuration of Figure 16.1 has the following characteristics: A_{v0} = 100 kV/V, R_i = 10 MΩ, R_o = 10Ω, and voltage swing $v_{o\ \max} = \pm 13.5$ V. Perform the following tasks:

1. Determine resistor values R_1 and R_2 to realize a closed-loop gain $A_v = -8$ and a closed-loop input resistance $R_i = 100$ kΩ.Verify with Multisim.
2. Calculate the input voltage v_{sM} that causes the output voltage saturation. Confirm with Multisim.

3. For $v_s = 100$ mV, calculate the current i_1 through the resistor R_1 and the voltage v_2 across the resistor R_2. Verify with Multisim.

16.2.2.1 Solutions to Problem 2

1. For the inverting configuration, the closed-loop input resistance R_i is equal to R_1; thus, the closed-loop gain A_v is given by

$$A_v = -\frac{R_2}{R_i} = -8 \text{ V/V} \Rightarrow R_2 = 800 \text{ k}\Omega. \tag{16.9}$$

The Multisim simulation of Figure 16.5 confirms the gain of $A_v = -8$ with the calculated value of R_2, showing an output voltage $v_o = -799.96$ mV for an input voltage $v_s = 100$ mV.

2. The voltage swing of the operational amplifier is given as $v_{o\ max} = \pm 13.5$ V. For the inverting configuration, the closed-loop gain A_v is -8. The magnitude of the input voltage v_{sM} that causes output voltage saturation can be calculated using the following formula:

$$|v_{sM}| = \frac{|v_{0\ max}|}{A_v} = 1.68 \text{ V}. \tag{16.10}$$

The Multisim simulation of Figure 16.6 confirms saturation for $v_s = 1.7$ V.

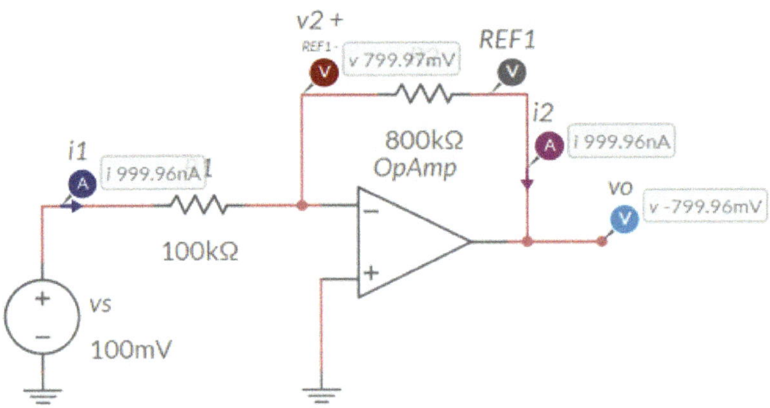

Figure 16.5 Multisim simulation verifying calculated values for R_1 and R_2.

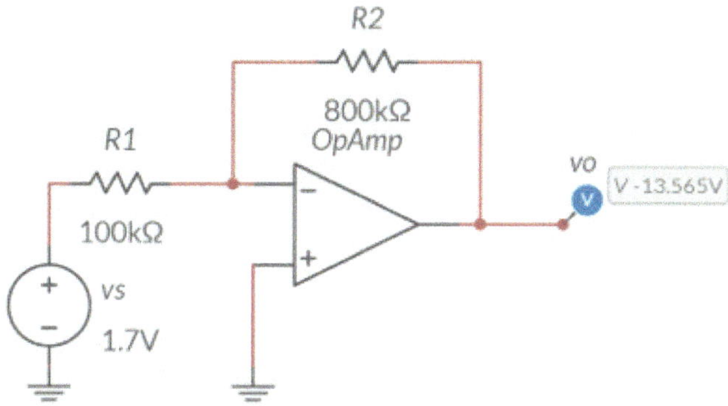

Figure 16.6 Inverting operational amplifier in saturation..

3. For the inverting configuration, the input voltage (v_s) is 100 mV. The current i_1 circulating through resistor R_1 and the voltage v_2 across the resistor R_2 can be calculated using Ohm's law:

$$i_1 = \frac{v_s}{R_1} = 1\ \mu\text{A} \tag{16.11}$$

$$v_2 = i_1 R_2 = 800mV. \tag{16.12}$$

To confirm these calculations, we perform a Multisim simulation with the given values of R_1, R_2, and v_s and verify that the results match the calculated values (Figure 16.5).

16.3 Lab Procedure 1

An operational amplifier in the inverting configuration of Figure 16.1 has the following characteristics: A_{v0} = 100 kV/V, R_i = 10 MΩ, R_o = 5Ω, voltage swing $v_{o\ \max}$ = ±15 V, R_1 = 2 kΩ, and R_2 = 8 kΩ. Perform the following tasks:

1. Calculate the magnitude of the input voltage v_{sM} that causes the saturation of the output voltage v_o. Verify with Multisim.
2. Determine analytically the closed-loop input resistance R_i. Verify with Multisim.
3. Calculate the closed-loop output resistance R_o. Verify with Multisim.

16.4 Lab Procedure 2

The operational amplifier in the inverting configuration of Figure 16.1 has the following characteristics: A_{v0} = 80 kV/V, R_i = 12 MΩ, R_o = 20Ω, and voltage swing $v_{o\ \max} = \pm 12.5$ V.

1. Determine resistor values R_1 and R_2 to realize a closed-loop gain $A_v = -10$ and a closed-loop input resistance $R_i = 150$ kΩ. Verify with Multisim.
2. Calculate the input voltage v_{sM} that causes the output voltage saturation. Confirm with Multisim.
3. For $v_s = 10$ mV, calculate the current i_1 through the resistor R_1 and the voltage v_2 across the resistor R_2. Verify with Multisim.

17

Measuring the Slew Rate of Operational Amplifiers

17.1 Student Learning Outcomes

- Define the *slew rate* (SR) of an operational amplifier.
- In Multisim, measure the slew rate of an operational amplifier using a unity gain configuration.
- Apply the formula to calculate the maximum frequency of the input signal that will not cause distortion due to slew rate limitations.

17.2 Prelab

The *slew rate* (*SR*) indicates an operational amplifier's capability to change its output voltage v_o in response to input voltage variations. It is a crucial parameter to consider when selecting an operational amplifier for high-speed applications.

Manufacturers typically specify the slew rate by applying a step signal (v_s) with an amplitude of 10 V to the op amp's input and measuring the rate of change between 10% and 90% of the output voltage's amplitude (Figure 17.1). The slew rate (SR) is mathematically represented as the derivative of the output voltage with respect to time and can be calculated using the following formula:

$$\mathrm{SR} = \frac{dv_o(t)}{dt} = \frac{v_{o90\%} - v_{o10\%}}{t_{90\%} - t_{10\%}}. \tag{17.1}$$

For instance, for the μA741C op amp, the specified slew rate is 0.5 V/μs.

The slew rate is typically measured in a unity gain configuration, also known as a *voltage follower*, as it allows the determination of *SR* independently of the voltage gain. The unit of measurement for *SR* is typically V/μs (volts per microsecond). The value of *SR* signifies an op amp's ability to deliver an output voltage that faithfully represents the input signal. Over a certain frequency range, the slew rate remains relatively constant. For instance, the μA741C op amp specifies a slew rate of 0.5 V/μs. One can calculate the maximum frequency (f_M) of the input signal that would not produce a distorted output due to slew rate limitations as follows:

$$f_M = \frac{SR}{2\pi V_{oM}}. \tag{17.2}$$

Here, V_{oM} is the peak of the output voltage, coinciding with the output voltage swing. In the case of the μA741C, the maximum output voltage swing is specified as ± 13.5 V achievable with a ± 15 V power supply.

From the formula, a lower output voltage peak allows for a higher maximum frequency of the input voltage that would not cause distortion. This formula can be applied to any periodic input signal, such as sine waves, square waves, and triangular waves, although specific applications may influence the op amp's performance, potentially limiting f_M.

17.2.1 Problem 1

To illustrate the application of these concepts, let us consider the operational amplifier μA741C in the configuration shown in Figure 17.1. It is a general-purpose device with a dual supply voltage of $V_{cc} = \pm 15$ V and $R_L = 2$ kΩ. A step signal with an amplitude of 10 V, duty cycle 50%, and frequency of 4 kHz is applied at the positive input terminal. Perform the following tasks.

1. By conducting a Multisim simulation, determine the *slew rate* (*SR*) of the op amp.
2. Analyze the input terminal potential difference $v_i = v_+ - v_-$ in response to step wave input.

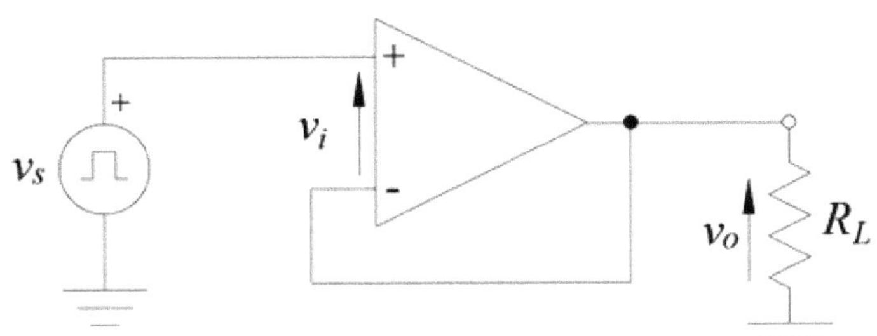

Figure 17.1 Slew rate measurement circuit.

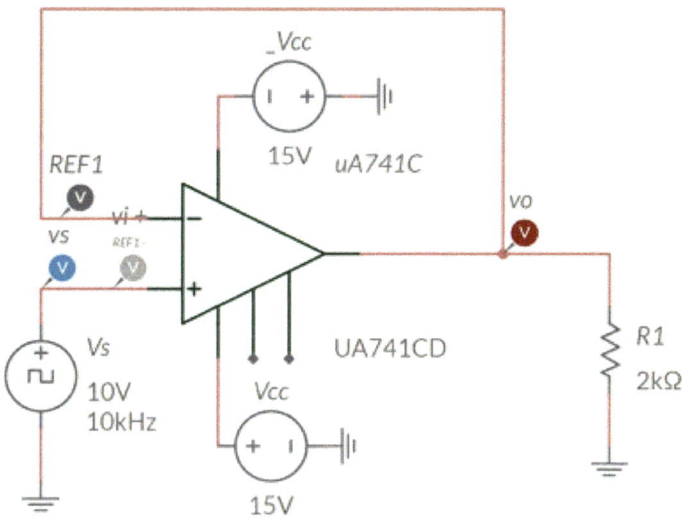

Figure 17.2 Multisim simulation of *voltage follower* circuit to determine the *slew rate* of an op amp.

17.2.1.1 Solutions to Problem 1

1. A simulation *voltage follower* circuit is constructed in Multisim (Figure 17.2).

Figure 17.3 graphically displays the input and output voltages of the *voltage follower* configuration over time. Analysis of the output waveform gives a slew rate:

$$S = \frac{v_{o90\%} - v_{o10\%}}{t_{90\%} - t_{10\%}} = 0.5\ \text{V}/\mu\text{s}. \tag{17.3}$$

This confirms the specified 0.5 V/μs slew rate for the μA741C.

The plots in Figure 17.3 demonstrate that the output voltage (v_o) is not a faithful representation of the input square wave due to the slew rate limitations of the μA741C.

2. The graph in Figure 17.3 reveals that, upon applying the step function, the potential difference v_i across the input terminals (+ and −) of the op amp is not zero, resulting in the absence of the virtual ground. This finite voltage across the input terminals arises due to the slower evolution of the output voltage compared to the input signal. As the output voltage (v_o) linearly increases to its final value, the voltage v_i across the input terminals decreases to zero.

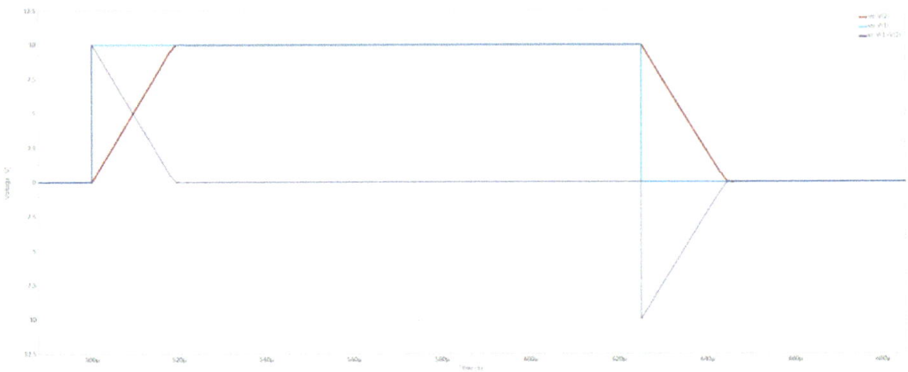

Figure 17.3 Plots of input and output voltages of the *voltage follower* configuration over time.

17.3 Lab Procedure 1

The operational amplifier LM324A in the configuration of Figure 17.1 is a general-purpose device with a dual supply voltage V_{cc} = ±15 V and R_L = 1 kΩ. A step signal with an amplitude of 10 V, duty cycle 50%, and frequency of 5 kHz is applied to the positive input terminal. Perform the following tasks.

1. By conducting a Multisim simulation, determine the *slew rate* (*SR*) of the op amp.
2. Analyze the input terminal's potential difference $v_i = v_+ - v_-$ in response to step wave input.

Part III

AC Circuits

18

Measurement of Parameters and Phase Differences of Sinusoids

18.1 Student Learning Outcomes

- Identify and calculate major parameters of a sinusoid.
- Calculate and measure the phase difference between sinusoids.

18.2 Prelab

A sinusoidal waveform, represented by $x(t)$, is characterized as a *periodic function*. This means that there exists a positive number T_0 such that the waveform $x(t)$ repeats itself after a time interval of T_0, or $x(t) = x(t + T_0)$. The period T of the sinusoid is defined as the smallest value of T_0, representing the time it takes for the waveform to complete one full cycle before repeating.

The following equation describes a sinewave:

$$x\,(t) = X_p \sin(\omega t + \varphi), \tag{18.1}$$

where

- X_p is the peak value.
- ω is the angular frequency measured in radians per second; $f = \frac{\omega}{2\pi}$ is the frequency measured in Hertz.
- $T = \frac{1}{f}$ is the period, measured in seconds.
- The phase angle φ, associated with a sinusoidal waveform, is typically measured in radians. However, it is also common to express φ in degrees as an accepted "abuse of notation" (e.g., $\sin(\omega t + \frac{\pi}{2}) = \sin(\omega t + 90°)$). This allows for convenient interchangeability between radians and degrees when describing phase angles in sinusoidal functions.

If the angular frequency ω approaches zero ($\omega \to 0$), the period T of the sinusoidal waveform becomes infinitely large ($T \to \infty$). In this case, the waveform oscillations slow down significantly, and the waveform appears as a steady, unchanging value equal to $X_p \sin(\varphi)$.

18.2.1 Mean value

We define the *mean* or *average* value $\langle x(t)\rangle$ of a signal $x(t)$ over the interval [0, *T*] as follows:

$$\langle x(t)\rangle = \frac{1}{T}\int_0^T x(t)\, dt\,. \tag{18.2}$$

The average value represents the constant value that, if sustained over the interval [0, *T*], would result in the same area under the curve as the waveform *x*(*t*).

We do note that the *mean* is a linear operator, such that $\langle ax_1(t) + bx_2(t)\rangle = a\langle x_1(t)\rangle + b\langle x_2(t)\rangle$, for any two waveforms and any constants *a* and *b*. For sinewaves $\langle x(t)\rangle = 0$ for any value of *f* and φ, as the areas above and below the time-axis are equal and of opposite sign, thus the net area over a period is zero.

We note that the *mean* is a linear operator, satisfying the property $\langle ax_1(t) + bx_2(t)\rangle = a\langle x_1(t)\rangle + b\langle x_2(t)\rangle$.

This means that for any two waveforms $x_1(t)$ and $x_2(t)$ and any constants *a* and *b*, the mean of their linear combination is equal to the linear combination of their means.

For sinusoidal waveforms, regardless of the values of frequency *f* and phase angle φ, we always have $\langle x(t)\rangle = 0$. This is because the areas above and below the time-axis are equal in magnitude and opposite in sign, resulting in a net area over a period of zero.

18.2.2 Root mean square, or *rms*

We define the *root mean square*, or *rms*, value of a waveform *x*(*t*) as follows:

$$X_{\text{rms}} = \sqrt{\frac{1}{T}\int_0^T x^2(t)\, dt}. \tag{18.3}$$

For sinusoidal waveforms, the *rms* value can be simplified as $X_{\text{rms}} = \frac{X_p}{\sqrt{2}}$, where X_p represents the peak amplitude of the sinusoid. This relationship allows us to directly calculate the *rms* value using the peak amplitude.

18.2.3 Phase difference between sinusoids

To determine the phase difference between sinusoids, certain conditions must be met:

1. The sinusoids must have the same frequency, denoted as *f*.
2. The amplitude of both sinusoids must be positive.
3. Both sinusoids must be written as either sine or cosine waves.

To satisfy these conditions and calculate the phase difference, we can use the following trigonometric identities:

$$\cos(\omega t) = \cos(\omega t \pm 180°) \tag{18.4}$$

$$\sin(\omega t) = \sin(\omega t \pm 180°) \tag{18.5}$$

$$\cos(\omega t) = \sin(\omega t + 90°) \tag{18.6}$$

$$\sin(\omega t) = \cos(\omega t - 90°). \tag{18.7}$$

Given $x_1(t) = X_1 \sin(\omega t + \alpha)$ and $x_2(t) = X_2 \sin(\omega t + \beta)$, if the phase difference $\Delta\varphi = (\alpha - \beta)$ is greater than zero, then $x_1(t)$ leads $x_2(t)$. This means that any point on the waveform $x_1(t)$ occurs earlier in time than the corresponding point on the waveform $x_2(t)$. In Figure 18.1, the sinewave $x_1(t)$ lags the sinewave $x_2(t)$ by $\pi/4$ radians (or 45°), which means that there is a phase difference $\Delta\varphi$ of $\pi/4$ radians between the two sinusoidal waveforms.

If the phase difference is zero, the sinewaves are said to be *in phase*. This means that they reach their minima and maxima at the same time.

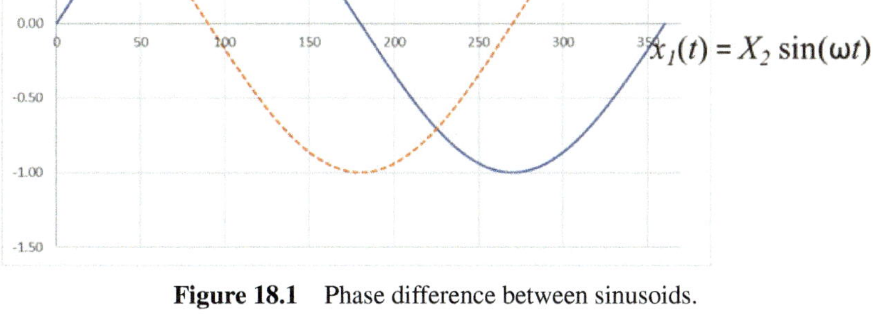

Figure 18.1 Phase difference between sinusoids.

From a plot, we can determine the phase difference between sinusoids by estimating the time difference between corresponding points on both waveforms, such as zero crossings, maxima, or minima. The relationship between the time difference Δt and the phase difference $\Delta\varphi$ is given by eqn (18.8).

$$\Delta\varphi = \frac{360°}{T}\Delta t = \omega\Delta t, \tag{18.8}$$

where $\Delta\varphi$ is the phase difference in degrees, Δt is the time difference in seconds, and T is the period of the sinusoids in seconds.

This equation allows us to relate the time difference between points on the waveforms to the corresponding phase difference, using the period T of the sinusoids.

18.2.4 Problem 1

Given the two sinusoids $i_1(t) = \sin(1000t + 45°)$ A and $i_2(t) = -4\cos(1000t + 30°)$ A, determine:

1. The phase difference $\Delta\varphi$ between $i_1(t)$ and $i_2(t)$.
2. The angular frequency ω, the frequency f and the period T.
3. Peak and *rms* values.

Report all results in Table 18.1.

18.2.4.1 Solutions to Problem 1

1. We cannot directly compare the phase angles of the two sinewaves because they have different forms (i.e., sine and cosine) and one is negative. To determine the phase difference between the sinusoids, we can use the trigonometric identities between the sine and cosine functions.

$$\begin{aligned} i_2(t) &= -4\cos(1000t + 30°) = 4\cos(1000t + 30° - 180°) \\ &= 4\sin(1000t - 150° + 90°)\ \text{A}. \end{aligned} \tag{18.9}$$

Therefore, $\Delta\varphi = (\alpha - \beta) = 45° - (-60°) = 105° > 0$, which indicates that $i_1(t)$ leads $i_2(t)$ by $105°$.

2. The angular frequency ω can be calculated as $\omega = 2\pi f = 1000\ \text{rad}\cdot\text{s}^{-1}$; hence the frequency is $f = \omega/2\pi = 159.2$ Hz, and the period is $T = 1/f = 6.28$ ms.

For $i_1(t)$, the peak value is 1 A, and the *rms* value is $1/\sqrt{2} = 0.707$ A. For $i_2(t)$, the peak value is 4 A, and the *rms* value is $4/\sqrt{2} = 2.828$ A.

Reporting the results in Table 18.1.

Table 18.1 Summary of calculations

$\Delta\varphi$ (°)	ω (rad · s^{-1})	f (Hz)	T (ms)	$i_{1\ peak}$ (A)	$i_{2\ peak}$ (A)	$i_{1\ rms}$ (A)	$i_{2\ rms}$ (A)
105	1,000	159.2	6.28	1	4	0.7	2.8

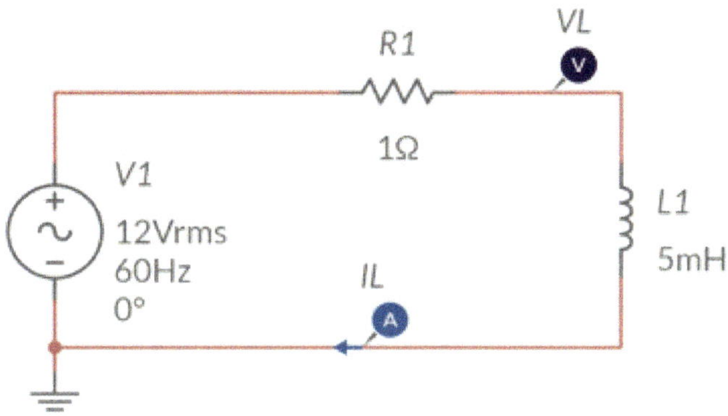

Figure 18.2 Ohmic-inductive circuit.

18.2.5 Problem 2

In the circuit of Figure 18.2, a series ohmic-inductive circuit is supplied by a sinusoidal voltage source $v_1\ (t) = \sqrt{2}{\cdot}12 \cdot\ \sin{(376.8t)}\ \text{V}$.

1. Through Multisim simulation, estimate the period T and the frequency f of the voltage $v_L(t)$ across the inductor and the current $i_L(t)$ through the inductor. Verify that the frequencies of $v_L(t)$ and $i_L(t)$ are the same as that of the voltage supply.
2. Through Multisim simulation, estimate the time delay Δt of the current $i_L(t)$ with respect to the voltage $v_L(t)$, and the phase difference $\Delta\varphi = \frac{360°}{T}\Delta t$ in degrees.

18.2.5.1 Solutions to Problem 2

1. The Multisim simulation will provide us with the necessary waveforms to analyze. By examining the waveform of $v_L(t)$ (Figure 18.3), we can measure the period T, which represents the time taken for the waveform to complete one full cycle (e.g., for one peak to the other peak). The period T equals 16.687 ms.

The reciprocal of the period gives us the frequency f, which indicates the number of cycles per unit time. The frequency is 59.954 Hz. We can compare the measured frequency of $v_L(t)$ with the given frequency of the voltage

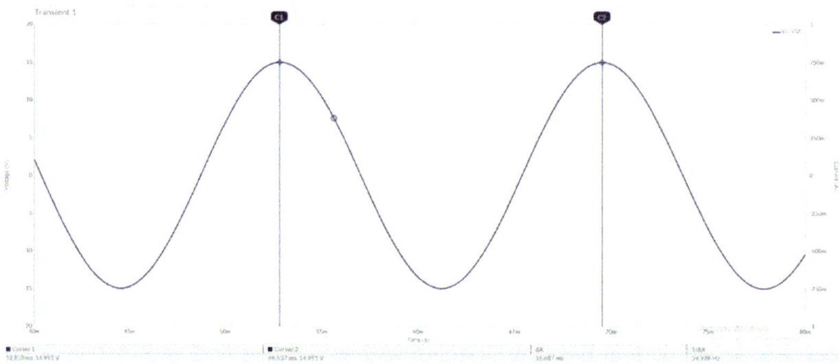

Figure 18.3 Plot of the voltage $v_L(t)$ as a function of time.

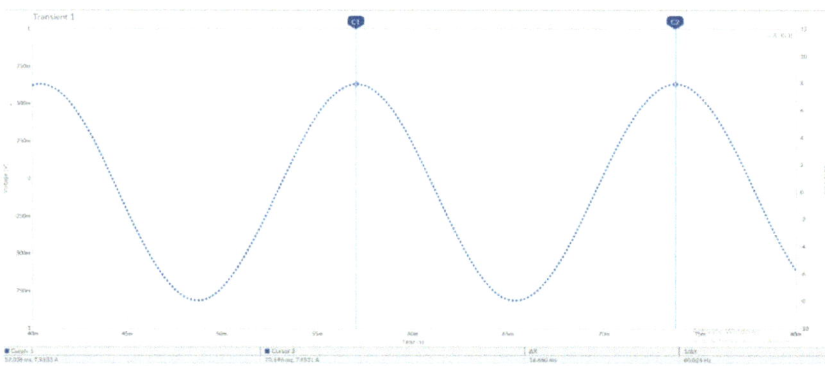

Figure 18.4 Plot of the current $i_L(t)$ as a function of time.

source, which is 60 Hz, to verify that they closely match. The period T, and the frequency f of the current $i_L(t)$ flowing through the inductor can be estimated through its plot as a function of time (Figure 18.4).

Based on the plot above, the period T and the frequency f of $i_L(t)$ are respectively 16.66 ms and 60.024 Hz, which closely match the system frequency of 60 Hz.

By conducting the Multisim simulation and analyzing the waveforms of the current $i_L(t)$, we can estimate the time delay Δt of the current $i_L(t)$ with respect to the voltage $v_L(t)$ (Figure 18.5).

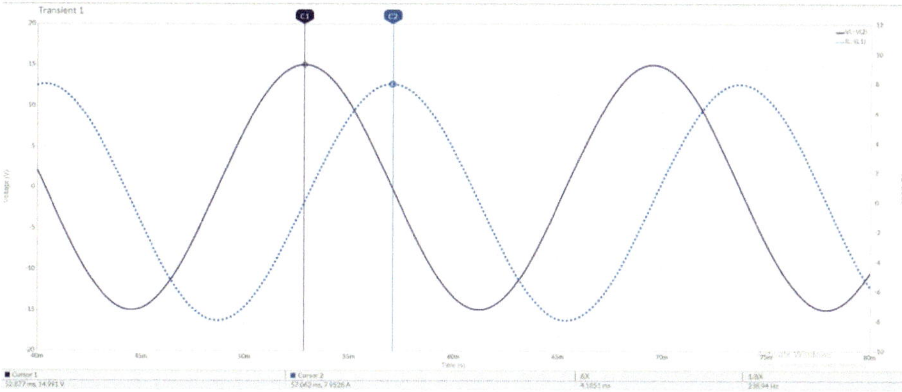

Figure 18.5 Plot of the voltage $v_L(t)$ and the current $i_L(t)$ as a function of time showing the time delay.

Based on the plot below, the time delay of $i_L(t)$ with respect to the voltage $v_L(t)$ is Δt = 4.18 ms. The phase difference $\Delta\varphi$ is calculated with the following formula:

$$\Delta\varphi \cong \frac{360^\circ}{T}\Delta t = \frac{360^\circ}{0.0166} \cdot 4.18\times10^{-3} = 90.6^\circ. \quad (18.10)$$

The value $\Delta\varphi$ measured with Multisim simulation closely matches the calculated value of the phase difference.

$$\Delta\varphi \cong \frac{X_L}{R} = \frac{1.8}{1} = 89.9^\circ. \quad (18.11)$$

18.3 Lab Procedure 1

In the circuit of Figure 18.6, an ohmic-inductive circuit is supplied by a voltage source with an *rms* value of 24 V and a frequency f = 60 Hz:

1. Estimate the period T and the frequency f of the voltage $v_L(t)$ across the inductor and the current $i_L(t)$ through the inductor. Verify that the frequencies of $v_L(t)$ and $i_L(t)$ are the same as that of the voltage supply.

AQ3 2. Estimate the time delay Δt of the current $i_L(t)$ with respect to the voltage $v_L(t)$, and the phase difference $\Delta\varphi = \frac{360^\circ}{T}\Delta t$ in degrees. Record results in Table 18.2 for further analysis and comparison.

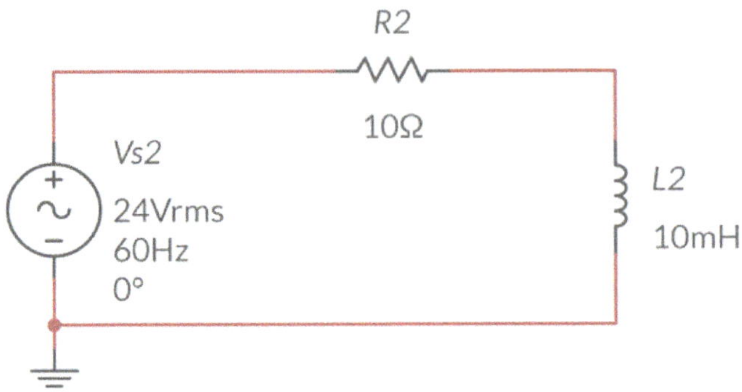

Figure 18.6 Ohmic-inductive circuit.

Table 18.2 Summary of calculations

	T (s)	f (Hz)	Δt (s)	$\Delta\varphi$ (°)
$v_L(t)$				
$i_L(t)$				

18.3.1 Lab Procedure 2

In the circuit of Figure 18.7, an ohmic-capacitive circuit is supplied by a voltage source with an *rms* value of 12 V and a frequency f = 60 Hz:

1. Estimate the period T and the frequency f of the voltage $v_C(t)$ across the capacitor and the current $i_C(t)$ through the capacitor. Verify that the frequencies of $v_C(t)$ and $i_C(t)$ are the same as that of the voltage supply. Record results in Table 18.3 for further analysis and comparison.

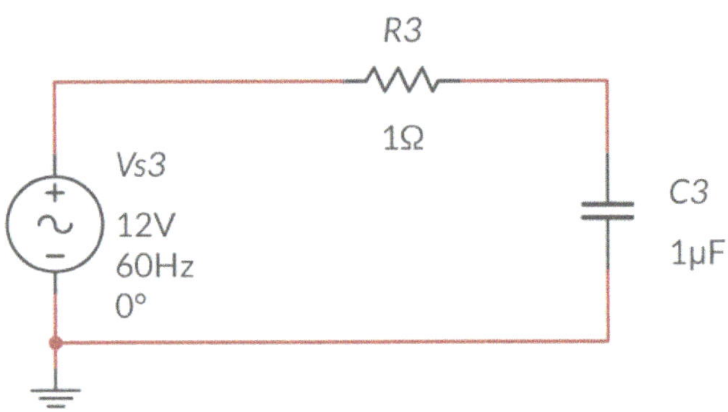

Figure 18.7 Ohmic-capacitive circuit.

Table 18.3 Summary of calculations

	T (s)	f (Hz)	Δt (s)	$\Delta\varphi$ (°)
$v_c(t)$				
$i_c(t)$				

2. Estimate the time delay Δt of the voltage $v_C(t)$ with respect to the current $i_C(t)$, and the phase difference $\Delta\varphi \cong \frac{360°}{T}\Delta t$ in degrees. Record results in Table 18.3 for further analysis and comparison.

19

Capacitive Reactance, Inductive Reactance, and Impedance

19.1 Student Learning Outcomes

- Demonstrate the ability to define the *impedance* $\overline{Z}\,(j\omega)$ of inductors and capacitors as a function of the frequency.
- Evaluate and compare the calculated capacitive and inductive *reactances* in electrical circuits with simulated data.
- Analyze capacitive and inductive *reactances* as a function of the frequency.

19.2 Prelab

In general, the *impedance* is defined as the ratio of an exponential voltage to the corresponding current. The impedance reflects the resistance to the flow of current and is measured in Ohms. Specifically, considering sinusoidal waves as a linear combination of complex exponentials,[1] we can express the complex impedance $\overline{Z}$ as the ratio of *phasors* representing sinusoidal voltages to currents.

$$\overline{Z} = \frac{\overline{V}}{\overline{I}} = R + jX = Z\angle\varphi_Z, \tag{19.1}$$

where *R* is the resistance; X is the *reactance*; $Z = \sqrt{R^2 + X^2}$ is the magnitude of the impedance, and $\varphi_Z = tg^{-1}(\frac{X}{R})$ is the phase of the impedance. By representing sinusoidal quantities as phasors, we simplify the analysis of electrical circuits operating with sinusoidal signals. The complex impedance $\overline{Z}$ captures both the magnitude and phase relationship between the voltage and current phasors, providing valuable insights into the circuit behavior at different frequencies.

[1]For example, $\sin x = \frac{e^{jx} - e^{-jx}}{2j}$.

If a current $i(t) = \sqrt{2} \cdot I \sin(\omega t)$, represented in the phasor domain by $\overline{I} = I \angle 0°$, circulates through a resistor *R*, the corresponding phasor of the voltage will be $\overline{V} = I \cdot R \angle 0°$. Therefore, the impedance for sinusoidal currents can be calculated as follows:

$$\overline{Z} = \frac{I \cdot R \angle 0°}{I \angle 0°} = R\angle 0°. \tag{19.2}$$

Hence, the impedance for sinusoidal currents in a resistor is equal to the resistance *R*.

If the same sinusoidal current $i(t) = \sqrt{2} \cdot I \sin(\omega t)$ flows through an inductor *L*, the voltage $v_L(t)$ across the inductor can be determined in the time domain as follows:

$$\begin{aligned} v_L(t) &= L\frac{di_L(t)}{dt} = L \cdot \sqrt{2} \cdot I\frac{d}{dt}\sin(\omega t) = \sqrt{2} \cdot \omega \cdot L \cdot I\cos(\omega t) \\ &= \sqrt{2} \cdot \omega \cdot L \cdot I \sin(\omega t + 90°). \end{aligned} \tag{19.3}$$

In the phasor domain, we represent the sinusoidal voltage $v_L(t)$ by $\overline{V}_L = \sqrt{2} \cdot \omega \cdot L \cdot I\angle 90°$, where ω is the angular frequency.

The associated complex impedance can be calculated as:

$$\overline{Z}_L = \frac{\omega LI \angle 90°}{I \angle 0°} = \omega L\angle 90° = j\omega L = jX_L, \tag{19.4}$$

where *j* is the imaginary unit ($\sqrt{-1}$); X_L is defined as the *inductive reactance,* measured in Ohms.

If a sinusoidal voltage $v_C(t) = \sqrt{2} \cdot V \cdot sin(\omega t)$ is applied across a capacitor *C*, the corresponding current $i_C(t)$ through the capacitance can be calculated in the time domain as follows:

$$\begin{aligned} i_C(t) &= C\frac{dv_c(t)}{dt} = \sqrt{2} \cdot V \cdot C \cdot \frac{d}{dt}\sin(\omega t) = \sqrt{2}\omega CV\cos(\omega t) \\ &= \sqrt{2}\omega CV\sin(\omega t + 90°). \end{aligned} \tag{19.5}$$

In the phasor domain, we represent the sinusoidal current $i_C(t)$ by $\overline{I}_C = \sqrt{2}\omega CV\angle 90°$, where ω is the angular frequency.

The associated complex impedance can be calculated as follows:

$$\overline{Z}_C = \frac{V\angle 0°}{\omega CV\angle 90°} = \frac{1}{\omega C}\angle -90° = -j\frac{1}{\omega C} = -jX_C. \tag{19.6}$$

X_C is defined as the *capacitive reactance*, measured in Ohms.

Based on the definitions of impedance, the phasor voltage–current relations for each component are as follows:

For the resistor *R*:

$$\overline{V}_R = I\angle\varphi \cdot R\angle 0^\circ = IR\angle\varphi. \tag{19.7}$$

The voltage phasor $\overline{V}_R$ across a resistor is proportional to the current flowing through it and is in phase with it.

For the inductor *L*:

$$\overline{V}_L = j\omega L \cdot \overline{I}_L = \omega L\angle 90^\circ \cdot I_L\angle\varphi = \omega L I_L\angle(\varphi + 90^\circ). \tag{19.8}$$

The voltage phasor $\overline{V}_L$ across an inductor leads the current circulating through it by 90°.

$$\overline{V}_C = -j\frac{1}{\omega C}\overline{I}_C = \frac{1}{\omega C}\angle{-90^\circ} \cdot I_C\angle\varphi = \frac{1}{\omega C}I_C\angle\left(\varphi - 90^\circ\right). \tag{19.9}$$

The voltage phasor $\overline{V}_C$ across a capacitor lags the current circulating through it by 90°.

19.2.1 Problem 1

In the circuit of Figure 19.1, the sinusoidal voltage source is $v_s\left(t\right) = \sqrt{2} \cdot 6\sin(628t)$ supplies a capacitor *C* = 2 ţF.

1. Calculate the reactance X_C of the capacitor at the frequency *f* = 100 Hz.
2. Determine the complex impedance $\overline{Z}_C$ of the capacitor at the frequency *f* = 100 Hz.
3. Confirm the calculated value of the reactance X_C of the capacitor using Multisim simulation; you can use the readings of the *root mean square* (*rms*) values of voltage and current to calculate the reactance with the

Figure 19.1 Capacitive reactance.

formula $X_C = \frac{V_{c\ \mathrm{rms}}}{I_{c\ \mathrm{rms}}}$. The utilization of Multisim simulation offers a practical and effective means to accurately assess and analyze the voltage and current across specific components. By capturing the *rms* values, a comprehensive understanding of the electrical behavior within the circuit can be obtained.

4. Calculate the capacitive reactance X_C at the frequencies 100 kHz, 200 kHz, 300 kHz, . . . , 1 MHz.
5. In correspondence with each value of the capacitive reactance, determine the *rms* values of the current through the capacitor using the formula $I_C = \frac{V_{c\ \mathrm{rms}}}{X_c}$. Confirm your calculated *rms* values at each frequency with Multisim simulation.

19.2.1.1 Solutions to Problem 1

1. The capacitive reactance at the frequency f = 100 Hz can be calculated using the following formula:

$$X_C = \frac{1}{\omega C} = \frac{1}{628 \cdot 2 \cdot 10^{-6}} = 796.2\ \Omega. \tag{19.10}$$

2. The impedance $\overline{Z}_C$ of the circuit at the frequency f = 100 Hz can be calculated using the following formula:

$$\overline{Z}_C = -j\frac{1}{\omega C} = -j796.2\ \Omega. \tag{19.11}$$

3. The Multisim simulation provides the following *rms* values of voltage and current at f = 100 Hz (Figure 19.2).

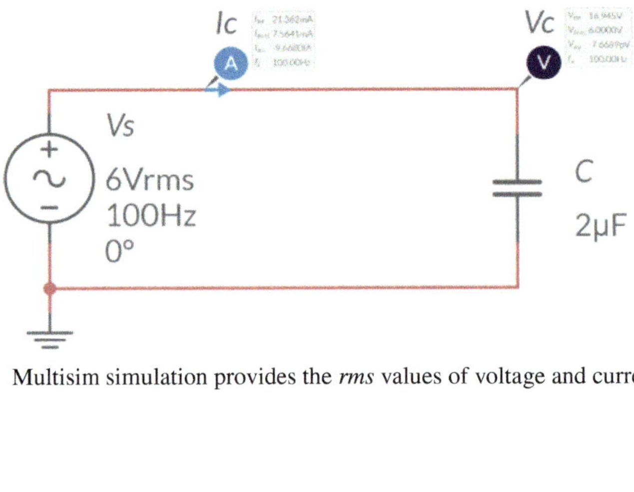

Figure 19.2 Multisim simulation provides the *rms* values of voltage and current at 100 Hz.

Table 19.1 Summary of calculations

f (kHz)	X_C (Ω)	$I_{C\ rms}$ (A)
100	0.796	7.54
200	0.398	15.08
300	0.265	22.62
400	0.199	30.16
500	0.159	37.70
600	0.133	45.24
700	0.114	52.78
800	0.099	60.32
900	0.088	67.86
1000	0.080	75.40

The reactance X_C of the capacitor can be calculated using the following formula:

$$X_c = \frac{V_{c\ \mathrm{rms}}}{I_{c\ \mathrm{rms}}} = \frac{6\ \mathrm{V}}{7.56 \cdot 10^{-3}\mathrm{A}} = 793.7\ \Omega. \tag{19.12}$$

The above value of X_C confirms the calculated value.

4. The values of the capacitive reactance X_C are listed in Table 19.1.

5. The *rms* values of the current I_C through the capacitor are listed in Table 19.1. The Multisim simulation confirms the calculated *rms* values at each frequency (Figure 19.2 shows results for f = 100 Hz).

19.2.2 Problem 2

In the circuit given in Figure 19.3, the sinusoidal current source is $i_s(t) = \sin(628t)$ supplies an inductor L = 10 mH.

Figure 19.3 Inductive reactance.

1. Calculate the reactance X_L of the inductor at the frequency f = 100 Hz.
2. Determine the complex impedance $\overline{Z}_L$ of the inductor at the frequency f = 100 Hz.
3. Determine the phasor $\overline{V}_L$ of the voltage across the inductor L at the frequency f = 100 Hz.
4. Confirm the calculated value of the reactance X_L of the inductor using Multisim simulation; you can use the readings of the *root mean square* (*rms*) values of voltage and current to calculate the reactance with the formula $X_L = \frac{V_{L\ \mathrm{rms}}}{I_{L\ \mathrm{rms}}}$.
5. Calculate the inductive reactance X_L at the frequencies 200 Hz, 300 Hz, ..., 1 kHz.
6. In correspondence with each value of the inductive reactance, determine the *rms* values of the voltage across the inductor using the formula $V_L = X_L \cdot I_L$. Confirm the calculated *rms* values at each frequency with Multisim simulation.

19.2.2.1 Solutions to Problem 2

1. The inductive reactance X_L at the frequency f = 100 Hz can be calculated with the following formula:

$$X_L = \omega L = 628 \cdot 10 \cdot 10^{-3} = 6.3\ \Omega. \tag{19.13}$$

2. The complex impedance $\overline{Z}_L$ of the inductor at the frequency f = 100 Hz can be calculated with the following formula:

$$\overline{Z}_L = j\omega L = 6.28\ \Omega. \tag{19.14}$$

3. The phasor $\overline{V}_L$ of the voltage across the inductor L at 100 Hz can be calculated with the following formula:

$$\overline{V}_L = \overline{I}_s \cdot j\overline{X}_L = 0.707\angle 0^\circ \cdot 6.28\angle 90^\circ = 4.4\angle 90^\circ\ \mathrm{V}. \tag{19.15}$$

4. The Multisim simulation provides the following *rms* values of voltage and current at f = 100 Hz.

The inductive reactance can be calculated with the following formula:

$$X_L = \frac{V_{L\ \mathrm{rms}}}{I_{L\ \mathrm{rms}}} = \frac{4.4}{0.7} = 6.3\ \Omega, \tag{19.16}$$

which confirms the calculated value.

5. The values of the inductive reactance X_L at different values of frequency are listed in Table 19.2.

Table 19.2 Summary of calculations

f (Hz)	X_L (Ω)	$V_{L\ rms}$ (V)
200	12.6	8.9
300	18.8	13.3
400	25.1	17.8
500	31.4	22.2
600	37.7	26.7
700	44.0	31.1
800	50.3	35.5
900	56.5	40.0
1000	62.8	44.4

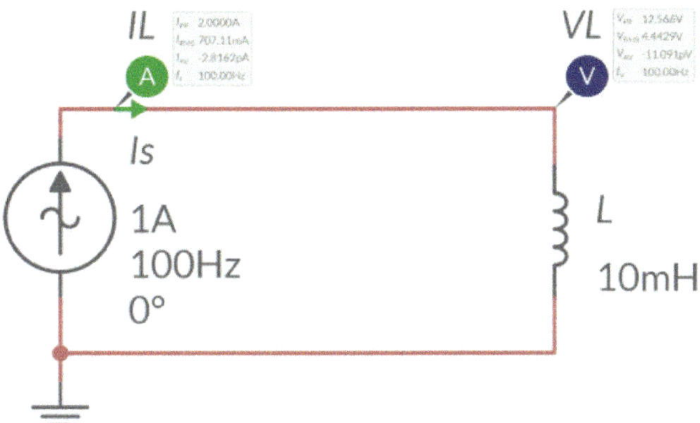

Figure 19.4 Multisim simulation providing *rms* values of voltage and current at 100 Hz.

6. The calculated *rms* values of the voltage V_L across the inductor are listed in Table 19.2. The Multisim simulation confirms the calculated *rms* values of V_L at each frequency (Figure 19.4 shows the *rms* value for f = 100 Hz).

19.3 Lab Procedure 1

In the circuit given in Figure 19.5, the sinusoidal voltage source $v_s(t) = \sqrt{2} \cdot 4 \cdot \sin(628t)$ supplies the series of a resistor R_1 = 470 Ω and a capacitor C_1 = 0.22 μF.

1. Through the utilization of Multisim simulation, determine the *rms* values of the voltages $v_{C1}(t)$ and $v_{R1}(t)$ across the capacitor C_1 and resistor R_1, respectively. These measurements will be performed across a range of frequencies as indicated in Table 19.3.

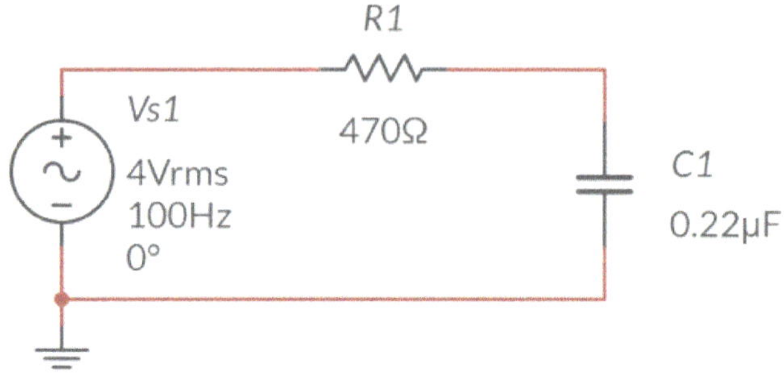

Figure 19.5 Ohmic-capacitive circuit with ac voltage generator, and resistor and capacitor in series.

2. By means of Multisim simulation, measure the *rms* values of the current $i(t)$ through the capacitor C_1 for the frequencies listed in Table 19.3. Through the utilization of Multisim simulation, measure the *rms* values of the current $i(t)$ flowing through the capacitor C_1. This analysis will be conducted for various frequencies as specified in Table 19.3. By performing these simulations, it becomes possible to accurately assess and quantify the current behavior through the capacitor across the specified frequency range. Multisim provides a reliable platform for conducting such measurements, allowing for precise and reproducible results in determining the *rms* current values.
3. For each frequency specified in Table 19.3, calculate the capacitive reactance X_{C1} of the capacitor C_1. These calculated values must be recorded in Table 19.3 to facilitate further analysis and comparison.
4. For each frequency, calculate the capacitive reactance X_{C1} through the ratio of the measurements of the *rms* values of voltage and current. Record values in Table 19.3.
5. For each frequency specified in Table 19.3, calculate the capacitive reactance X_{C1} by taking the ratio of the measurements of the *rms* values of voltage and current, by measuring them using Multisim. These calculated values should be recorded in Table 19.3 to provide a comprehensive understanding of the capacitive behavior at different frequencies.
6. For each frequency listed in Table 19.4, calculate the total complex impedance $\overline{Z}_T = R_1 - jX_{C1}$, its magnitude Z_T, and phase φ. These values must be recorded in Table 19.4 to facilitate the analysis and understanding of the impedance characteristics at different frequencies.
7. For each frequency specified in Table 19.4, calculate the magnitude Z_T of the total impedance by taking the ratio of the measurements of

Table 19.3 Summary of readings and calculations

f (Hz)	$V_{C1\ rms}$ (V)	$V_{R1\ rms}$ (V)	I_{rms} (mA)	$X_{C1} = \frac{1}{\omega C_1}$ (kΩ)	$X_{C1} = \frac{V_{C1\ \mathrm{rms}}}{I_{\ \mathrm{rms}}}$ (kΩ)
100					
200					
300					
400					
500					
600					
700					
800					
900					
1000					

Table 19.4 Summary of calculations

f (Hz)	$\overline{Z}_T = R_1 - jX_{C1}$ (kΩ)	$Z_T = \sqrt{R_1^2 + X_{C1}^2}$ (kΩ)	$\varphi = tg^{-1}\left(-\frac{X_{C1}}{R_1}\right)$	$Z_T = \frac{V_{s1}}{I_{\ \mathrm{rms}}}$ (kΩ)
100				
200				
300				
400				
500				
600				
700				
800				
900				
1000				

the *rms* values of voltage and current. These calculated values must be recorded in Table 19.4 for comparison with the impedance values obtained through other calculations to verify the consistency and accuracy of the measurements and calculations.

8. Calculate the frequency f_q of the generator at which $X_{C1} = R_1$; set up a Multisim simulation with the given components and parameters to verify that the calculated value of f_q is correct.

19.4 Lab Procedure 2

In the circuit given in Figure 19.6, consisting of a sinusoidal current source $i_{s1}(t) = \sqrt{2} \cdot 2 \cdot \sin(628t)$ supplying a series combination of a resistor R_2 = 470 Ω and an inductor L_2 = 20 mH, use Multisim simulation for measurements.

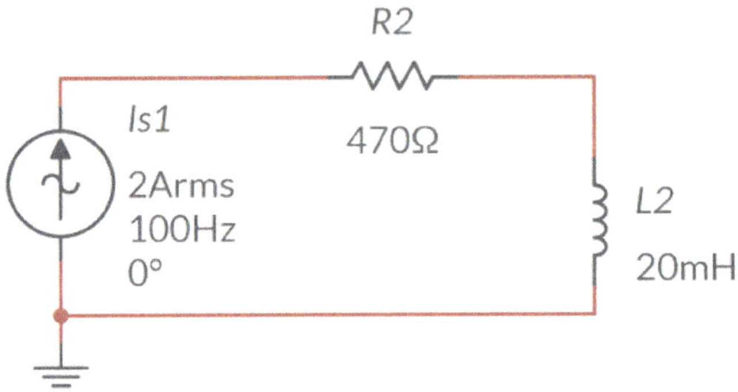

Figure 19.6 Ohmic-inductive circuit with an ac current generator, and resistor and inductor in series.

1. Measure the *rms* values of the voltages $v_{L2}(t)$ and $v_{R2}(t)$, across the inductor L_2 and the resistor R_2, respectively, for the frequencies specified in Table 19.5.
2. Measure the *rms* values of the current $i(t)$ through the inductor L_2 for the frequencies specified in Table 19.5.
3. Calculate the inductive reactance X_{L2} for the frequencies specified in Table 19.5. These values must be recorded in the corresponding column of Table 19.5. This allows for a comprehensive analysis of the inductive reactance across different frequencies and helps understand the impact of frequency on the behavior of the inductor in the circuit.
4. For the frequencies specified in Table 19.5, calculate the inductive reactance X_{L2} by taking the ratio of the measurements of the *rms* values of voltage and current. Record values in Table 19.5. Record the calculated inductive reactance values in the corresponding column in Table 19.5 for comparison with the values obtained through other calculations to verify the consistency and accuracy of the measurements and calculations.
5. For each frequency specified in Table 19.6, calculate the total complex impedance $\overline{Z}_T = R_2 + jX_{L2}$, its magnitude Z_T, and phase φ. These values must be recorded in Table 19.6 to facilitate the analysis and understanding of the impedance characteristics at different frequencies.
6. For each frequency specified in Table 19.6, calculate the magnitude Z_T of the total impedance by taking the ratio of the measurements of the *rms* values of the voltage $v(t)$ across the series combination and the current. These values must be recorded in Table 19.6 for comparison with the values obtained through other calculations to verify the consistency and accuracy of the measurements and calculations.

Table 19.5 Summary of readings and calculations

f (Hz)	$V_{L2\ rms}$ (V)	$V_{R2\ rms}$ (V)	I_{rms} (mA)	$X_{L2} = \omega L_2$ (kΩ)	$X_{L2} = \frac{V_{L2rms}}{I_{rms}}$ (kΩ)
100					
200					
300					
400					
500					
600					
700					
800					
900					
1000					

Table 19.6 Summary of calculations

f (Hz)	$\overline{Z}_T = R_2 + jX_{L2}$ (kΩ)	$Z_T = \sqrt{R_2^2 + X_{L2}^2}$ (kΩ)	$\varphi = tg^{-1}\left(\frac{X_{L2}}{R_2}\right)$	$Z_T = \frac{V_{rms}}{I_{rms}}$ (kΩ)
100				
200				
300				
400				
500				
600				
700				
800				
900				
1000				

7. Calculate the frequency f_q of the generator at which $X_{L2} = R_2$; set up a Multisim simulation with the given components and parameters to verify that the calculated value of f_q is correct.

20

Nonsinusoidal Waveforms

20.1 Student Learning Outcomes

- Apply measurement techniques to determine the *mean* and *rms* values of specific nonsinusoidal waveforms, such as square wave, triangular wave, and sinewave with a superimposed DC voltage.
- Utilize theoretical knowledge to calculate the *mean* and *rms* values of nonsinusoidal waveforms based on waveform characteristics.

20.2 Prelab

20.2.1 Square wave

An ideal square wave exhibits two symmetrical vertical ramps of same height, separated by a defined period of time called the *pulse width* t_w (Figure 20.1). The *duty cycle D*, often expressed as a percentage, represents the ratio of the

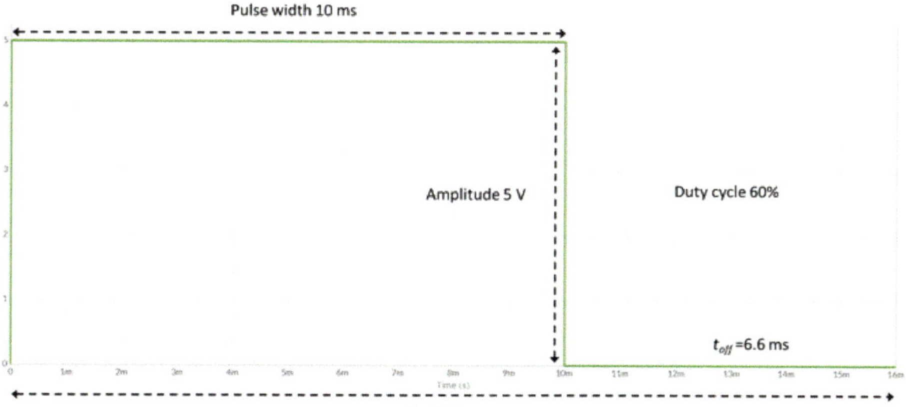

Figure 20.1 Ideal square wave.

pulse width to the period T.

$$D = \frac{t_w}{T} \times 100. \tag{20.1}$$

The *duty cycle* quantifies the duration of the active high portion of the square wave relative to the entire period.

The *mean* and *root mean square* (*rms*) values for a square wave can be determined using the same general expressions as defined for sinusoids. They can be calculated using the following formulae:

$$V_{\text{ave}} = \frac{1}{T}\int_0^T v(t)\,dt = \frac{1}{T}\int_0^{t_w} V\,dt = \frac{V \cdot t_w}{T} \tag{20.2}$$

$$V_{\text{rms}} = \sqrt{\frac{1}{T}\int_0^T v^2(t)\,dt} = \sqrt{\frac{1}{T}\int_0^{t_w} V^2 dt} = V\sqrt{\frac{t_w}{T}}. \tag{20.3}$$

20.2.2 Triangular waveform

A triangular waveform is characterized by its composition of two ramps, each possessing equal slopes but with opposite signs (Figure 20.2). The duration of the downward slope, known as the *fall time*, corresponds precisely to half of the waveform's period. Like other waveforms, the *mean* and *root mean square* (*rms*) values can be determined using standard formulae employed for waveform analysis. However, by exploiting the inherent symmetry of the waveform, it is possible to streamline the calculation process. Specifically,

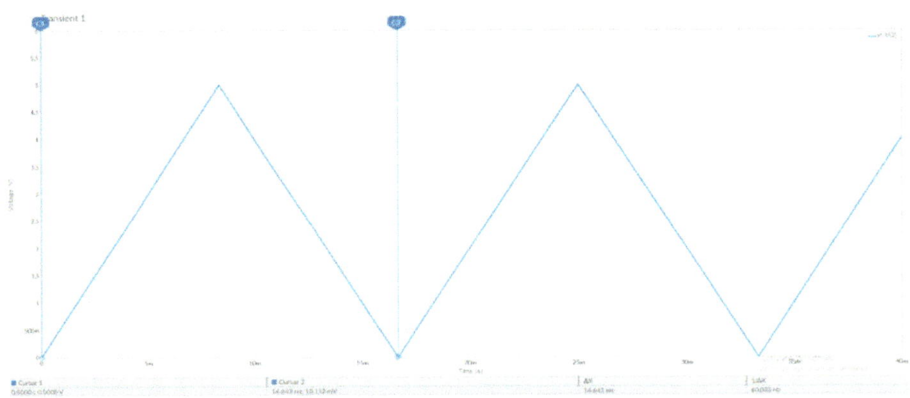

Figure 20.2 Triangular waveform.

the analysis focuses on only one-quarter of the waveform's period, (i.e., from $t = 0$ to $t = T/4$). By restricting the integration to this interval, the calculations can be performed efficiently while accurately capturing the waveform's characteristics.

The integrand function $v(t)$ valid in the interval $t = 0$ and $t = T/4$ can be expressed as follows:

$$v(t) = \frac{V_p}{\frac{T}{4}} t, \tag{20.4}$$

where V_p is the amplitude of the triangular waveform, and T is the period.

The *mean* value can be calculated as follows:

$$V_{\text{ave}} = \frac{4}{T} \int_0^{\frac{T}{4}} v(t)\, dt = \frac{4}{T} \int_0^{\frac{T}{4}} \frac{4V_p}{T} t\, dt = \frac{V_p}{2}. \tag{20.5}$$

The *rms* value can be calculated as follows:

$$V_{\text{rms}} = \sqrt{\frac{4}{T} \int_0^{\frac{T}{4}} v^2(t)\, dt} = \sqrt{\frac{4}{T} \left(\frac{4V_p}{T}\right)^2 \int_0^{\frac{T}{4}} t^2 dt} = \sqrt{\frac{64{V_p}^2}{T^3} \cdot \left.\frac{t^3}{3}\right|_0^{\frac{T}{4}}}$$
$$= \frac{V_p}{\sqrt{3}}. \tag{20.6}$$

20.2.3 Sine wave with a superimposed DC voltage

When a direct current (DC) quantity V_0 is superimposed onto a sinusoidal waveform, the resulting waveform $v(t)$ is a sinusoid that appears to be shifted up or down, "riding" the DC level (Figure 20.3).

The equation describing this resulting waveform is expressed as follows:

$$v(t) = V_0 + V_p \cdot \sin(\omega t). \tag{20.7}$$

The waveform $v(t)$ exhibits a maximum amplitude equal to the sum of the DC quantity and the peak amplitude of the sinusoid (i.e., $V_0 + V_P$). It also has a minimum amplitude equal to the difference between the DC quantity and the peak amplitude of the sinusoid ($V_0 - V_P$).

The *mean* value of the waveform is the DC component and is given by

$$V_{\text{ave}} = \frac{1}{T} \int_0^T v(t)\, dt = \frac{1}{T} \int_0^T (V_0 + V_P \sin\omega t)\, dt = V_0. \tag{20.8}$$

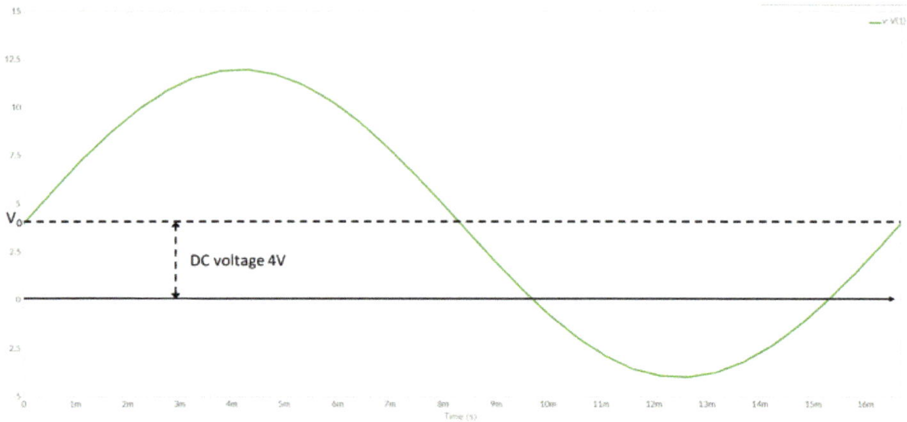

Figure 20.3 Sinusoidal waveform with a superimposed DC voltage.

The *root mean square* (*rms*) value of the waveform can be calculated using the following equation:

$$V_{\text{rms}} = \sqrt{\frac{1}{T}\int_0^T v^2\left(t\right)dt} = \sqrt{\frac{1}{T}\int_0^T \left(V_0 + V_P\sin\omega t\ \right)^2 dt} = \sqrt{{V_0}^2 + \frac{V_P^2}{2}}. \tag{20.9}$$

The validity of V_{rms} can be verified when $V_0 = 0$ (i.e., the waveform $v\left(t\right)$ is solely determined by the sinusoidal component), which yields $V_{\text{rms}} = \frac{V_P}{\sqrt{2}}$.

20.2.4 Problem 1

The circuit shown in Figure 20.4 includes a square wave generator, also known as a *Clock* source, operating at a frequency of 60 Hz. The square wave generator produces a square waveform with an amplitude of 5 V and a duty cycle of 50%.

1. Determine the *mean* and the *rms* values of the square wave using theoretical calculations and validate the results through simulation using Multisim.

20.2.4.1 Solutions to Problem 1

The *mean* value of the square wave with a 50% duty cycle can be calculated as follows:

$$V_{\text{ave}} = \frac{V \cdot t_w}{T} = \frac{5 \cdot 0.5T}{T} = 2.5\ \text{V}. \tag{20.10}$$

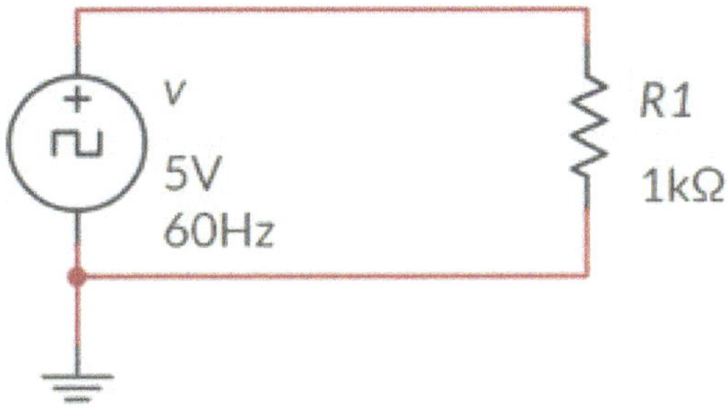

Figure 20.4 Resistive circuit with square wave generator.

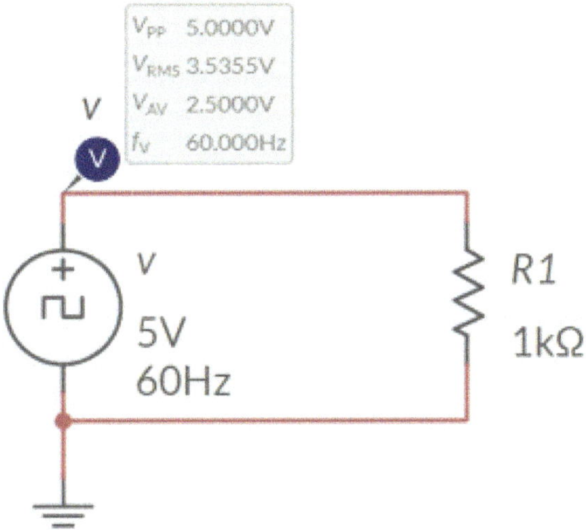

Figure 20.5 Multisim simulation showing the *mean* and the *rms* values of a square wave.

The *rms* value of the square wave with a 50% duty cycle can be calculated as follows:

$$V_{\text{rms}} = V\sqrt{\frac{t_w}{T}} = 3.54 \text{ V}. \tag{20.11}$$

Multisim simulation confirms the calculated values (Figure 20.5).

20.2.5 Problem 2

The circuit shown in Figure 20.6 includes a triangular waveform generator operating at a frequency of 60 Hz. The triangular wave generator produces a waveform with an amplitude of 5 V and a *fall time* of 8.33 ms.

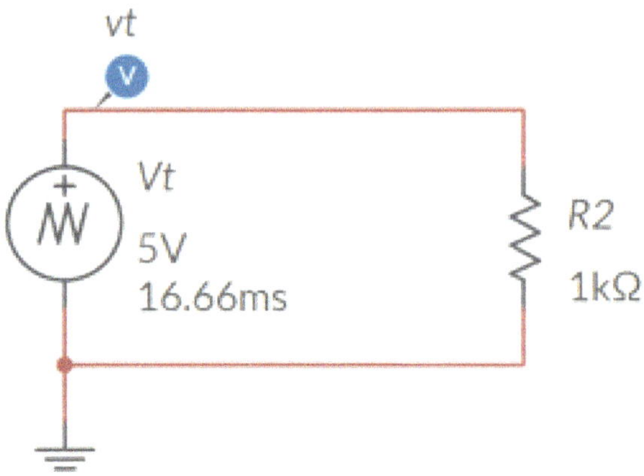

Figure 20.6 Resistive circuit with triangular waveform generator.

1. Determine the *mean* and the *rms* values of the triangular waveform using theoretical calculations and validate the results through simulation using Multisim.

20.2.5.1 Solutions to Problem 2

The *mean* value of the triangular waveform can be calculated as follows:

$$V_{\text{ave}} = \frac{V_P}{2} = 2.5\ \text{V}. \tag{20.12}$$

The *rms* value of the triangular waveform can be calculated as follows:

$$V_{\text{rms}} = \frac{V_P}{\sqrt{3}} = 2.9\ \text{V}. \tag{20.13}$$

Multisim simulation confirms the calculated values (Figure 20.7).

20.2.6 Problem 3

The circuit depicted in Figure 20.8 consists of a sinusoidal waveform generator with a peak voltage V_P of 8 V and a frequency f of 60 Hz. Additionally, a DC voltage with an amplitude V_0 of 4 V is superimposed onto the sinusoidal waveform.

1. Determine the *mean* and the *rms* values of the resulting voltage waveform $v_1(t)$ across the resistor using theoretical calculations and validate the results through simulation using Multisim simulation.

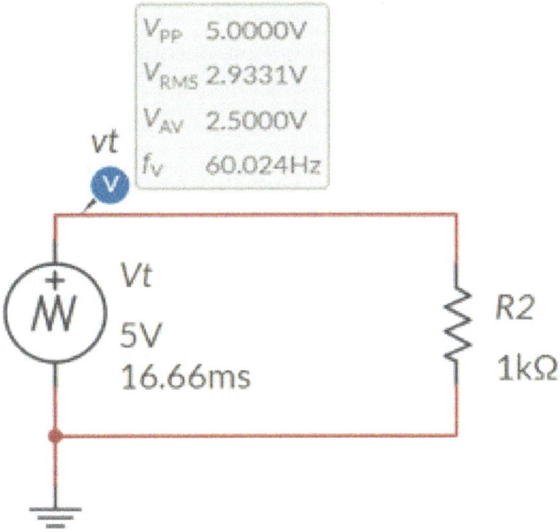

Figure 20.7 Multisim simulation showing the *mean* and the *rms* values of the triangular square wave.

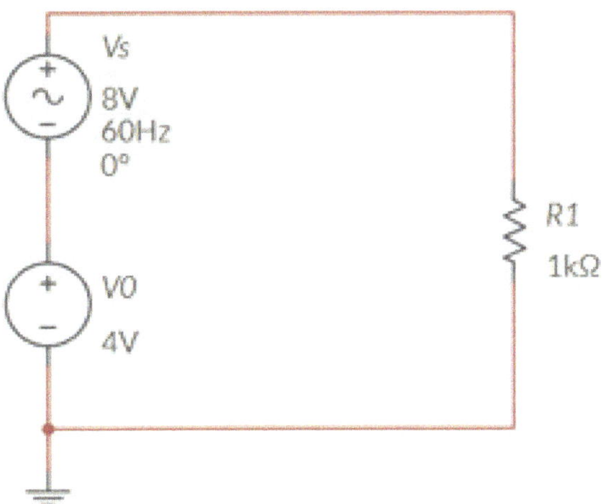

Figure 20.8 Resistive circuit with a sinusoidal wave generator with a superimposed DC voltage.

2. Determine the maximum and minimum values of the resulting voltage waveform $v_1(t)$ across the resistor using theoretical calculations and validate the results through Multisim simulation.

20.2.6.1 Solutions to Problem 3

1. The *mean* value of $v_1(t)$ can be calculated as follows:

$$V_{\text{ave}} = V_0 = 4 \text{ V}. \tag{20.14}$$

The *rms* value of $v_1(t)$ can be calculated as follows:

$$V_{\text{rms}} = \sqrt{{V_0}^2 + \frac{V_P^2}{2}} = 6.9 \text{ V}. \tag{20.15}$$

Multisim simulation confirms the calculated values (Figure 20.9).

2. The maximum and minimum values of $v_1(t)$ can be calculated as follows:

$$V_{1\,\max} = V_0 + V_P = 12V \tag{20.16}$$

$$V_{1\,\min} = V_0 - V_P = -4V. \tag{20.17}$$

Multisim simulation confirms the calculated values (Figure 20.10).

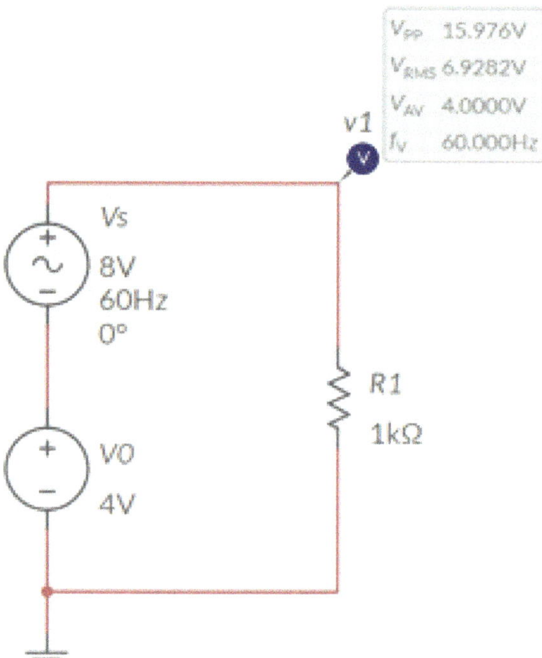

Figure 20.9 Multisim simulation showing the *mean* and the *rms* values of the resulting waveform.

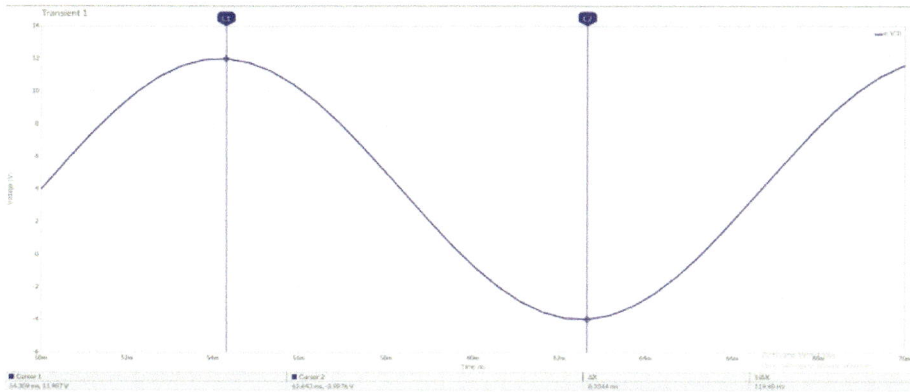

Figure 20.10 Multisim simulation showing maximum and minimum values of the resulting waveform $v_1(t)$.

20.3 Lab Procedure 1

The circuit shown in Figure 20.4 includes a square wave generator, also known as a *Clock* source, operating at a frequency of 60 Hz. The square wave generator produces a square waveform with an amplitude of 5 V and a duty cycle of 50%.

The circuit illustrated in Figure 20.11 incorporates a square wave generator that operates at a frequency of 120 Hz. The square wave generator produces a square waveform with an amplitude of 10 V and a duty cycle of 80%. This configuration allows us to explore the characteristics of the square waveform and compare theoretical calculations with simulated results.

1. Determine the *mean* and the *rms* values of the square wave using theoretical calculations and validate the results through simulation

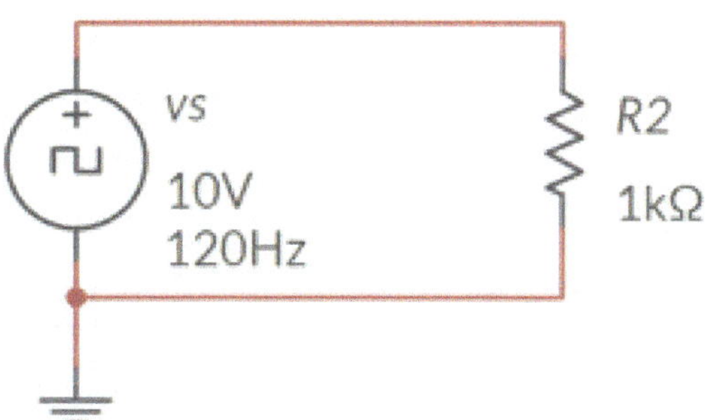

Figure 20.11 Resistive circuit with square wave generator.

Table 20.1 Summary of readings and calculations

	Calculated values		Simulated values	
Duty cycle	**mean** (V)	**rms** (V)	**mean** (V)	**rms** (V)
80%				
20%				

using Multisim. Record results in Table 20.1 for further analysis and comparison.

2. Repeat the above calculations using a duty cycle of 20%. Record results in Table 20.1 for further analysis and comparison.

20.4 Lab Procedure 2

The circuit shown in Figure 20.12 includes a triangular waveform generator operating at a frequency of 50 Hz. The triangular wave generator produces a waveform with an amplitude of 12 V and a *fall time* of 10 ms.

1. Determine the *mean* and the *rms* values of the triangular waveform using theoretical calculations and validate the results through Multisim simulation. Record results in Table 20.2 for further analysis and comparison.
2. Repeat the above calculations using a *fall time* of 10 ms. Record results in Table 20.2 for further analysis and comparison.

Figure 20.12 Resistive circuit with triangular waveform generator.

Table 20.2 Summary of readings and calculations

	Calculated values		Simulated values	
Fall time (ms)	**mean** (V)	**rms** (V)	**mean** (V)	**rms** (V)
10				
5				

21

Resistive Circuits with AC Sources

21.1 Student Learning Outcomes

- Determine the voltage across and the current through resistors in different operating conditions, both theoretically and through simulations using Multisim.
- Determine the *instantaneous* power and the *mean* power absorbed in resistive circuits, both theoretically and through simulations using Multisim.

21.2 Prelab

In AC circuits, resistors still have resistance, which remains the same as in DC circuits. The *impedance* of a resistor in an AC circuit is equal to its resistance and is purely a real number. This means that the resistance of a resistor remains unchanged regardless of the frequency of the AC signal. The Ohm's law still holds true, where the voltage across a resistor is proportional to the current passing through it. Other formulae and principles commonly used in DC circuits, such as power calculations, also apply to AC circuits with resistors.

In AC circuits, the voltage and current may not be in phase due to the presence of *reactance* in addition to resistance. However, in a resistive circuit, the voltage and current remain in phase. This means that the waveforms of voltage and current reach their peak and zero points simultaneously.

In resistive AC circuits, power is dissipated solely in the form of heat due to the resistance. Resistors are passive elements in electrical circuits, for which the *passive sign* convention is adopted (Figure 21.1).

This convention signifies that a passive element cannot generate or supply more energy than it receives. For any instant of time, eqn (21.1) must hold

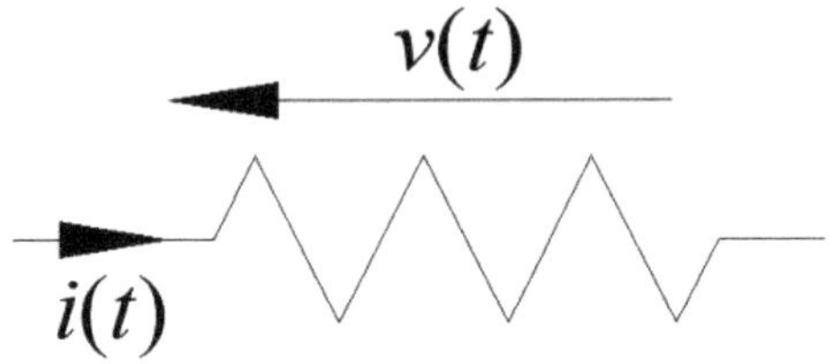

Figure 21.1 Passive sign convention for a resistor.

true, ensuring the passive nature of the component.

$$w_{\text{abs}}(t) = \int_{-\infty}^{t} p_{abs}(t')dt' \geq 0, \ \ \forall t, \tag{21.1}$$

where $p_{\text{abs}}(t)$ is the power absorbed by the component.

When an alternating current (AC) voltage $v(t) = V_p\cos(\omega t + \varphi)$ is applied across a resistor *R*, the resulting AC steady-state current *i*(*t*) through the resistor is described by eqn (21.2).

$$i(t) = \frac{V_p\cos(\omega t + \varphi)}{R}. \tag{21.2}$$

It is noteworthy that the voltage and current possess the same phase angle φ, while their peak values differ by the factor *R*.

The instantaneous steady-state power $p_R(t)$ absorbed by the resistor can be obtained by multiplying the voltage and current waveforms, as given in eqn (21.3). For simplicity, we may assume $\varphi = 0$.[1]

$$p_R(t) = v(t)\,i(t) = \frac{V_p^2}{R}\cos^2(\omega t) = \frac{V_p^2}{2R} + \frac{V_p^2}{2R}\cos(2\omega t)\ . \tag{21.3}$$

Eqn (21.3) reveals that the power waveform consists of a constant term and a sinusoidal time-varying term. The sinusoid has a zero *mean* value, resulting in the average power *P* coinciding with the constant term, as expressed in eqn (21.4).

$$P = \langle p_R(t)\rangle = \frac{1}{2}\frac{V_p^2}{R} = \frac{V_{\text{rms}}^2}{R} = I_{\text{rms}}^2 R = V_{\text{rms}} I_{\text{rms}}. \tag{21.4}$$

---

[1]To develop the expression of $p_R(t)$, we can use the trigonometric identity $\cos^2 x = \frac{1}{2} + \frac{1}{2}\cos(2x)$.

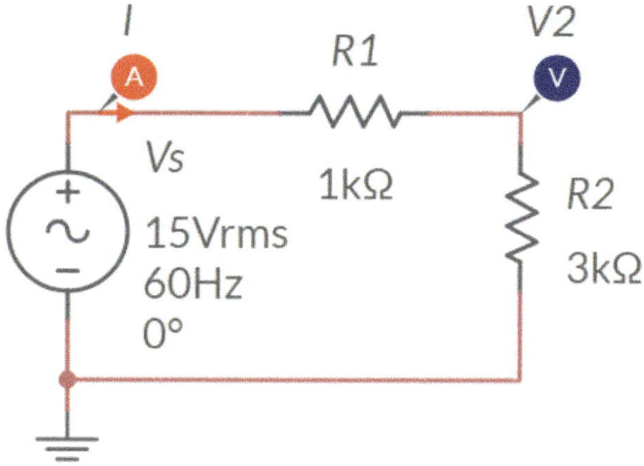

Figure 21.2 Series resistive circuit with an AC generator.

21.2.1 Problem 1

In relation to the circuit shown in Figure 21.2, the following tasks are to be performed utilizing both theoretical calculations and Multisim simulations:

1. Conduct a Multisim simulation to plot the sinusoidal current waveform $i(t)$ and the corresponding voltage waveform $v_2(t)$ across resistor R_2. Ensure that the plots clearly demonstrate the phase relationship between the current and voltage signals.
2. Calculate the *rms* values, I_{rms} and V_{2rms}, of the current flowing through resistor R_2 and the voltage across resistor R_2 using the appropriate theoretical formulae. Simulate the circuit in Multisim and measure the *rms* values of the current and voltage using the simulation tools. Compare these measured values with the theoretical calculations obtained earlier.
3. Calculate the average power P_2 absorbed by the resistor R_2 and the average power P_s supplied by the AC generator.

21.2.1.1 Solutions to Problem 1

1. The plot of $v_2(t)$ and $i(t)$ obtained with Multisim simulation is presented in Figure 21.3.

The plot demonstrates that the current and voltage are in phase, indicating that they reach peak values and zero-crossings at the same time. This observation aligns with the expected behavior of current and voltage for a resistor in an AC circuit.

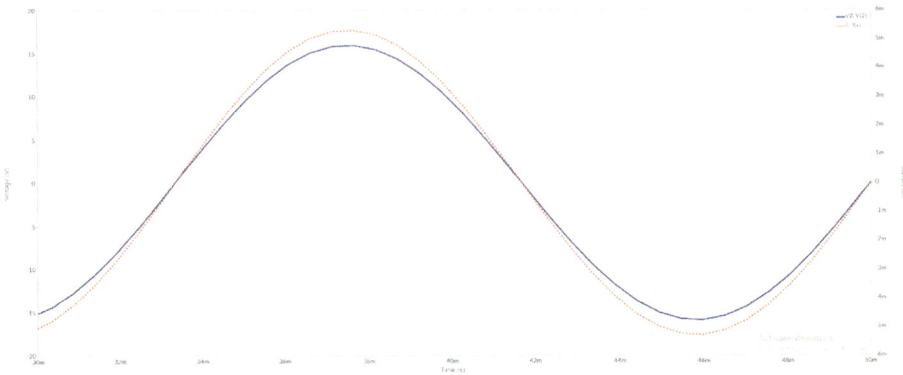

Figure 21.3 Plot of $v_2(t)$ and $i(t)$ obtained with Multisim simulation.

2. The *rms* value of $v_2(t)$ cane be calculated as follows:

$$\overline{V}_2 = \overline{V}_s \frac{R_2}{R_1 + R_2} = 15\angle 0^\circ \cdot 0.75 = 11.25\angle 0^\circ \text{ V}. \tag{21.5}$$

Therefore, V_{2rms} = 15 V.

The *rms* value of $i(t)$ cane be calculated as follows:

$$\overline{I} = \frac{\overline{V}_2}{R_2} = \frac{11.25\angle 0^\circ}{3 \text{ k}\Omega} = 3.75 \angle 0^\circ \text{ mA}. \tag{21.6}$$

Therefore, I_{rms} = 3.75 mA.

The theoretical and simulated values match, confirming the accuracy and reliability of the theoretical formulae in determining the *rms* values of the current and voltage (Figure 21.4).

3. The average power P_2 absorbed by the resistor R_2 can be calculated as follows:

$$P_2 = I_{\text{rms}}^2 R_2 = 42.2 \text{ mW}. \tag{21.7}$$

The average power P_s supplied by the AC voltage generator can be calculated as follows:

$$P_s = V_{\text{rms}} I_{\text{rms}} = 56.2 \text{ mW}. \tag{21.8}$$

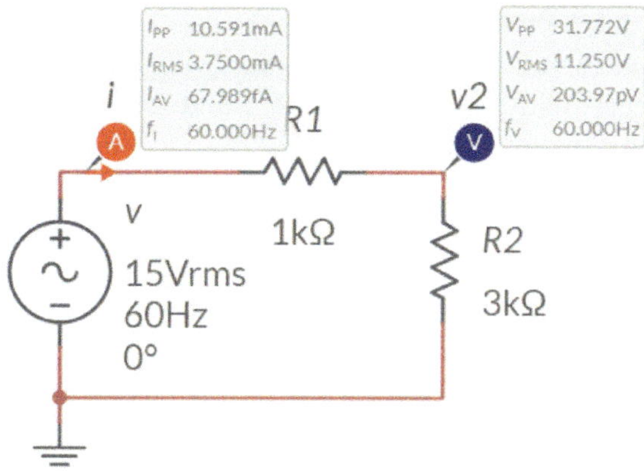

Figure 21.4 Multisim simulation providing *rms* values of voltage and current.

21.2.2 Problem 2

The given circuit (Figure 21.5) consists of two iso-frequency voltage sources, $v_1(t) = 10 \cdot \sin(376.8t)$ and $v_1(t) = 10 \cdot \sin(376.8t + \varphi)$ connected to a resistor $R = 2\ \mathrm{k}\Omega$. Analyze the circuit using Multisim simulation and calculate/measure various quantities for different phase angles φ (0°, 45°, –45°). The following tasks need to be performed:

1. Calculate the *rms* value of the current $i(t)$ circulating through R. Measure the *rms* value of the current $i(t)$ using Multisim and compare it with the calculated value.
2. Calculate the *rms* value of the voltage $v(t)$ across R. Measure the *rms* value of the of the voltage $v(t)$ using Multisim and compare it with the calculated value.
3. Calculate the average power absorbed by the resistor R for $\varphi = \pm 45°$.

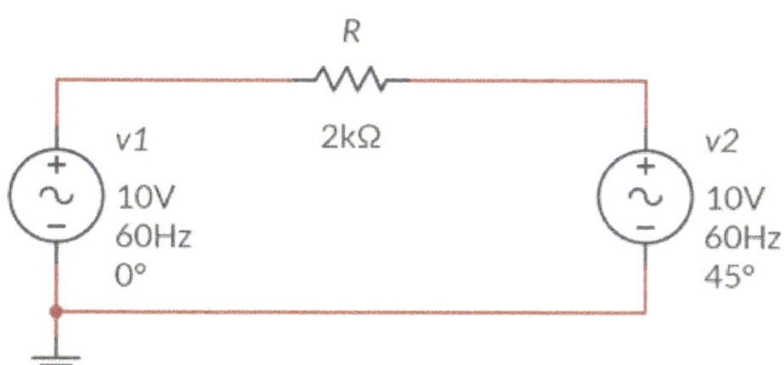

Figure 21.5 Resistive circuit with two AC iso-frequency generators of same amplitude.

21.2.2.1 Solutions to Problem 2

1. The *rms* value of the current $i(t)$ circulating through R can be calculated in correspondence with φ = 0°, φ = 45°, and φ = –45° by applying Kirchhoff's voltage law as follows:

$$10\angle 0^\circ - I\angle\alpha\cdot R - 10\angle\varphi = 0, \tag{21.9}$$

$$I_{\mathrm{rms}}\angle\alpha = \frac{10\angle 0^\circ - 10\angle\varphi}{\sqrt{2}R}. \tag{21.10}$$

For $\varphi = 0^\circ$, $I_{rms} = 0$; for $\varphi = \pm 45^\circ$, I_{rms} = 2.71 mA.

The current flowing through the resistor is determined by the voltage difference across the resistor and the resistance value. Even if the two generators have the same amplitude, the current will not be zero if their voltages are not in phase (i.e., $\varphi \neq 0$). Multisim simulation confirms the calculated *rms* values of the current (Figure 21.6).

2. The *rms* value of the voltage the voltage $v(t)$ across resistor R can be calculated for each phase angle $\varphi = 0^\circ$, $\varphi = 45^\circ$and $\varphi = -45^\circ$ by applying Ohm's law:

$$V_{\mathrm{rms}}\angle\beta = I_{\mathrm{rms}}\angle\alpha\cdot R. \tag{21.11}$$

For $\varphi = 0^\circ$, $V_{rms} = 0$; for $\varphi = \pm 45^\circ$, V_{rms} = 5.4 V.

Multisim simulation confirms the calculated *rms* values of the voltage (Figure 21.6).

3. The average power P dissipated by the resistor R for $\varphi = \pm 45^\circ$ can be calculated with the following formula:

$$P = I_{\mathrm{rms}}^2 R = 14.7mW. \tag{21.12}$$

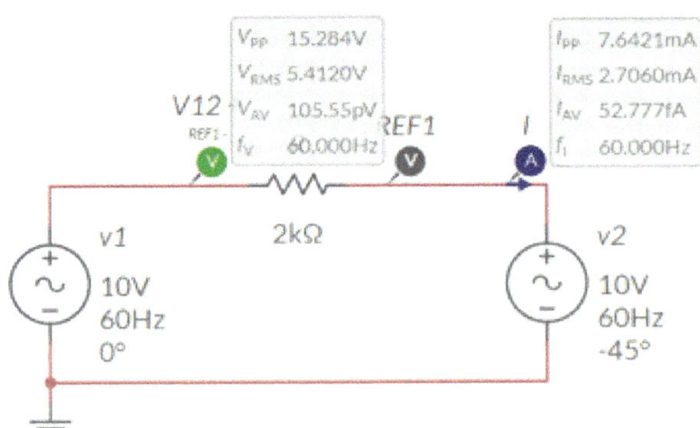

Figure 21.6 Multisim simulation providing *rms* values of voltage and current for $\varphi = -45^\circ$.

21.3 Lab Procedure 1

In relation to the circuit shown in Figure 21.7, the following tasks are to be performed utilizing both theoretical calculations and Multisim simulations:

1. Conduct a Multisim simulation to plot the sinusoidal current waveform $i_1(t)$ and the corresponding voltage waveform $v_5(t)$ across resistor R_5. Ensure that the plots clearly demonstrate the phase relationship between the current and voltage signals.
2. Calculate the *rms* values, I_{rms} and V_{5rms}, of the current flowing through resistor R_5 and the voltage across resistor R_5 using the appropriate theoretical formulae. Simulate the circuit in Multisim and measure the *rms* values of the current and voltage using the simulation tools. Compare these measured values with the theoretical calculations obtained earlier. Record values in Table 21.1.
3. Calculate the average power P_5 absorbed by the resistor R_5 and the average power P_{s1} supplied by the AC generator. Record values in Table 21.1.

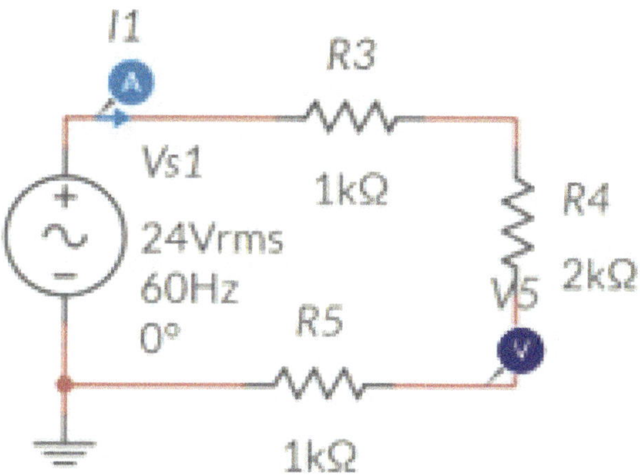

Figure 21.7 Resistive circuit with an AC voltage generator.

Table 21.1 Summary of readings and calculations

	Calculated	Measured
I_{rms} (mA)		
V_{5rms} (V)		
P_5 (mW)		
P_{s1} (mW)		

21.4 Lab Procedure 2

The given circuit (Figure 21.8) consists of two iso-frequency voltage sources, $v_1(t) = 12 \cdot \sin(376.8t)$ and $v_1(t) = 8 \cdot \sin(376.8t + \varphi)$ connected to two resistors. Analyze the circuit using Multisim simulation and calculate/measure various quantities for different phase angles φ (0°, 30°, –30°). The following tasks need to be performed:

1. Calculate the *rms* value of the current $i(t)$ circulating through resistor R. Measure the *rms* value of $i(t)$ using Multisim and compare it with the calculated value. Record value in Table 21.2.
2. Calculate the *rms* value of the voltage $v(t)$ across resistor R. Measure the *rms* value of the voltage $v(t)$ using Multisim and compare it with the calculated value. Record values in Table 21.2.
3. Calculate the average power absorbed by R.

These tasks will provide insights into the behavior of the circuit and allow for comparisons between the theoretical calculations and the measurements obtained from the Multisim simulation.

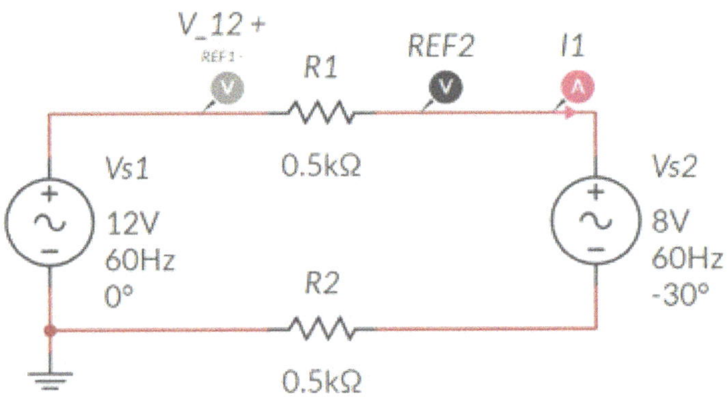

Figure 21.8 Resistive circuit with two AC iso-frequency generators of different amplitude.

Table 21.2 Summary of readings and calculations

φ (deg)	Calculated I_{rms} (mA)	Measured I_{rms} (mA)	Calculated V_{rms} (V)	Simulated V_{rms} (V)	Calculated P (mW)
0					
30					
–30					

22

Purely Capacitive Circuits with Sinusoidal and Non-sinusoidal Sources

22.1 Student Learning Outcomes

- Determine the voltage across and the current through capacitors in different operating conditions, both theoretically and through simulations using Multisim.
- Analyze the phase relationship between the voltage and current in purely capacitive circuits, both theoretically and through simulations using Multisim.
- Understand that the active power is zero in purely capacitive circuits with AC sources.

22.2 Prelab

Capacitors are passive elements for which the *passive sign* convention is adopted (Figure 22.1).

Let us consider an AC steady-state voltage across a capacitor, denoted as $v_C\left(t\right) = V_p sin\left(\omega t\right)$. The corresponding current through the capacitor can be determined using eqn (22.1).

$$i_C\left(t\right) = C\frac{dv_c\left(t\right)}{dt} = CV_p\frac{d}{dt}\sin\left(\omega t\right) = \omega CV_p\cos\left(\omega t\right) = I_p\sin\left(\omega t + 90°\right). \tag{22.1}$$

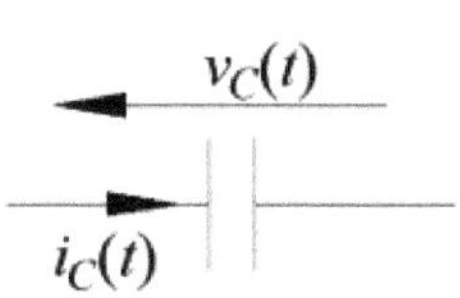

Figure 22.1 Passive sign convention for a capacitor.

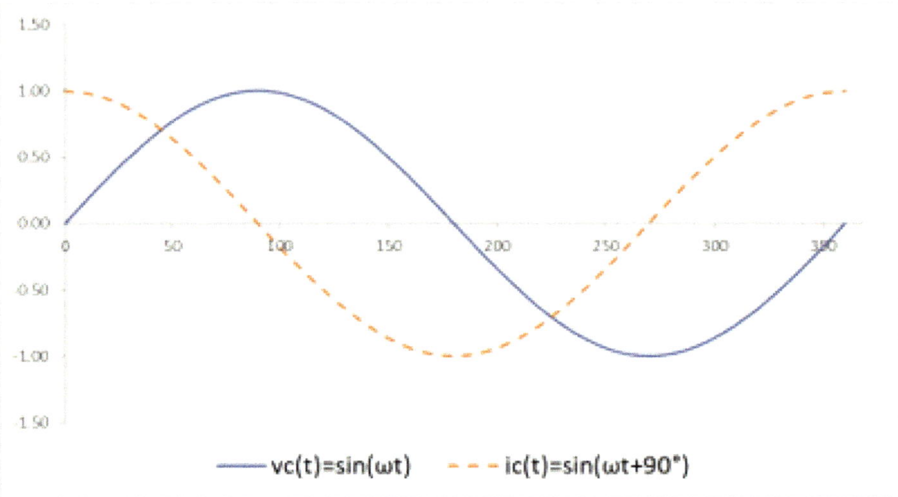

Figure 22.2 Expressions of $v_C\,(t)$ and $i_C\,(t)$ for a capacitor as a function of time.

The expressions of $v_C\,(t)$ and $i_C\,(t)$ as a function of time are given in Figure 22.2.

The interesting aspect is that the current through the capacitor leads the voltage across it by 90°, or that the current waveform reaches its peak value 90° ahead of the voltage waveform. To understand this phase relationship, it helps to consider the behavior of a capacitor in response to a sinusoidal voltage. When the voltage across a capacitor changes, the capacitor responds by charging or discharging to maintain the voltage balance. In the case of a sinusoidal voltage, the rate of change of the voltage is highest when the voltage is crossing zero. This is where the current through the capacitor is at its maximum value. As the voltage increases, the current begins to decrease, reaching zero at the peak of the voltage waveform. Then, as the voltage decreases, the current begins to flow in the opposite direction, reaching its negative peak when the voltage is at its negative peak.

The instantaneous steady-state power absorbed by the capacitor, denoted $p_C(t)$, is given by eqn (22.2).

$$p_C(t) = V_p I_p \sin(\omega t)\sin(\omega t + 90^\circ). \tag{22.2}$$

By applying the trigonometric identity shown in eqn (22.3):

$$\sin\alpha\,\sin\beta = \frac{1}{2}\left[\cos(\,\alpha - \beta) - \cos(\alpha + \beta)\right], \tag{22.3}$$

we can rewrite eqn (22.2) as follows:

$$p_C(t) = \frac{1}{2}\omega C V_p^2 \sin(2\omega t). \tag{22.4}$$

We observe that the instantaneous power oscillates between positive and negative values. The positive power corresponds to energy being stored in the capacitor during the charging phase, while the negative power represents the discharge phase when energy is returned to the circuit. However, since the positive and negative power values are equal over a complete cycle, the average power over that cycle is zero. This is because the energy stored and released by the capacitor in each cycle cancels out over time. Therefore, on average, a capacitor neither absorbs nor delivers net power to the circuit. This characteristic of zero average power absorption by a capacitor is an important distinction from resistive elements, where power is continuously dissipated in the form of heat. Capacitors primarily store and release energy in the form of electric fields, rather than converting it into other forms of energy.

22.2.1 Problem 1

In relation to the circuit shown in Figure 22.3, where the sinusoidal voltage source $v_s(t)$ has a peak of 1 V, the following tasks are to be performed utilizing both theoretical calculations and Multisim simulations:

1. Determine the expressions of the instantaneous current $i_C(t)$ and instantaneous power $p_C(t)$ of the capacitor C.
2. Conduct a Multisim simulation to plot the sinusoidal voltage waveform $v_c(t)$ across the capacitor C and the corresponding current waveform $i_c(t)$. Ensure that the plots clearly demonstrate the phase relationship between the current and voltage signals; estimate their phase difference $\Delta\varphi$.

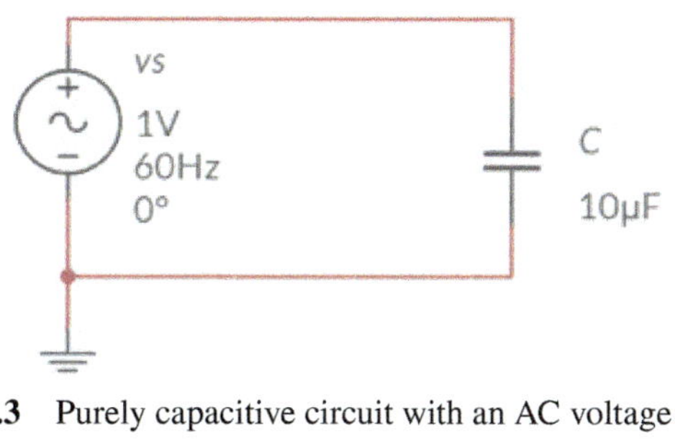

Figure 22.3 Purely capacitive circuit with an AC voltage generator.

22.2.1.1 Solutions to Problem 1

1. The expressions of $i_C(t)$ and $p_C(t)$ can be calculated as follows:

$$i_C\left(t\right) = \omega C V_p \sin\left(\omega t + 90°\right) = 3.76\sin\left(376.8t + 90°\right)\ \text{mA}\ , \tag{22.5}$$

$$p_C(t) = \frac{1}{2}\omega C V_p^2 \left[\sin(\ 2\omega t)\right] = 1.88\sin\left(753.6t\right)\ \text{mW}. \tag{22.6}$$

2. The plot of the instantaneous voltage $v_C(t)$ and current $i_C(t)$ obtained with Multisim simulation is presented in Figure 22.4.

The plot clearly shows that the current and voltage waveforms are not in phase, indicating that they reach their peak values and zero-crossings at different times. This phase difference between the current and voltage waveforms can be evaluated using eqn (22.7), which relates the phase difference $\Delta\varphi$ to the time difference Δt; Δt can be approximately evaluated by measuring it between corresponding points on the current and voltage waveforms on the plot created with Multisim.

$$\Delta\varphi = \frac{360°}{T}\Delta t = \omega\Delta t = 89.1°. \tag{22.7}$$

Comparing the calculated phase difference to the theoretical value of 90°, we can see that the measured value closely matches the expected theoretical phase difference.

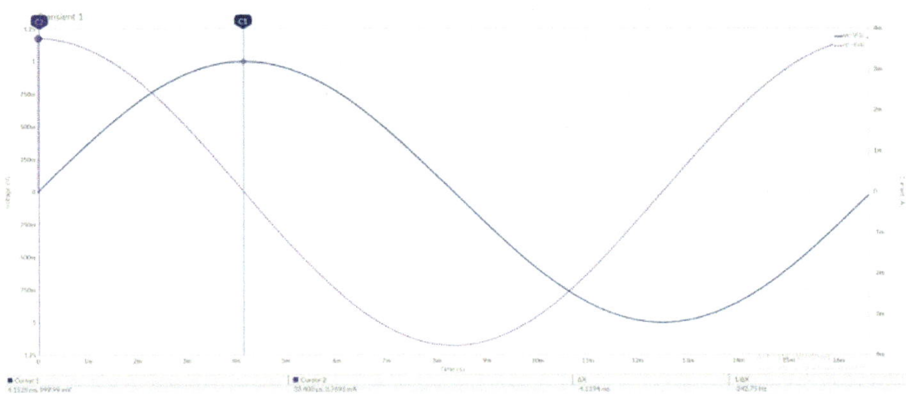

Figure 22.4 Plot of the sinusoidal current and voltage of the capacitor.

22.2.2 Problem 2

In relation to the circuit shown in Figure 22.5, where the triangular wave voltage generator $v_s(t)$ has a peak V_p = 2 V, a period T = 1 ms, and a *fall time TF* = T/2 = 0.5 ms, the following tasks are to be performed utilizing both theoretical calculations and Multisim simulations:

1. Determine the expression of the instantaneous current $i_C(t)$ through the capacitor.
2. Conduct a Multisim simulation to plot the voltage waveform $v_c(t)$ across the capacitor C and the corresponding current waveform $i_c(t)$.

22.2.2.1 Solutions to Problem 2

1. The piecewise expression of the instantaneous triangular voltage waveform $v_s\,(t)$ applied to the capacitor C is expressed over intervals of time as follows:

$$v_s\,(t) = \frac{V_p}{\frac{T}{2}}t = 4 \cdot 10^3 t, \qquad 0 \leq t \leq \frac{T}{2} \tag{22.8}$$

$$v_s\,(t) = -\frac{I_p}{\frac{T}{2}}t = -4 \cdot 10^3 t, \quad \frac{T}{2} \leq t \leq 0. \tag{22.9}$$

The instantaneous current $i_C(t)$ through the capacitor C can be determined using the basic relationship between current and voltage and can also be

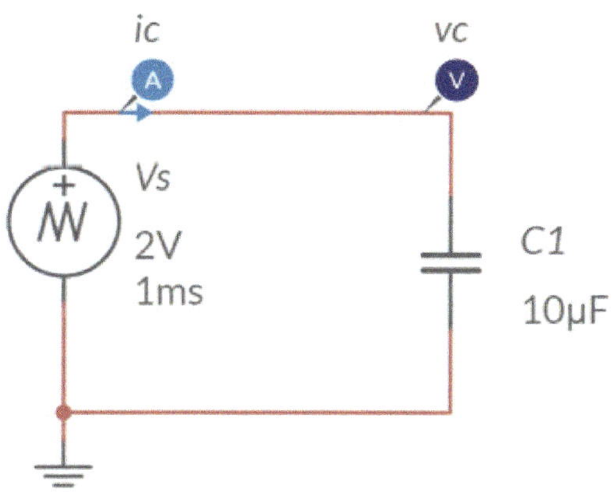

Figure 22.5 Purely capacitive circuit with triangular wave generator.

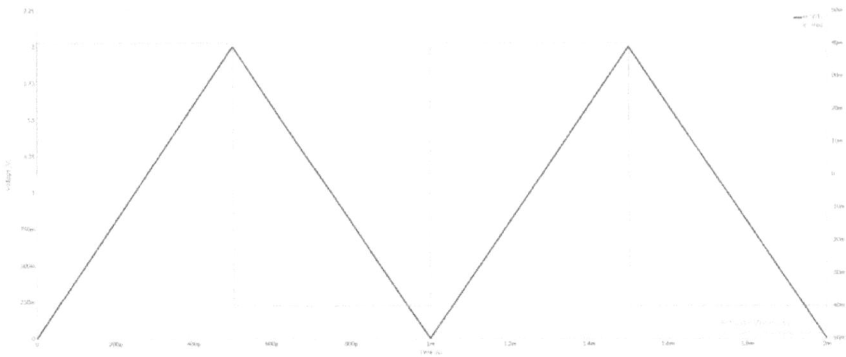

Figure 22.6 Plot of the instantaneous current and voltage of the capacitor.

described as a piecewise function:

$$i_C(t) = C\frac{dv_s(t)}{dt} = 10\ \mu\text{F} \cdot 4 \cdot 10^3 = 40\ \text{mA}, \qquad 0 \le t \le \frac{T}{2} \quad (22.10)$$

$$i_C(t) = C\frac{dv_s(t)}{dt} = -10\ \mu\text{F} \cdot 4 \cdot 10^3 = -40\ \text{mA}, \quad \frac{T}{2} \le t \le 0. \quad (22.11)$$

The current $i_C(t)$ through the capacitor is a square waveform.

2. The plot of the waveforms $v_c(t)$ and $i_c(t)$ is provided in Figure 22.6.

By examining the plot created with Multisim, we can confirm the theoretical values for the current $i_C(t)$.

22.3 Lab Procedure 1

In relation to the circuit shown in Figure 22.7, where the sinusoidal voltage source $v_s(t)$ has a peak of 2 V and a phase angle of 30°, the following tasks are to be performed utilizing both theoretical calculations and Multisim simulations:

1. Determine the expressions of the instantaneous current $i_C(t)$ and instantaneous power $p_C(t)$ of the capacitor C. Record expressions in Table 22.1.
2. Conduct a Multisim simulation to plot the sinusoidal voltage waveform $v_c(t)$ across the capacitor C and the corresponding current waveform $i_c(t)$. Ensure that the plots clearly demonstrate the phase relationship between the current and voltage signals; estimate their phase difference $\Delta\varphi$. Record value in Table 22.1.

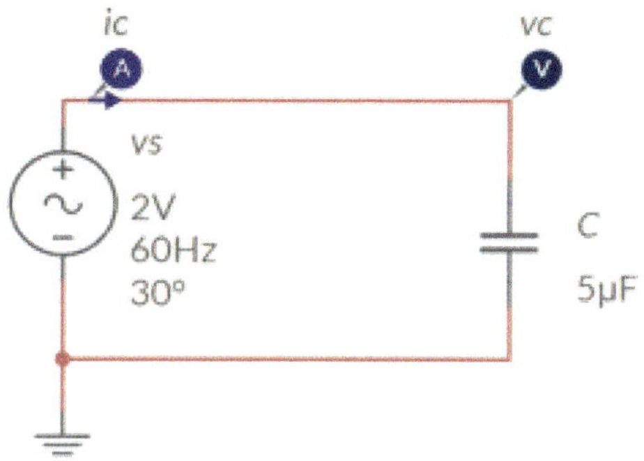

Figure 22.7 Purely capacitive circuit with an AC voltage generator.

Table 22.1 Summary of readings and calculations

$v_C(t)$	$p_C(t)$	$\Delta\varphi$

22.4 Lab Procedure 2

In relation to the circuit shown in Figure 22.8, where the triangular wave voltage generator $v_s(t)$ has a peak V_p = 1 V, a period T = 2 ms, and a *fall time TF* = T/2 = 1 ms, the following tasks are to be performed utilizing both theoretical calculations and Multisim simulations:

1. Determine the expression of the instantaneous current $i_C(t)$ through the capacitor.
2. Conduct a Multisim simulation to plot the voltage waveform $v_c(t)$ across the capacitor C and the corresponding current waveform $i_c(t)$.

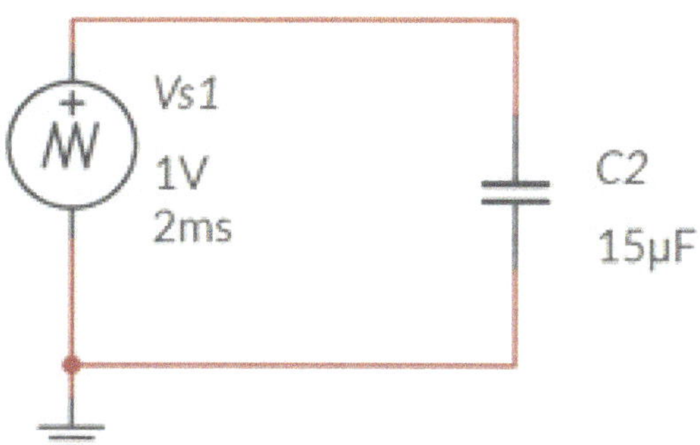

Figure 22.8 Purely capacitive circuit with triangular wave generator.

23

Purely Inductive Circuits with Sinusoidal and Non-sinusoidal Sources

23.1 Student Learning Outcomes

- Determine the voltage across and the current through inductors in different operating conditions, both theoretically and through simulations using Multisim.
- Analyze the phase relationship between the voltage and current in purely inductive circuits, both theoretically and through simulations using Multisim.
- Understand that the active power is zero in purely inductive circuits with AC sources.

23.2 Prelab

Inductors are passive elements for which the *passive sign* convention is adopted (Figure 23.1).

Let us consider an AC steady-state current through an inductor, denoted as $i_L(t) = I_p sin(\omega t)$. The corresponding voltage across the inductor can be determined using eqn (23.1).

$$v_L(t) = L\frac{di_L(t)}{dt} = LI_p\frac{d}{dt}\sin(\omega t) = \omega LI_p\cos(\omega t) = V_p\sin(\omega t + 90°)\,. \tag{23.1}$$

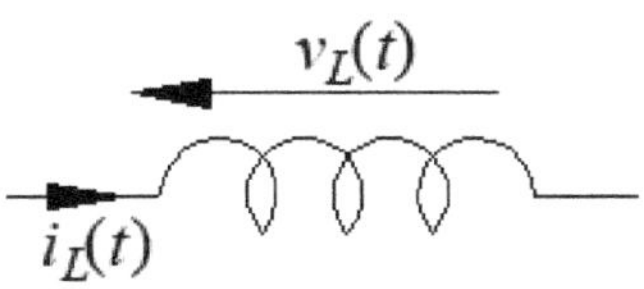

Figure 23.1 Passive sign convention for an inductor.

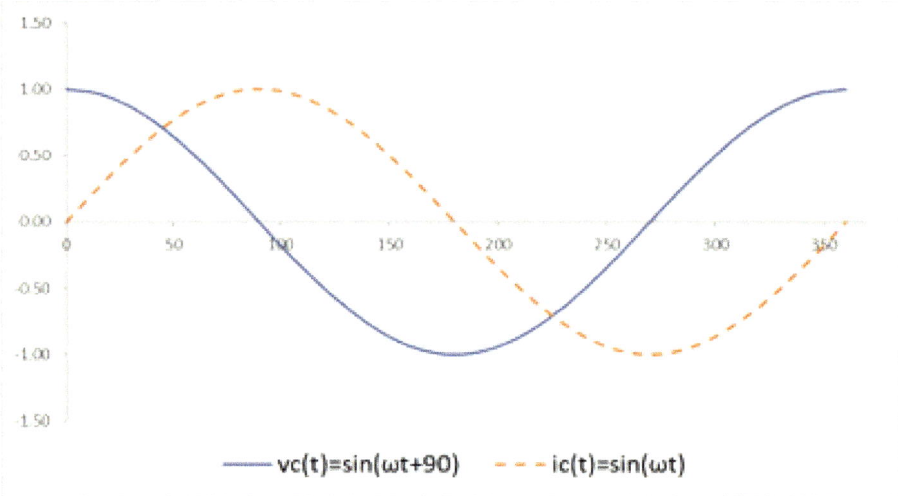

Figure 23.2 Expressions of $v_L\ (t)$ and $i_L\ (t)$ for an inductor as a function of time.

The expressions of $v_C\ (t)$ and $i_C\ (t)$ as a function of time are given in Figure 23.2.

Eqn (23.1) shows that the voltage $v_L\ (t)$ leads the current $i_L\ (t)$ by 90°. This means that at any given point in time, the voltage waveform reaches its peak value 90° (or π/2 radians) ahead of the current waveform. As the current is increasing or decreasing, the voltage across the inductor is at its maximum positive or negative value. The peak values of the voltage and current differ by the factor ωL, where ω is the angular frequency (i.e., $2\pi f$) and L is the inductance of the inductor. This factor represents the relationship between the rate of change of current and the resulting voltage across the inductor.

The instantaneous steady-state power absorbed by the inductor, denoted $p_L(t)$, is given by eqn (23.2).

$$p_L(t) = V_p I_p \sin(\ \omega t + 90°)\sin(\ \omega t). \tag{23.2}$$

By applying the trigonometric identity shown in eqn (23.3):

$$\sin a\ \sin\beta\ = \frac{1}{2}\left[\cos(\ a - \beta) - \cos(\ a + \beta)\right], \tag{23.3}$$

we can rewrite eqn (23.2) as follows:

$$p_L\ (t) = \frac{1}{2}\omega L I_p^2\left[\sin(\ 2\omega t)\right]. \tag{23.4}$$

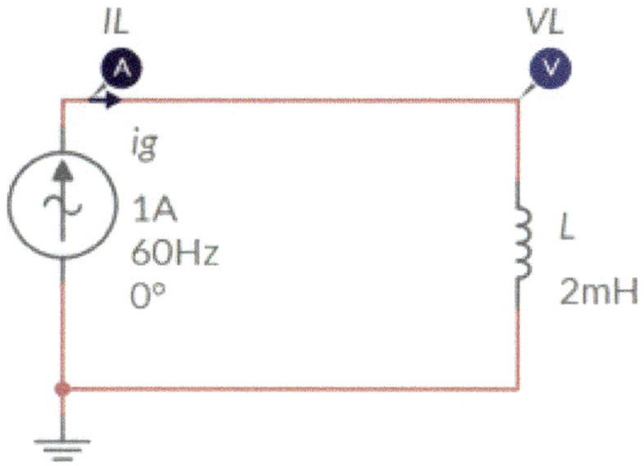

Figure 23.3 Purely inductive circuit with AC current generator.

Eqn (23.4) shows that the instantaneous power oscillates between positive and negative values during each cycle. Since the integral of $\sin(2\omega t)$ over one complete cycle is zero, the average power in an inductor is also zero. This means that, on average, an inductor neither absorbs nor dissipates power. The energy stored in the inductor during one half-cycle is released back to the circuit during the other half-cycle, resulting in no net power transfer over time. This highlights the *reactive* nature of inductors, where they store and release energy rather than convert it into usable power.

23.2.1 Problem 1

In relation to the circuit shown in Figure 23.3, where the sinusoidal current source $i_g(t) = \sin(\omega t)$ has a peak of 1 A, the following tasks are to be performed utilizing both theoretical calculations and Multisim simulations:

1. Determine the expressions of the instantaneous voltage $v_L(t)$ and instantaneous power $p_L(t)$ of the inductor L.
2. Conduct a Multisim simulation to plot the sinusoidal current waveform $i_L(t)$ through the inductor L and the corresponding voltage waveform $v_L(t)$. Ensure that the plots clearly demonstrate the phase relationship between the current and voltage signals; estimate their phase difference $\Delta\varphi$.

23.2.1.1 Solutions to Problem 1

1. The expressions of $v_L(t)$ and $p_L(t)$ can be calculated as follows:

$$v_L(t) = \omega L I_p \sin(\omega t + 90^\circ) = 753 \sin(376.8t + 90^\circ) \text{ mV} \quad (23.5)$$

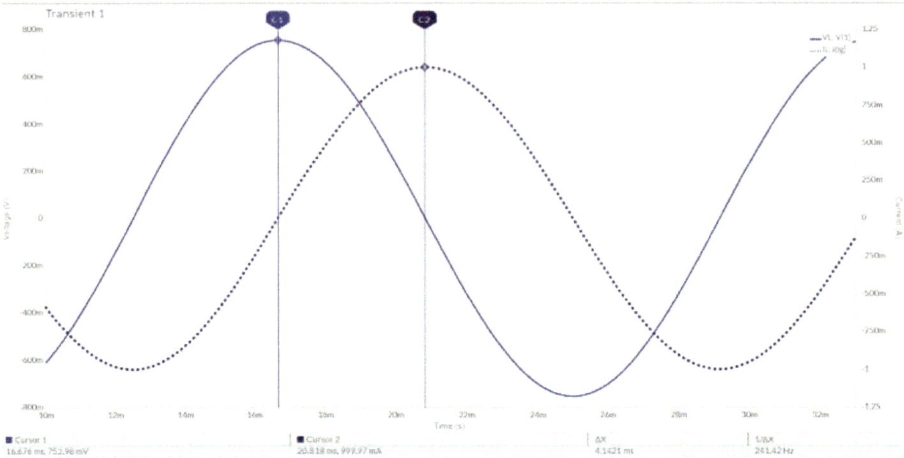

Figure 23.4 Plot of the instantaneous current and voltage of the inductor.

$$p_L(t) = \frac{1}{2}\omega L I_p^2 \left[\sin(\, 2\omega t)\right] = 376.8\sin(\, 753.6t)\ \text{mW}. \qquad (23.6)$$

2. The plot of the instantaneous voltage $v_L\ (t)$ and current $i_L\ (t)$ obtained with Multisim simulation is presented in Figure 23.4.

The plot clearly shows that the current and voltage waveforms are not in phase, indicating that they reach their peak values and zero-crossings at different times. This phase difference between the current and voltage waveforms can be evaluated using eqn (23.7), which relates the phase difference $\Delta\varphi$ to the time difference Δt; Δt can be approximately evaluated by measuring it between corresponding points on the current and voltage waveforms on the plot created with Multisim.

$$\Delta\varphi = \frac{360^\circ}{T}\Delta t = \omega\Delta t = 89.8^\circ. \qquad (23.7)$$

Comparing the calculated phase difference to the theoretical value of 90°, we can see that the measured value closely matches the expected theoretical phase difference.

23.2.2 Problem 2

In relation to the circuit shown in Figure 23.5, where the triangular wave current generator $i_g(t)$ has a peak I_p = 2 mA, a period T = 1 ms, and a *fall*

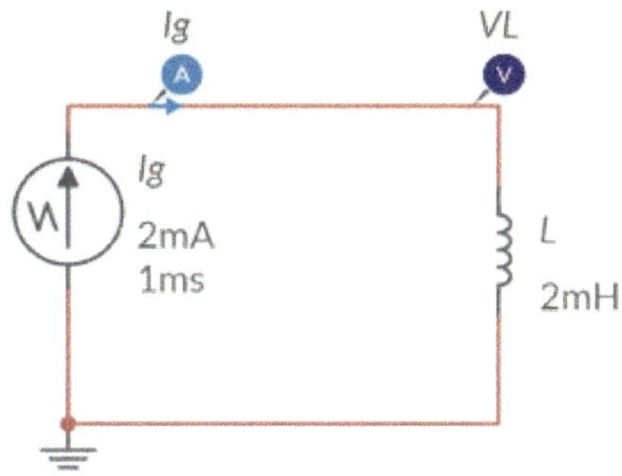

Figure 23.5 Purely inductive circuit with triangular wave generator.

time TF = *T*/2 = 0.5 ms, the following tasks are to be performed utilizing both theoretical calculations and Multisim simulations:

1. Conduct a Multisim simulation to plot the current waveform $i_g(t)$ through the inductor and the corresponding voltage waveform $v_L(t)$ across the inductor.
2. Determine the expression of the instantaneous voltage $v_L(t)$ across the inductor *L*.

23.2.2.1 Solutions to Problem 2

1. The plot of the waveforms $v_L(t)$ and $i_L(t)$ is provided in Figure 23.6.

Figure 23.6 Plot of the instantaneous triangular current and voltage of the inductor.

The piece-wise expression of the triangular current waveform $i_g(t)$ through the inductor L is expressed over intervals of time as:

$$i_g(t) = \frac{I_p}{\frac{T}{2}}t = 4t \text{ A}, \qquad 0 \le t \le \frac{T}{2} \tag{23.8}$$

$$i_g(t) = -\frac{I_p}{\frac{T}{2}}t = -4t \text{ A}, \quad \frac{T}{2} \le t \le 0. \tag{23.9}$$

2. The instantaneous voltage $v_L(t)$ across the inductor L can be determined using the basic relationship between current and voltage and can also be described as a piecewise function:

$$v_L(t) = L\frac{di_L(t)}{dt} = 2 \text{ mH} \cdot 4 = 8 \text{ mV}, \qquad 0 \le t \le \frac{T}{2} \tag{23.10}$$

$$v_L(t) = L\frac{di_L(t)}{dt} = -2 \text{ mH} \cdot 4 = -8 \text{ mV}, \quad \frac{T}{2} \le t \le 0. \tag{23.11}$$

The voltage $v_L(t)$ of the inductor is a square waveform.

23.3 Lab Procedure 1

In relation to the circuit shown in Figure 23.7, where the sinusoidal current source $i_g(t) = \sin(\omega t)$ has a peak of 2 A and a phase angle of 45°, the following tasks are to be performed utilizing both theoretical calculations and Multisim simulations:

1. Conduct a Multisim simulation to plot the sinusoidal voltage waveform $v_L(t)$ through the inductor L and the corresponding current waveform $i_L(t)$. Ensure that the plots clearly demonstrate the phase relationship

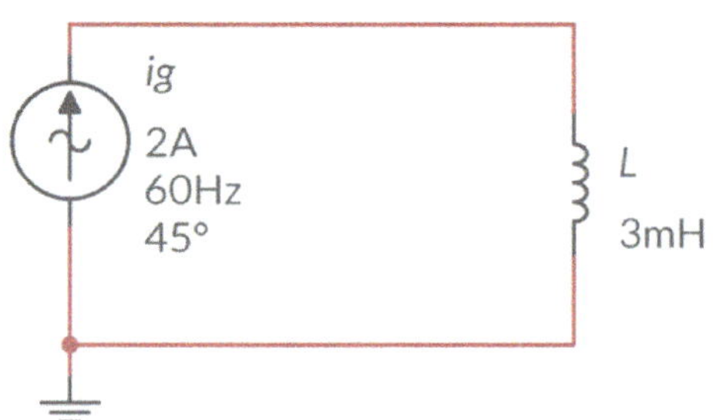

Figure 23.7 Purely inductive circuit with an AC current generator.

Table 23.1 Summary of calculations

$\Delta\varphi$	$v_L(t)$	$p_L(t)$

between the current and voltage signals; estimate their phase difference $\Delta\varphi$. Record value in Table 23.1.

2. Determine the expressions of the instantaneous voltage $v_L(t)$ and of the instantaneous power $p_L(t)$ of the inductor L.

23.4 Lab Procedure 2

In relation to the circuit shown in Figure 23.8, where the triangular wave current generator $i_g(t)$ has a peak of 1 mA, a period T = 1 ms, and a *fall time* $TF = T/2 = 0.5$ ms, the following tasks are to be performed utilizing both theoretical calculations and Multisim simulations:

1. Conduct a Multisim simulation to plot the current waveform $i_g(t)$ through the inductor and the corresponding voltage waveform $v_L(t)$ across the inductor.
2. Determine the expression of the instantaneous voltage $v_L(t)$ across the inductor L.

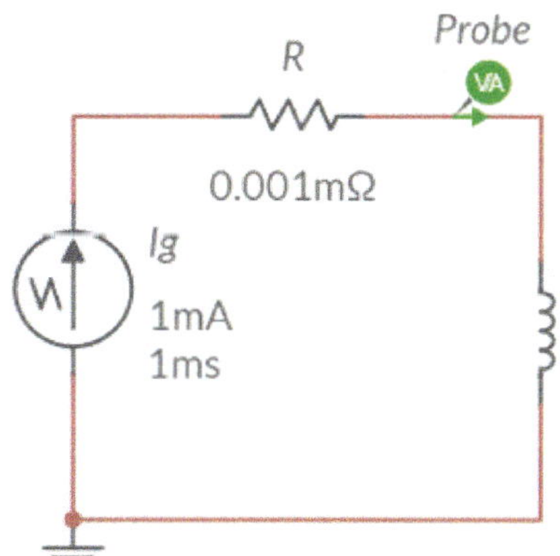

Figure 23.8 Purely inductive circuit with triangular wave current generator.

24

Sinusoidal Steady-state Analysis in the Phasor Domain

24.1 Student Learning Outcomes

- Recall the definition of impedance for resistors, inductors, and capacitors and the voltage and current relationships in ohmic-inductive and ohmic-capacitive circuits in the phasor domain.
- Apply phasor analysis techniques to determine the voltage, current, and reactive power in ohmic-inductive and ohmic-capacitive circuits.
- Utilize Multisim to calculate and measure the phase angle difference between voltages and currents in simulated ohmic-inductive and ohmic-capacitive circuits.

24.2 Prelab

The definitions of the impedance $\overline{Z}\,(j\omega)$ for resistor, inductor, and capacitor are summarized in Figure 24.1.

Circuit component	$\bar{Z}(j\omega) = \frac{\bar{V}}{\bar{I}}$
R	R
L	$j\omega L$
C	$-j\frac{1}{\omega C}$

Figure 24.1 Impedance $\overline{Z}\,(j\omega)$ for circuit elements.

24.2.1 Resistor

In the phasor domain, the relationship between voltage and current in resistors can be expressed as:

$$\overline{V} = R\overline{I}. \tag{24.1}$$

Since phasors are complex numbers, eqn (24.1) implies that $\left|\overline{V}\right| = R\left|\overline{I}\right|$ and $\varphi_v = \varphi_i$, indicating that the waveforms of voltage and current are in phase.

The mean or *real* power *P* absorbed by the resistor can be calculated using the following expression:

$$P = \langle p_R\,(t)\rangle = \frac{1}{2}\frac{V_p^2}{R} = \frac{V_{\text{rms}}^2}{R} = I_{\text{rms}}^2 R. \tag{24.2}$$

This equation demonstrates that the power absorbed by the resistor is always greater than (or equal to) zero, since all the terms are nonnegative values. This indicates that the resistor acts as a passive element, dissipating energy rather than generating it.

24.2.2 Inductor

For inductors in the phasor domain, the voltage–current relationship can be expressed as:

$$\overline{V} = j\omega L\overline{I} = jX_L\overline{I}. \tag{24.3}$$

In this equation, X_L represents the inductive reactance, and *L* is the value of the inductance.

Eqn (24.3) implies that the magnitude of the voltage phasor, $\left|\overline{V}\right|$, is equal to ωL times the magnitude of the current phasor, $\left|\overline{I}\right|$. Additionally, the phase angle of the voltage across the inductor can be expressed as $\varphi_v = \varphi_i + 90°$, indicating that the voltage waveform leads the current waveform by 90°.

The reactive power Q_L, measured in volt-amps reactive (VAR), represents the power associated with the reactive component of the current flowing through an inductor. The reactive power can be calculated using the equation:

$$Q_L = \frac{1}{2}X_L I^2 = \frac{1}{2}\frac{V_p^2}{\omega L} = \frac{V_{\text{rms}}^2}{\omega L}. \tag{24.4}$$

In this equation, X_L represents the inductive reactance, *I* is the current phasor, V_p is the peak voltage.

The equation shows that the reactive power absorbed by an inductor is always greater than (or equal to) zero due to the square terms. This means that the energy is being stored and released by the inductor in each AC cycle without any net power dissipation.

24.2.3 Capacitor

In the phasor domain, the relationship between voltage $\overline{V}$ and current $\overline{I}$ in a capacitor C can be described by the equation:

$$\overline{V} = -j\frac{1}{\omega C}\overline{I} = jX_C\overline{I}. \tag{24.5}$$

In this equation, X_C represents the capacitive reactance, and C is the value of the capacitance.

From eqn (24.5), we can infer that the magnitude of the voltage phasor ($|\overline{V}|$) is equal to $1/\omega C$ times the magnitude of the current phasor ($|\overline{I}|$). Furthermore, we can deduce that $\varphi_v = \varphi_i - 90°$, where φ_v is the phase angle of the voltage phasor and φ_i is the phase angle of the current phasor. This indicates that the waveforms of voltage and current in a capacitor are out-of-phase by 90 degrees, with the current leading the voltage.

The reactive power in VAR absorbed by the capacitor is: The reactive power Q_C, measured in volt-amps reactive (VAR), associated with the capacitor can be calculated using the equation:

$$Q_C = \frac{1}{2}X_C I_p^2 = -\frac{1}{2}\omega C V_p^2 = -\omega C V_{\text{rms}}^2 \tag{24.6}$$

Here, I_p represents the peak current, and V_p is the peak voltage.

Eqn (24.6) shows that the reactive power absorbed by the capacitor is always less than (or equal to) zero. The negative sign in the equation indicates that the reactive power associated with the capacitor is of the opposite sign compared to inductors. It signifies that the capacitor acts as a reactive power source, providing reactive power back to the circuit. The negative sign does not imply a loss of power but rather reflects the direction of power flow.

24.2.4 Problem 1

In the given circuit shown in Figure 24.2, the AC sinusoidal voltage source is defined as $v_g\,(t) = \sqrt{2} \cdot 12\sin(376.8t + 45°)$.

1. Determine the phasor $\overline{I}$ of the current through R, and its expression in the time domain. Additionally, simulate the circuit using Multisim and compare the results with the calculated values.

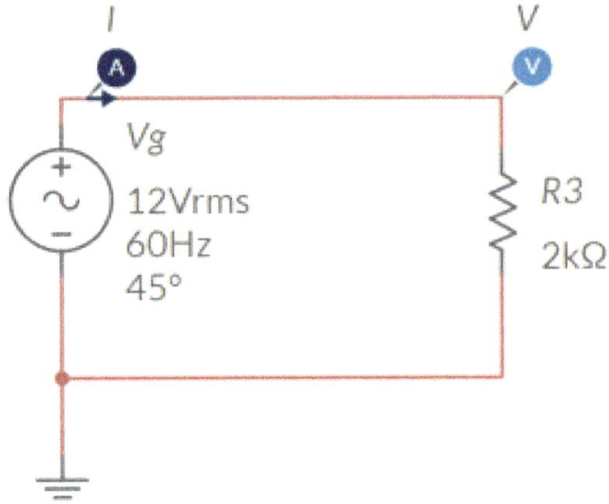

Figure 24.2 Purely resistive circuit with an AC voltage generator.

2. Plot the instantaneous current and voltage of the resistor and estimate the phase difference $\Delta\varphi$ with a Multisim simulation.
3. Calculate the real power (i.e., average power) P_3 absorbed by the resistor R_3.

24.2.4.1 Solutions to Problem 1

1. To determine the phasor of the current through the resistor R and its expression in the time domain, we can apply Ohm's Law and divide the phasor voltage $\overline{V}_s = 12\angle 45°\text{V}$ by the resistance R:

$$\overline{I} = \frac{12\angle 45°V}{2000\ \Omega} = 6\angle 45°\text{mA}. \tag{24.7}$$

Since the voltage and current are in phase for a resistor, the phasor of the current has the same phase angle as the voltage.

In the time domain, the expression of the current $i\,(t)$ is

$$i\,(t) = \sqrt{2} \cdot 6\sin\left(376.8t + 45°\right)\ \text{mA}\,. \tag{24.8}$$

Multisim simulation confirms the calculated value of the current (Figure 24.3).

2. The plot of $v(t)$ and $i(t)$ obtained with Multisim simulation is presented in Figure 24.4.

The plot demonstrates that the current and voltage are in phase, indicating that they reach peak values and zero-crossings at the same time. This observation

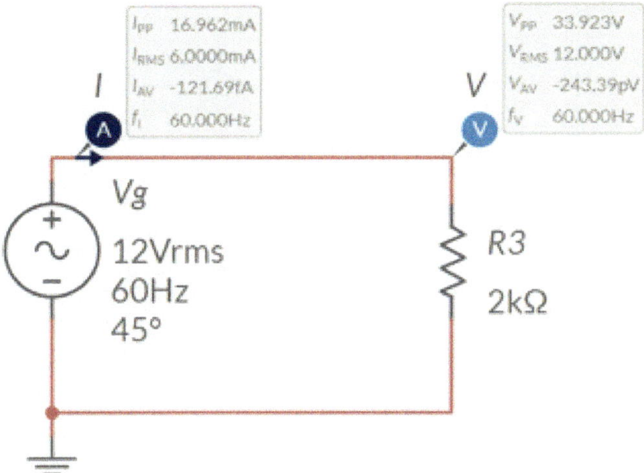

Figure 24.3 Multisim simulation confirming the calculated value of current.

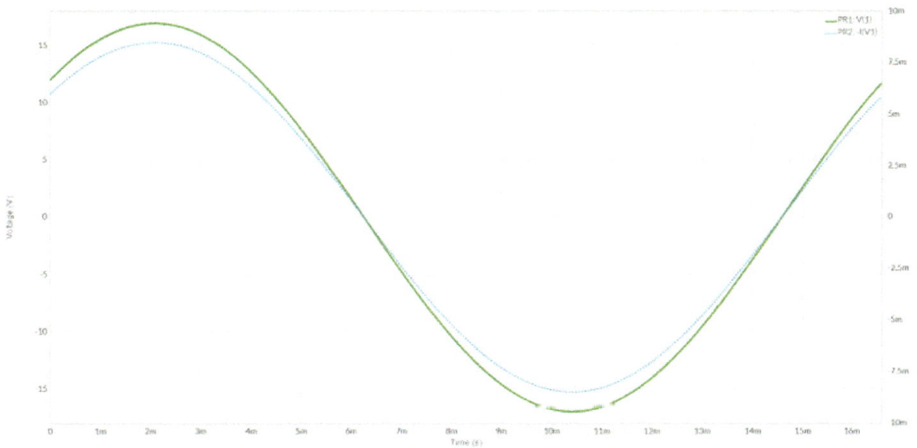

Figure 24.4 Plot of the instantaneous sinusoidal current and voltage of the resistor.

aligns with the expected behavior of current and voltage for a resistor an AC circuit.

3. The real power P_3 absorbed by the resistor R_3 can be calculated as follows:

$$P_3 = \frac{V_{\text{rms}}^2}{R} = 72\,\text{mW}. \tag{24.9}$$

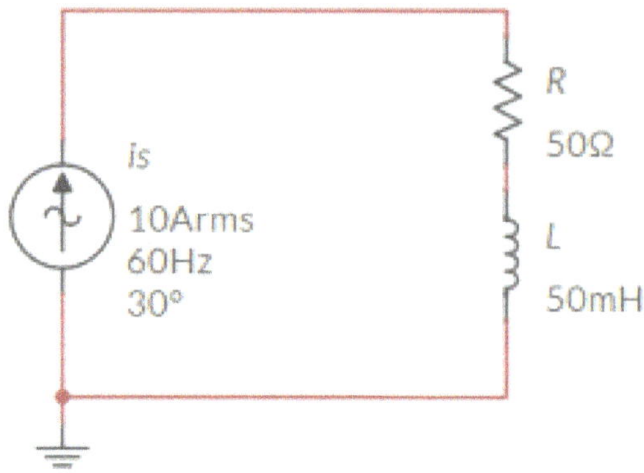

Figure 24.5 Ohmic-inductive circuit with an AC current generator.

24.2.5 Problem 2

Consider the circuit shown in Figure 24.5, where a sinusoidal current source $i_s\,(t) = \sqrt{2}\cdot 10\,\sin(376.8t + 30°)$ supplies an ohmic-inductive circuit.

1. Determine the total impedance $\overline{Z}\,(j\omega)$ of the circuit considering both the resistive and inductive components. Simulate the circuit using Multisim to verify the calculated value.
2. Determine the phasor $\overline{V}$ of the voltage across the series combination of *R* and *L*; simulate the circuit using Multisim to measure the *rms* value of the voltage waveform and compare it with the calculated value. Additionally, calculate the phase difference $\Delta\varphi$ in degrees between voltage and current.
3. With Multisim, plot the instantaneous values of the current through and the voltage across the series combination of *R* and *L*. Estimate the phase difference $\Delta\varphi$ in degrees between voltage and current by analyzing the plot and compare it with the calculated value.
4. Calculate the real power *P* absorbed by the resistor.
5. Calculate the reactive power Q_L associated with the inductor.

24.2.5.1 Solutions to Problem 2

1. The total impedance is given by the following formula:

$$\overline{Z} = (R + j\omega L) = \left(50 + j2\pi\cdot 60\cdot 50\cdot 10^{-3}\right) = 53.4\angle 20.6°\ \Omega. \tag{24.10}$$

It is noteworthy that the argument of the total impedance in a circuit represents the phase relationship between the current and voltage. A positive

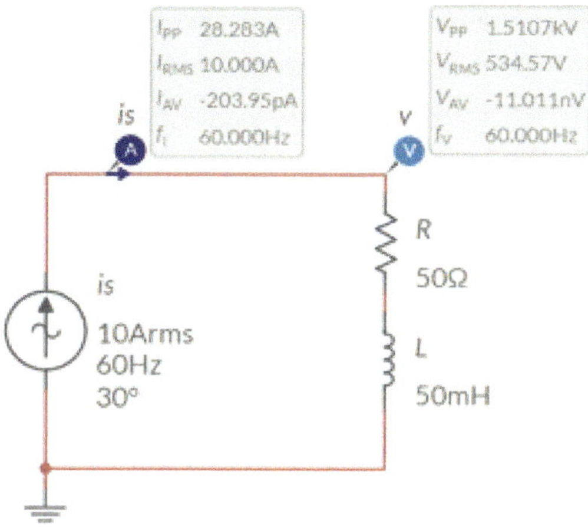

Figure 24.6 Multisim simulation providing *rms* values of voltage and current.

phase angle indicates that the voltage waveform leads the current waveform, while a negative phase angle indicates that the voltage waveform lags the current waveform.

To verify this calculation using Multisim, we can simulate the circuit and measure the *rms* values of the voltage across the series combination of *R* and *L* and the current circulating through it (Figure 24.6).

The magnitude of the impedance can be determined by applying Ohm's law dividing the measured *rms* values of voltage and current.

$$\left|\overline{Z}\right| = \frac{V_{\mathrm{rms}}}{I_{\mathrm{rms}}} = 53.4\ \Omega. \tag{24.11}$$

The theoretical and simulated values match, confirming the accuracy of the calculated value.

2. The phasor $\overline{V}$ of the voltage across the series combination of *R* and *L* can be determined with Ohm's law, by multiplying the current phasor $\overline{I}$ by the total impedance $\overline{Z}$.

$$\overline{V} = (R + j\omega L)\,\overline{I} = 53.4\angle 20.6^\circ \cdot 10\angle 30^\circ \text{ V} = 534.4\ \angle 50.6^\circ \text{ V}. \tag{24.12}$$

The *rms* value of the voltage waveform provided by the Multisim simulation of Figure 24.6 closely matches the calculated value.

The phase difference $\Delta\varphi$ between voltage and current can be calculated with the formula:

$$\Delta\varphi = \varphi_V - \varphi_{I_s} = 50.6^\circ - 30^\circ = 20.6^\circ. \tag{24.13}$$

Since $\Delta\varphi$, the phase difference between the current and voltage, is greater than zero, we can conclude that in the time domain, the voltage waveform leads the current waveform supplied by the generator. This behavior is expected in an ohmic-inductive circuit, where the presence of an inductor causes the current to lag the voltage. As a result, the voltage waveform reaches its peak value before the current waveform does. This observation aligns with the calculated value of the argument of the total impedance, which represents the phase angle between the current and voltage in the circuit. The agreement between the phase difference and the impedance calculation confirms the expected temporal relationship between current and voltage in the circuit.

3. The plot of the current and voltage through and across the series combination of *R* and *L* obtained with Multisim simulation is presented in Figure 24.7.

Analyzing the plot, we can identify the time difference Δt between corresponding points on the waveforms. The phase difference $\Delta\varphi$ can be estimated with the following formula:

$$\Delta\varphi = \frac{360}{0.016} \cdot 0.965 = 21.7.^\circ \tag{24.14}$$

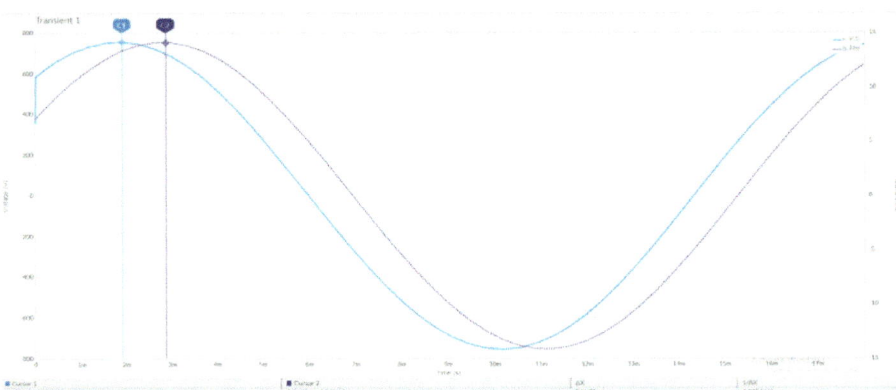

Figure 24.7 Plot of the instantaneous values of current and voltage across the series combination of *R* and *L*.

The result from the Multisim plot closely matches the calculated value.

4. The real power P absorbed by the resistor can be calculated using the formula that relates the real power absorbed by a resistive component to the current and resistance.

$$P{=}I_{\rm rms}^{2}R{=}10^{2}\cdot 50=5\ {\rm kW}. \tag{24.15}$$

5. The reactive power Q_L associated with the inductor can be calculated using the formula that relates the reactive power to the current and reactance. Q_L represents the energy storage and release by the inductive component.

$$Q_{L}=I_{\rm rms}^{2}X_{L}=10^{2}\cdot 18.8=1.8\ {\rm kVAR}. \tag{24.16}$$

24.3 Lab Procedure 1

In the given circuit shown in Figure 24.8, the AC sinusoidal voltage source is defined as $v_1\,(t) = \sqrt{2}\cdot 12\,\sin(376.8t + 45°)$.

1. Determine the phasor $\overline{I}_1$ and $\overline{I}_2$ of the currents respectively through R_1 and R_2. Simulate the circuit using Multisim and measure the *rms* values of the currents compare the results with the calculated values. Record values in Table 24.1.
2. With a Multisim simulation, plot the supply voltage and the total current and estimate their phase difference $\Delta\varphi$. Record values in Table 24.1.
3. Calculate the real power P absorbed by the two resistors in parallel. Record value in Table 24.1.

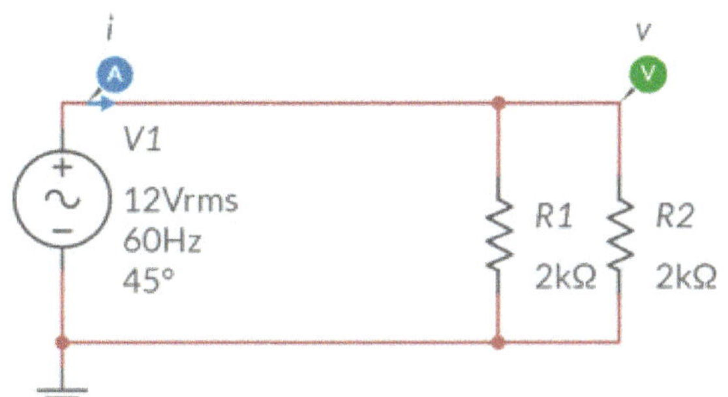

Figure 24.8 Purely resistive circuit with an AC voltage generator and resistors in parallel.

Table 24.1 Summary of readings and calculations

$\overline{I}_1 = I_{1\,\mathrm{rms}}\angle\varphi_1$	$\overline{I}_2 = I_{2\,\mathrm{rms}}\angle\varphi_2$	$I_{1\,rms}$	$I_{2\,rms}$	$\Delta\varphi$	P

24.4 Lab Procedure 2

Consider the circuit shown in Figure 24.9, where an AC voltage source $v_1(t) = \sqrt{2}\cdot 24\sin(376.8t + 25°)$ V supplies an ohmic-inductive circuit.

1. Determine the total impedance $\overline{Z}(j\omega)$ of the circuit. Use Multisim to verify the calculation. Record result in Table 24.2.
2. Determine the phasor $\overline{I}$ of the voltage through the series combination of *R* and *L*; simulate the circuit using Multisim to measure the *rms* value of the current waveform. Record results in Table 24.2.
3. Calculate the phase difference $\Delta\varphi$ in degrees between supply voltage and current. Record result in Table 24.2.
4. With Multisim simulation, plot the instantaneous values of the current through and the voltage across the series combination of *R* and *L*. Estimate the phase difference $\Delta\varphi$ in degrees between voltage and current by analyzing the plot and compare the estimation with the calculated value.
5. Calculate the real power *P* absorbed by the resistor and the reactive power Q_L associated with by the inductor. Record results in Table 24.2.

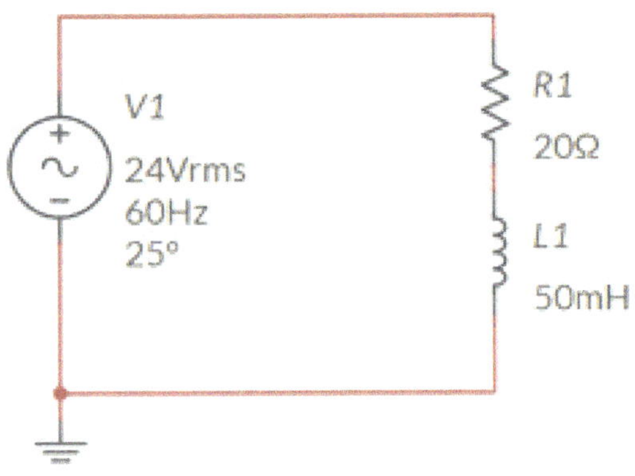

Figure 24.9 Ohmic-inductive circuit with AC voltage generator.

Table 24.2 Summary of readings and calculations

$\overline{Z} = Z\angle\varphi$	$\overline{I} = I_{\mathrm{rms}}\angle\varphi_I$	I_{rms} (measured)	$\Delta\varphi$	P	Q_L

25

Series Combination AC Circuits

25.1 Student Learning Outcomes

- Calculate the total impedance of AC series circuits using theoretical principles.
- Apply knowledge of AC circuits to calculate in the phasor domain and measure with Multisim simulations voltage drops across and currents through resistors, inductors, and capacitors connected in AC series circuits.
- Apply Multisim to simulate AC series circuits, verify theoretical calculations, and observe the behavior of voltage drops and currents.

25.2 Prelab

In a series combination of components in an electrical circuit, such as resistors, inductors, and capacitors, the key characteristic is that the same circulating current flows through all the components. This means that the current passing through one component is identical to the current passing through the other components in the series (Figure 25.1).

Since the current is the same in a series combination, the voltage drop across a component can be obtained by applying Ohm's law by multiplying the current flowing through it and its impedance. By summing up the individual voltage drops, we can and determine the total voltage across the series combination.

In the context of AC series circuits, the equivalent impedance $\overline{Z}_s$ of n impedances connected in series can be calculated using the following expressions:

$$\overline{Z}_s = \sum_{i=1}^{n} \overline{Z}_i = R_s + j\, X_s, \tag{25.1}$$

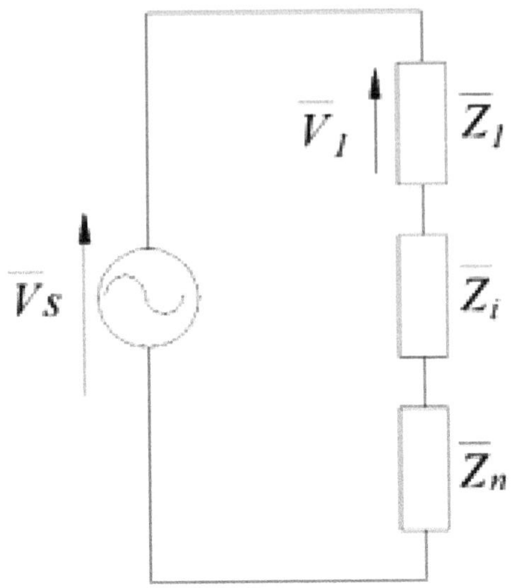

Figure 25.1 Impedances $\overline{Z}\,(j\omega)$ in series.

where

$$R_s = \sum_{i=1}^{n} R_i \tag{25.2}$$

$$X_s = \sum_{i=1}^{n} X_i. \tag{25.3}$$

In an AC series circuit, we can still apply the voltage divider rule to determine the phasor voltage $\overline{V}_i$ across a specific impedance $\overline{Z}_i$. The voltage divider rule states that the phasor voltage across a component is proportional to its impedance relative to the total impedance of the circuit.

$$\overline{V}_i = \overline{V}_s \frac{\overline{Z}_i}{\overline{Z}_s}. \tag{25.4}$$

Here, $\overline{V}_s$ represents the total voltage in the series circuit, and $\overline{Z}_s$ is the total impedance of the series combination.

25.2.1 Problem 1

In the given circuit shown in Figure 25.2, the AC voltage source is defined as $v_s\,(t) = \sqrt{2} \cdot 12\sin(376.8t + 30°)$. Perform the following tasks:

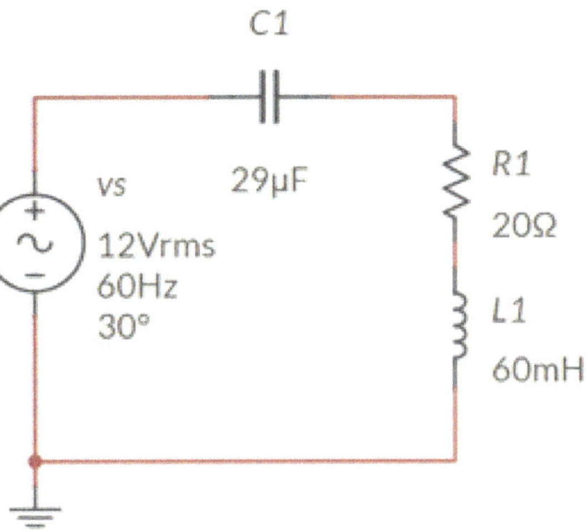

Figure 25.2 *R-L-C* series circuit.

1. Calculate the equivalent impedance $\overline{Z}_s$ of the *R-L-C* series combination. Use Multisim simulation to verify the calculated value.
2. Determine the phasor $\overline{I}$ of the current through the *R-L-C* series combination, and its instantaneous expression $i\,(t)$. Simulate the circuit using Multisim to measure the *rms* value of the current waveform and compare it with the calculated value.
3. Calculate the phase difference $\Delta\varphi = \varphi_I - \varphi_V$ between the current $\overline{I}$ flowing in the circuit and the generator voltage $\overline{V}_s$.
4. With Multisim, plot the instantaneous values of the current through and the voltage across the series combination. By analyzing the plot, estimate the phase difference $\Delta\varphi$ in degrees between voltage and current and compare it with the calculated value.
5. With Multisim, plot the instantaneous voltages $v_L(t)$ and $v_C(t)$ respectively across the inductor and the capacitor and estimate their phase difference $\Delta\varphi_{LC}$.

25.2.1.1 Solutions to Problem 1

1. The equivalent impedance $\overline{Z}_s$ of the *R-L-C* series combination can be calculated by summing up the single impedances using the following expressions:

$$\overline{Z}_{R1} = R_1 = 20\ \Omega, \tag{25.5}$$

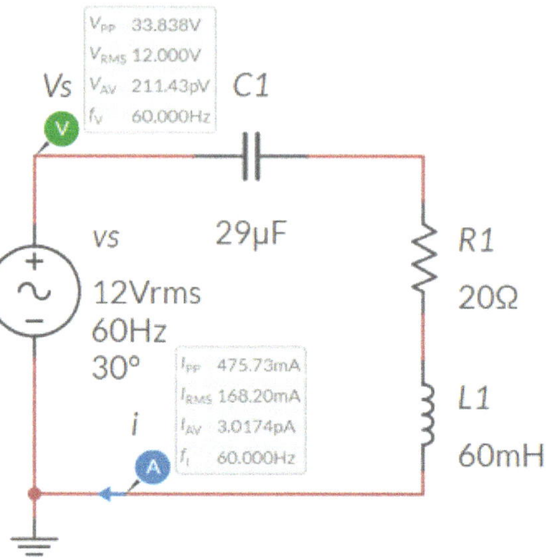

Figure 25.3 Multisim simulation providing *rms* values of voltage and current in the *R-L-C* series circuit.

$$\overline{Z}_{L1} = j\omega L = \ j22.6\ \Omega, \tag{25.6}$$

$$\overline{Z}_{C1} = -j\frac{1}{\omega C} = -j91.5\ \Omega, \tag{25.7}$$

$$\overline{Z}_s = \overline{Z}_R + \overline{Z}_L + \overline{Z}_C = 20 - j68.9\ \Omega = 71.7\angle{-73.8}°\Omega. \tag{25.8}$$

To verify the calculated value of the impedance using Multisim, we can simulate the circuit and measure the *rms* values of the voltage across the series combination and the current circulating through it (Figure 25.3).

The magnitude of the impedance can be determined by applying Ohm's law dividing the measured *rms* values of voltage and current.

$$\left|\overline{Z}_s\right| = \frac{V_{\text{rms}}}{I_{\text{rms}}} = 71.3\ \Omega. \tag{25.9}$$

The calculated value closely matches the Multisim simulated value, confirming its accuracy.

2. The phasor $\overline{I}$ of the current through the series combination can be determined with Ohm's law, by dividing the voltage phasor $\overline{V}_s$ by the total impedance $\overline{Z}_s$. The phasor of the voltage is $\overline{V}_s = 12\angle 30°\text{V}$, hence:

$$\overline{I} = \frac{\overline{V}_s}{\overline{Z}_s} = \frac{12\angle 30° V}{71.7\angle{-73.8}°\ \Omega} = 167.3\angle 103.8°\text{mA}. \tag{25.10}$$

The theoretical and Multisim simulated value closely match (Figure 25.3), confirming the accuracy of the calculated value.

The instantaneous expression of the current can be obtained as follows:

$$i\,(t) = \sqrt{2} \cdot 167.3 \sin\,(376.8t + 103.8^{\circ})\ \mathrm{mA}. \tag{25.11}$$

3. The phase difference $\Delta\varphi$ between the current $\overline{I}$ and the generator voltage $\overline{V}_s$ can be obtained with the following formula:

$$\Delta\varphi = \varphi_I - \varphi_V = 103.8^{\circ} - 30^{\circ} = 73.8^{\circ}. \tag{25.12}$$

Since $\Delta\varphi$ is greater than zero, we can conclude that in the time domain the current waveform leads the voltage waveform supplied by the generator. This result is consistent with the calculated value of the argument of the total impedance $\overline{Z}_s$, which represents the phase angle between the current and voltage in the circuit.

The capacitor's reactance contributes to the overall capacitive nature of the *R-L-C* series combination.

4. The plot of the instantaneous values of the current and the supply voltage obtained with Multisim simulation is presented in Figure 25.4.

Analyzing the plot, we can identify the time difference Δt between corresponding points on the waveforms.

The phase difference $\Delta\varphi$ can be estimated in degrees with the following formula:

$$\Delta\varphi = \frac{360}{0.0166} 3.4 \cdot 10^{-3} = 74^{\circ}. \tag{25.13}$$

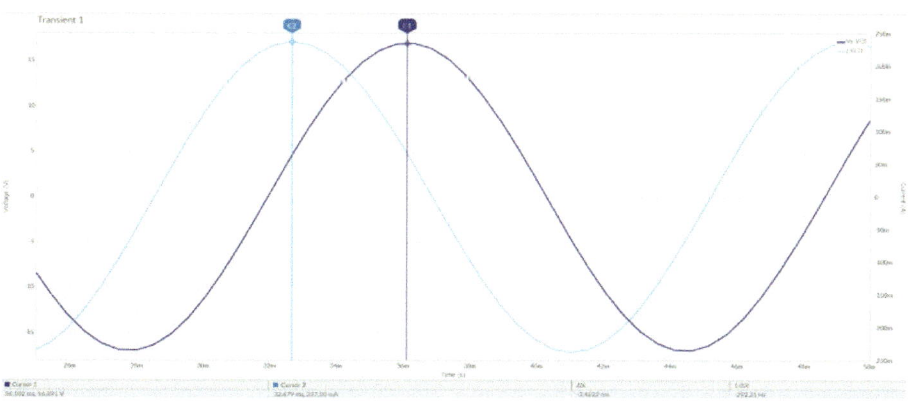

Figure 25.4 Plot of the instantaneous sinusoidal current and supply voltage in the series circuit.

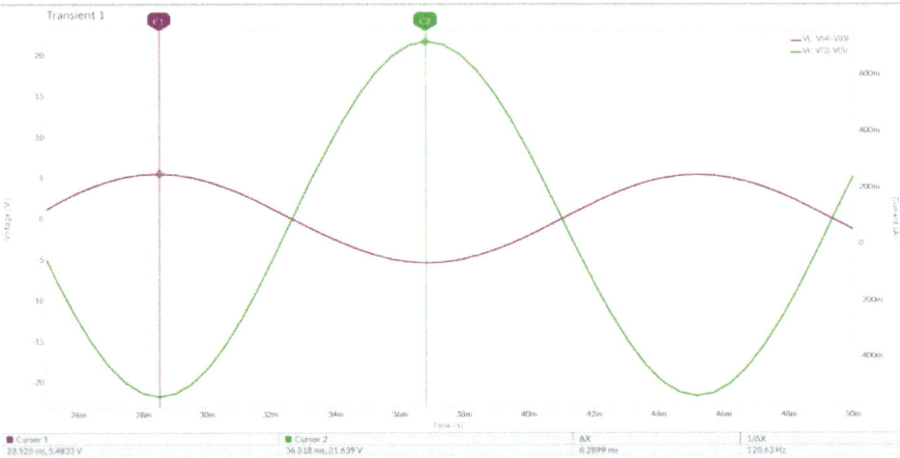

Figure 25.5 Plot of the instantaneous sinusoidal voltages across the inductor and the capacitor.

The result from the Multisim plot closely matches the calculated value.

5. The plot of $v_L(t)$ and $v_C(t)$ is given in Figure 25.4. The plot of $v_L(t)$ and $v_C(t)$ obtained with the oscilloscope in Multisim is presented in Figure 25.5.

Analyzing the plot, we can identify the time difference Δt between corresponding points on the waveforms.

The phase difference $\Delta\varphi_{LC}$ can be estimated in degrees with the following formula:

$$\Delta\varphi_{LC} = \frac{360}{0.0166} \cdot 8.3 \cdot 10^{-3} = 180°. \tag{25.14}$$

The two voltages are completely out-of-phase, or equivalently, in phase opposition of each other. This behavior arises from the inherent characteristics of the inductor and capacitor in an AC circuit.

25.2.2 Problem 2

In the given circuit shown in Figure 25.6, the AC voltage source is defined as $v_{s2}(t) = \sqrt{2} \cdot 12 \sin(376.8t)$. Perform the following tasks:

1. Calculate the equivalent impedance $\overline{Z}(j\omega)$ of the *R*-*C* series combination. Use Multisim simulation to verify the calculated value.

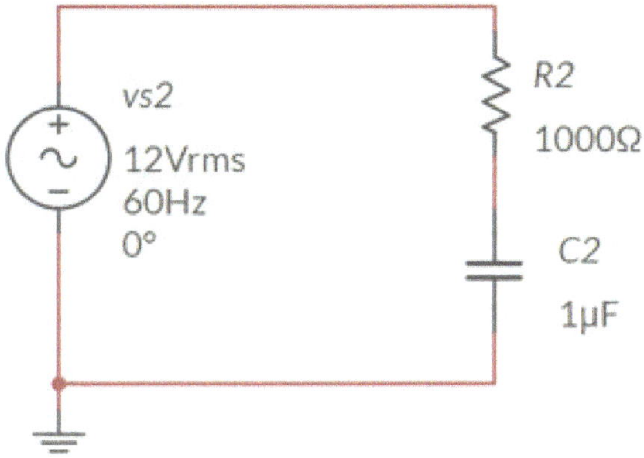

Figure 25.6 Ohmic-capacitive circuit with an AC voltage generator.

2. Determine the phasors $\overline{V}_R$ and $\overline{V}_C$, representing the voltages across R_2 and C_2, respectively. Verify that the calculated *rms* values of these voltages match the values obtained from Multisim simulation.
3. Calculate the frequency f_q at which the *rms* values of the voltages $v_R(t)$ and $v_C(t)$, across R_2 and C_2 respectively, are equal. Verify that the calculated frequency f_q aligns with the frequency observed in the Multisim simulation.

25.2.2.1 Solutions to Problem 2

1. The equivalent impedance $\overline{Z}$ of the *R*-*C* series combination can be calculated with the following formula:

$$\overline{Z} = R_2 - j\frac{1}{\omega C_2} = 1000 - j2653.9 = 2.8\angle{-69.3^\circ}\ \text{k}\Omega. \tag{25.15}$$

To verify the calculated value of the impedance using Multisim, we can simulate the circuit and measure the *rms* values of the supply voltage and the current circulating through the series combination (Figure 25.7).

The magnitude of the impedance can be determined by applying Ohm's law dividing the measured *rms* values of supply voltage and current.

$$\left|\overline{Z}\right| = \frac{V_{\text{rms}}}{I_{\text{rms}}} = \frac{12}{4.2 \cdot 10^{-3}} = 2.8\ \text{k}\Omega. \tag{25.16}$$

The theoretical value matches the simulated result, confirming its accuracy.

2. The phasors $\overline{V}_R$ and $\overline{V}_C$, representing the voltages across R_2 and C_2, can be calculated by applying the voltage divider formula:

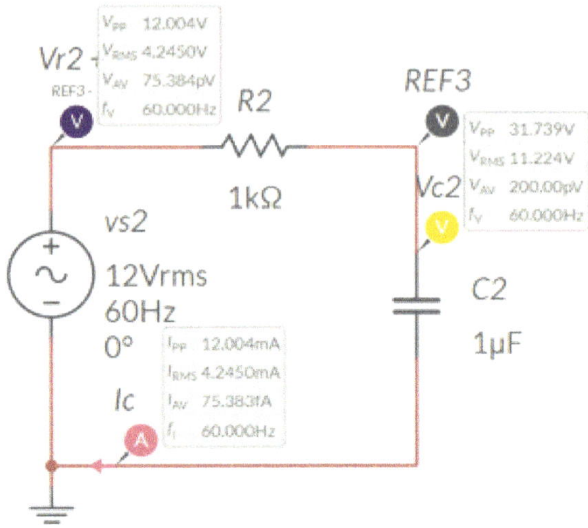

Figure 25.7 Multisim simulation providing *rms* values of supply voltage and current in the *R-C* series combination circuit.

$$\overline{V}_R = \overline{V}_s \frac{R_2}{\overline{Z}_s} = 12\angle 0^\circ \frac{1,000}{2,836\angle -69.3^\circ} = 4.2\angle 69.3^\circ \text{ V}. \tag{25.17}$$

$$\overline{V}_C = \overline{V}_s \frac{\overline{Z}_{C2}}{\overline{Z}_s} = 12\angle 0^\circ \frac{-j2,653.9}{2,836\angle -69.3^\circ} = 11.2\angle -20.7^\circ \text{ V} \tag{25.18}$$

The Multisim simulation results align with the calculated values, confirming their accuracy (Figure 25.7).

3. In the *R-C* series combination, the voltages $v_R(t)$ and $v_C(t)$ have the same *rms* value if the resistance and the capacitive reactance are equal. Thus, the corresponding frequency f_q can be determined by writing the following equation:

$$R_2 = X_{C2} = \frac{1}{2\pi f_q C_2}. \tag{25.19}$$

By solving the equation for f_q, we obtain f_q = 159.2 Hz.

The calculated value of the frequency f_q can be verified with Multisim simulation using a voltage source with a frequency set to f_q (Figure 25.8).

By observing the simulation results, we determine that the voltage drops across R_2 and C_2 are indeed equal, which confirms that the calculated value of f_q is correct.

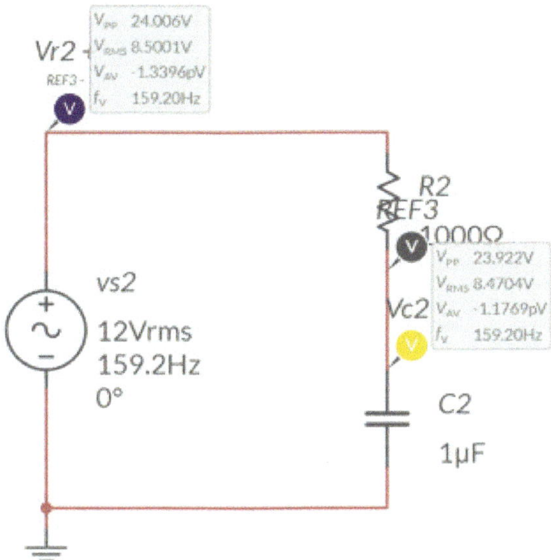

Figure 25.8 Multisim simulation showing voltages with same *rms* value at f_q = 159.2 Hz.

25.3 Lab Procedure 1

In the given circuit shown in Figure 25.9, the AC voltage source is defined as $v_{s3}\,(t) = \sqrt{2} \cdot 12\,\sin(376.8t)$. Perform the following tasks:

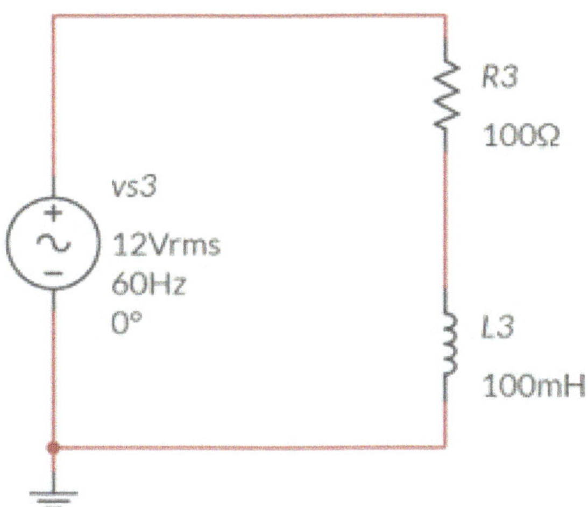

Figure 25.9 Ohmic-inductive circuit with an AC voltage generator.

Table 25.1 Summary of readings and calculations

$\overline{Z} = Z\angle\varphi$	$\overline{I} = I_{\text{rms}}\angle\varphi_I$	$i(t)$	$\Delta\varphi$	f_q	$\overline{Z}_q = Z_q\angle\varphi_q$	$\overline{I}_q = I_q\angle\varphi_{Iq}$

1. Calculate the equivalent impedance $\overline{Z}$ of the *R*-*L* series combination. Use Multisim simulation to verify the calculated value. Record values in Table 25.1.
2. Determine the phasor $\overline{I}$ of the current through the *R*-*L* series combination, and its instantaneous expression $i(t)$. Simulate the circuit using Multisim to measure the *rms* value of the current waveform and compare it with the calculated value. Record values in Table 25.1.
3. Calculate the phase difference $\Delta\varphi = \varphi_I - \varphi_V$ between the current $\overline{I}$ and the generator voltage $\overline{V}_{s3}$.
4. With Multisim, plot the instantaneous values of the current $i(t)$ through the series combination and the voltage source $v_{s3}(t)$. By analyzing the plot, estimate the phase difference $\Delta\varphi$ in degrees between voltage and current and compare it with the calculated value.
5. Calculate the frequency f_q at which the *rms* values of the voltages $v_R(t)$ and $v_L(t)$, across R_3 and L_3 respectively, are equal. Verify that the calculated frequency f_q aligns with the frequency observed in the Multisim simulation. Record value in Table 25.1.
6. Calculate the equivalent impedance $\overline{Z}_q$ at the frequency f_q and the phasor current $\overline{I}_q$ circulating through the *R*-*L* series combination. Verify that the calculated value of the *rms* for $\overline{I}_q$ is confirmed by the Multisim simulation. Record values in Table 25.1.

25.4 Lab Procedure 2

In the given circuit shown in Figure 25.10, the AC voltage source is defined as $v_{s4}(t) = \sqrt{2} \cdot 120 \sin(314t)$. Perform the following tasks:

1. Calculate the equivalent impedance $\overline{Z}$ of the *R*-*L* series combination. Use Multisim simulation to confirm the calculated value. Record values in Table 25.2.
2. Determine the phasor $\overline{I}$ of the current through the *R*-*L* series combination, and its instantaneous expression $i(t)$. Simulate the circuit using

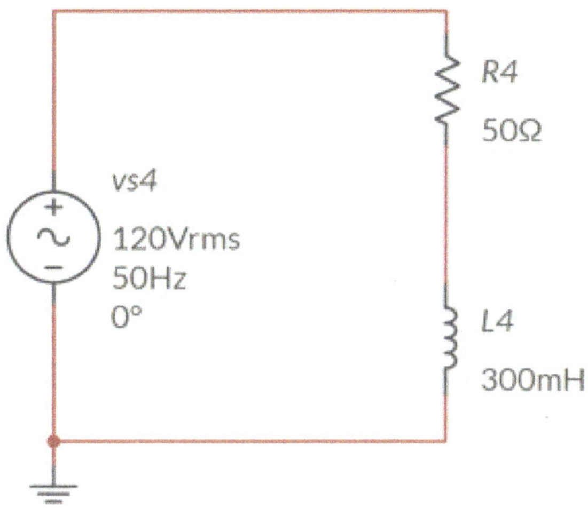

Figure 25.10 Ohmic-inductive circuit with an AC voltage generator.

Table 25.2 Summary of readings and calculations

$\overline{Z} = Z\angle\varphi$	$\overline{I} = I_{\text{rms}}\angle\varphi_I$	$i(t)$	$\Delta\varphi$	R_4

Multisim to measure the *rms* value of the current waveform and compare it with the calculated value. Record values in Table 25.2.

3. Calculate the phase difference $\Delta\varphi = \varphi_I - \varphi_V$ between the current $\overline{I}$ and the generator voltage $\overline{V}_{s4}$. Record values in Table 25.2.
4. With Multisim, plot the instantaneous values of the current $i(t)$ through the series combination and the voltage source $v_{s4}(t)$. By analyzing the plot, estimate the phase difference $\Delta\varphi$ in degrees between voltage and current and compare it with the calculated value.
5. Calculate the new value of the resistor R_4 that halves the *rms* value of the current. Use Multisim simulation to confirm the calculated value. Record value in Table 25.2.

26

Parallel Combination AC Circuits

26.1 Student Learning Outcomes

- Calculate the total impedance of AC parallel combination circuits using theoretical principles.
- Apply knowledge of AC circuits to calculate in the phasor domain and measure with Multisim simulations voltage drops across and currents through resistors, inductors, and capacitors connected in AC parallel combination circuits.
- Apply Multisim to simulate AC parallel combination, verify theoretical calculations, and observe the behavior of voltage drops and currents.

26.2 Prelab

In a parallel combination of impedances in an AC electrical circuit, each component is connected to the same voltage source. The voltage across each component is, therefore, identical, regardless of the individual component's impedance or characteristics (Figure 26.1).

It is noteworthy that while the voltage is the same across parallel components, the current flowing through each component may vary depending on their individual impedances.

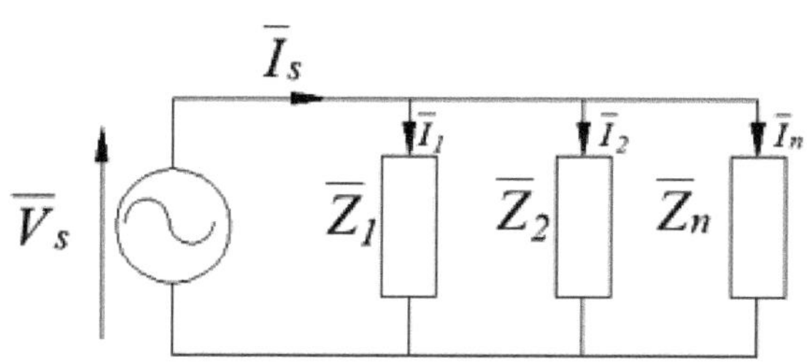

Figure 26.1 Impedances $\overline{Z}\,(j\omega)$ connected in a parallel combination.

The equivalent impedance $\overline{Z}_p$ of *n* impedances connected in parallel can be calculated as the reciprocal of the sum of the reciprocals of the individual impedances. Mathematically, it can be expressed as follows:

$$\frac{1}{\overline{Z}_p} = \sum_{i=1}^{n} \frac{1}{\overline{Z}_i}. \tag{26.1}$$

Alternatively, the equivalent impedance can be determined using the *admittance* $\overline{Y}$:

$$\overline{Y} = G + jB = \frac{1}{R} - j\frac{1}{\omega L} + j\omega C, \tag{26.2}$$

where G is termed *conductance* and B *susceptance*.

The parallel admittance $\overline{Y}_p$ is calculated as the sum of the individual admittances:

$$\overline{Y}_p = \sum_{i=1}^{n} \overline{Y}_i. \tag{26.3}$$

Then, the equivalent impedance is given by the reciprocal of the total admittance.

In an AC series circuit, we can still apply the current divider rule to determine the phasor current $\overline{I}_i$ flowing through a specific impedance $\overline{Z}_i$.

$$\overline{I}_i = \overline{I}_s \frac{\overline{Z}_p}{\overline{Z}_i}, \tag{26.4}$$

where $\overline{I}_s$ is the total current in the parallel circuit, and $\overline{Z}_p$ is the equivalent impedance of the parallel combination.

26.2.1 Problem 1

In the given circuit shown in Figure 26.2, the sinusoidal current source is defined as $i_s(t) = \sqrt{2} \cdot 3\sin(376.8t + 30°)$. Perform the following tasks:

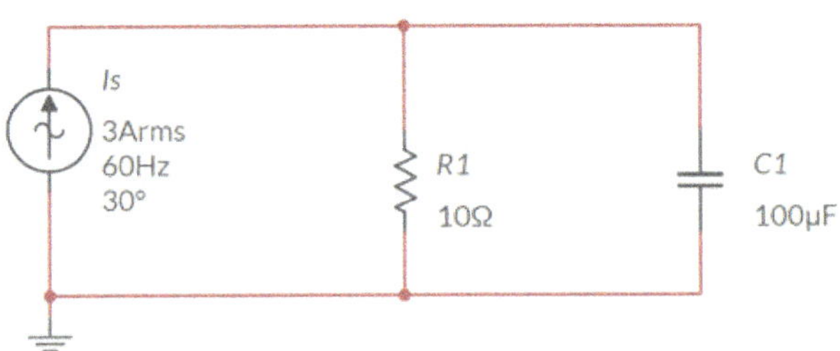

Figure 26.2 *R-C* parallel combination circuit.

1. Calculate the equivalent impedance $\overline{Z}$ of the *R-C* parallel combination. Use Multisim simulation to verify the calculated value.
2. Determine the phasor $\overline{V}$ of the voltage across the *R-C* parallel combination, and its instantaneous expression $v\,(t)$. Simulate the circuit using Multisim to measure the *rms* value of the voltage waveform and compare it with the calculated value.
3. Calculate the phase difference $\Delta\varphi = \varphi_V - \varphi_I$ between the phasor representing the voltage $\overline{V}$ across the *R-C* parallel combination and the phasor representing the generator current $\overline{I}_s$.
4. With Multisim, plot the instantaneous values of the current generator $i_s\,(t)$ and the voltage $v\,(t)$ across the parallel combination. By analyzing the plot, estimate the phase difference $\Delta\varphi$ in degrees between voltage and current waveforms and compare it with the calculated value.
5. Determine the phasor $\overline{I}_{R1}$ and $\overline{I}_{C1}$ representing respectively the currents through the resistor and the capacitor in the parallel combination. Simulate the circuit using Multisim to measure the *rms* values of the current waveforms and compare them with the calculated value.
6. Calculate the phase difference $\Delta\varphi_{I-V} = \varphi_{I_{c1}} - \varphi_V$ between the phasor $\overline{I}_{c1}$ representing the current through the capacitor and the phasor $\overline{V}$ representing the voltage across the capacitor.
7. With Multisim, plot the instantaneous current $i_{C1}\,(t)$ through the capacitor and the voltage *v*(*t*) across the capacitor. By analyzing the plot, estimate the phase difference $\Delta\varphi_{I-V}$ in degrees between voltage and current and compare it with the calculated value.
8. Calculate the frequency f_q at which the *rms* values of the currents $i_{R1}\,(t)$ and $i_{C1}\,(t)$, respectively through R_1 and C_1, are equal. Verify that the calculated frequency f_q aligns with the frequency observed in the Multisim simulation.

26.2.1.1 Solutions to Problem 1

1. The equivalent impedance $\overline{Z}$ of the *R-C* parallel combination can be calculated with the following formula:

$$\overline{Z} = \frac{R \cdot \left(-j\frac{1}{\omega C}\right)}{R - j\frac{1}{\omega C}} = 9.36 \angle -20.7°\Omega. \tag{26.5}$$

To verify the calculated value of the impedance using Multisim, we can simulate the circuit and measure the *rms* values of the current generator and of the voltage across the parallel combination (Figure 26.3).

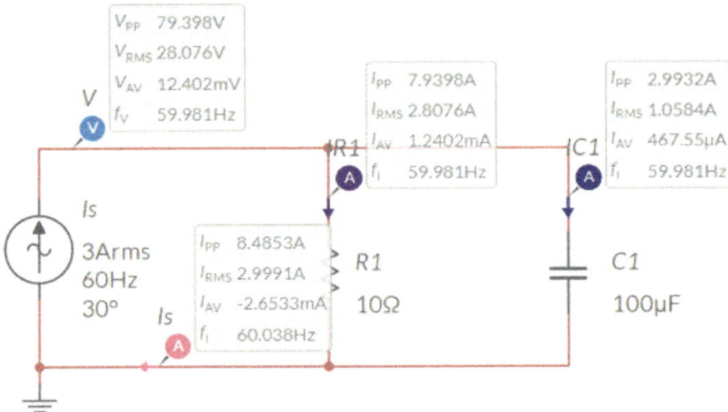

Figure 26.3 Multisim simulation providing *rms* values of supply voltage and current in the *R-C* parallel combination circuit.

The magnitude of the impedance can be determined by applying Ohm's law, which involves dividing the measured *rms* values of voltage across the parallel combination and the generator current.

$$\left|\overline{Z}\right| = \frac{V_{\mathrm{rms}}}{I_{\mathrm{rms}}} = \frac{28}{2.99} = 9.36\ \Omega. \tag{26.6}$$

The theoretical value matches the simulated result, confirming its accuracy.

2. The voltage phasor $\overline{V}$ across the *R-C* parallel combination can be calculated by using Ohm's law. Since the current phasor is given as $\overline{I}_s = 3\angle 30°\,\mathrm{A}$, we can use the following formula:

$$\overline{V} = \overline{I}_s \cdot \overline{Z} = 3\ \angle 30° \cdot 9.36\ \angle -20.7°= 28\ \angle 9.2°\mathrm{V}. \tag{26.7}$$

In the time domain, the expression of the voltage is given as follows:

$$v\left(t\right) = \sqrt{2} \cdot 28\sin\left(376.8t + 9.2°\right)\ \mathrm{V}\,. \tag{26.8}$$

The theoretical value matches the simulated result, confirming its accuracy (Figure 26.3).

3. The phase difference $\Delta\varphi$ between the generator current $\overline{I}_s$ and the voltage $\overline{V}$ across the *R-C* parallel can be obtained with the following formula:

$$\Delta\varphi = \varphi_I - \ \varphi_V = 30° - 9.2° = 20.8°. \tag{26.9}$$

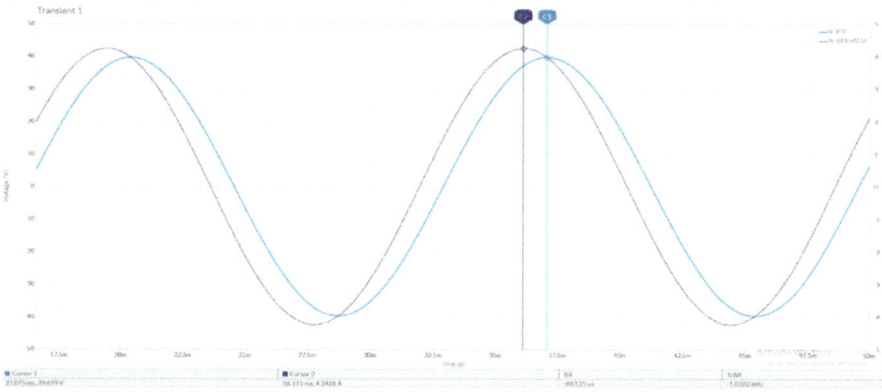

Figure 26.4 Plot of the instantaneous waveforms $i_s(t)$ and $v(t)$ obtained with Multisim simulation.

Since $\Delta\varphi$ is greater than zero, we can conclude that in the time domain, the current waveform leads the voltage waveform supplied by the generator. This result is consistent with the calculated value of the argument of the total impedance $\overline{Z}$, which represents the phase angle between the current and voltage in the circuit.

4. The plot of the instantaneous waveforms $i_s(t)$ and $v(t)$ obtained with Multisim simulation is presented in Figure 26.4.

Analyzing the plot, we can identify the time difference Δt between corresponding points on the waveforms.

The phase difference $\Delta\varphi$ can be estimated in degrees with the following formula:

$$\Delta\varphi = \frac{360}{0.0166} \cdot 963.2 \cdot 10^{-6} = 20.9^\circ. \tag{26.10}$$

The result from the Multisim plot closely matches the calculated value.

5. The phasors currents $\overline{I}_{R1}$ and $\overline{I}_{C1}$ can be calculated by applying Ohm's law:

$$\overline{I}_{R1} = \frac{\overline{V}}{R_1} = \frac{28\angle 9.2^\circ}{10} = 2.8\angle 9.2^\circ \text{A}, \tag{26.11}$$

$$\overline{I}_{C1} = \frac{\overline{V}}{-j\frac{1}{\omega C1}} = \frac{28\angle 9.2^\circ}{26.5\angle -90^\circ} = 1.05\angle 99.2^\circ \text{A}. \tag{26.12}$$

The Multisim simulation results align with the calculated values, confirming their accuracy (Figure 26.3).

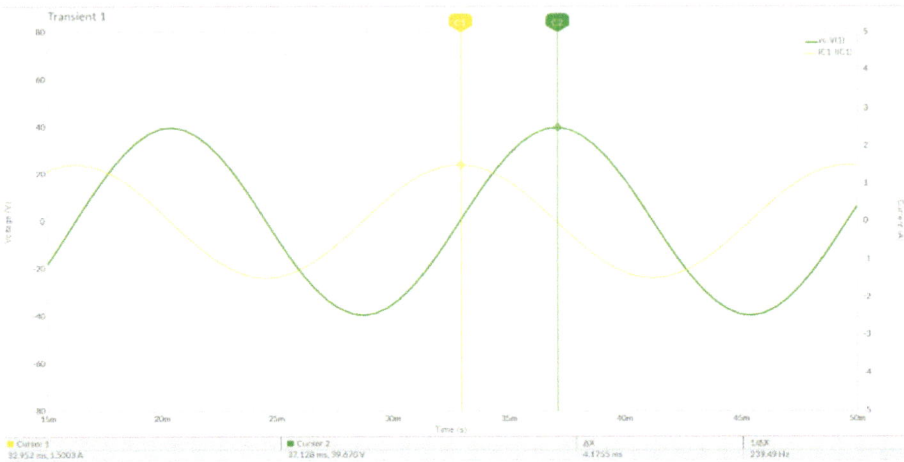

Figure 26.5 Plot of the current $i_{C1}(t)$ through the capacitor and the voltage $v(t)$ across the capacitor.

6. The phase difference $\Delta\varphi_{Ic1-V}$ between the phasor $\overline{I}_c$ and the phasor $\overline{V}$ can be obtained with the following formula:

$$\Delta\varphi_{I-V} = \varphi_{Ic1} - \varphi_V = 99.2^\circ - 9.2^\circ = 90^\circ. \quad (26.13)$$

7. The plot of $i_{C1}(t)$ and $v(t)$ obtained with Multisim simulation is presented in Figure 26.5. Analyzing the plot, we can identify the time difference Δt between corresponding points on the waveforms.

From the plot, the phase difference in degrees between $i_C(t)$ and $v(t)$ can be estimated with the following formula:

$$\Delta\varphi_{I-V} = \frac{360}{0.0166} \cdot 4.17 \cdot 10^{-3} = 90.5^\circ. \quad (26.14)$$

The result from the Multisim plot closely matches the calculated value.

8. In the branches of the *R*-*C* parallel combination, the currents $i_{R1}(t)$ and $i_{C1}(t)$ have the same *rms* value if the resistance and the capacitive reactance are equal. Thus, the corresponding frequency f_q can be determined by writing the following equation:

$$R_1 = X_{C1} = \frac{1}{2\pi f_q C_1}. \quad (26.15)$$

By solving the equation for f_q, we obtain f_q = 159.2 Hz.

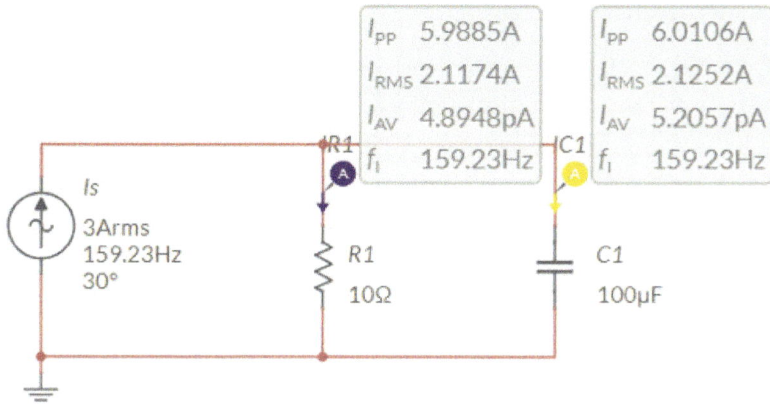

Figure 26.6 Equal *rms* values of the currents at f_q = 159.2 Hz.

The calculated value of the frequency f_q can be verified with Multisim simulation using a voltage source with a frequency set to f_q (Figure 26.6).

By observing the simulation results, we determine that the voltage drops across R_1 and C_1 are indeed equal, which confirms that the calculated value of f_q is correct.

26.2.2 Problem 2

In the presented circuit of Figure 26.7, a sinusoidal current source, denoted as $i_{s2}\,(t) = \sqrt{2} \cdot \sin(376.8t)$, supplies a parallel combination of a resistor and of an unknown inductor L_2. Perform the following tasks:

1. Determine the value of the inductance L_2 required to achieve an *rms* value of 936 mA for the current $i_{L2}\,(t)$ through the inductor. Performing a simulation using Multisim to validate this calculated value of L_2 and provide confirmation of its accuracy.
2. Calculate the equivalent impedance $\overline{Z}$ of the R_2-L_2 parallel combination. Validate this calculated value of $\overline{Z}$ performing a Multisim simulation.
3. Determine the phasor $\overline{V}_2$ representing the voltage across the R_2-L_2 parallel combination. Compare the calculated *rms* value and phase angle of $\overline{V}_2$ with the simulation results obtained from Multisim to ensure their consistency.
4. Determine the phasor $\overline{I}_{R2}$ representing the current flowing through the resistor R$_2$. The calculated *rms* value of $\overline{I}_{R2}$ must be cross-checked using Multisim simulation data.

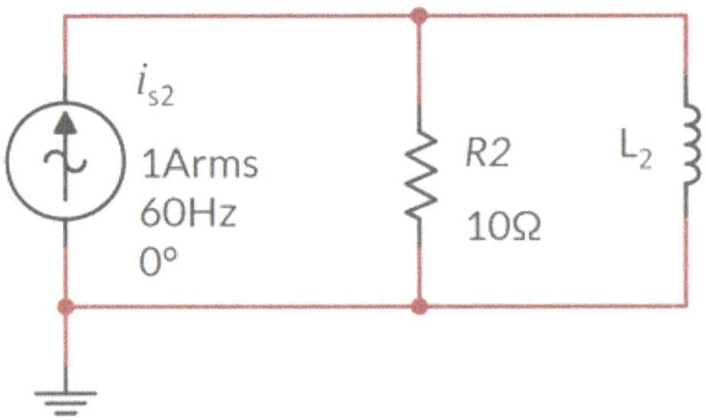

Figure 26.7 Ohmic-inductive parallel circuit with an AC current generator and an unknown inductor.

26.2.2.1 Solutions to Problem 2

1. The value of the inductance L_2 to achieve the desired *rms* current value of 936 mA can be determined by applying the current divider formula to determine $\overline{I}_{L2}$:

$$\overline{I}_{L2} = \overline{I}_{s2}\frac{R_2}{R_2 + j\omega L_2} = \frac{10}{10 + j376.8 \cdot L_2} = 936 \angle\varphi_{I_{L2}} \text{ mA}. \tag{26.16}$$

Eqn (26.16) is a complex equation that involves the unknown inductance L_2. To solve for L_2, we can equate the magnitudes of the complex numbers on both sides of the equation and focus solely on the real values. This approach allows us to obtain a numerical value that satisfies the given condition.

$$\left|\overline{I}_{L2}\right| = \frac{10}{\sqrt{10^2 + 376.8^2 \cdot L_2^2}} = 936 \text{ mA}. \tag{26.17}$$

Eqn (26.17) provides a calculated value for L_2 of 10 mH.

To verify the calculated value of the inductance L_2 using Multisim, we can simulate the circuit and measure the *rms* values of the current through it (Figure 26.8).

The calculated value of the inductance closely matches the Multisim simulated value, confirming its accuracy.

2. The equivalent impedance $\overline{Z}$ of the R_2-L_2 parallel combination can be calculated using the following formula:

$$\overline{Z} = \frac{R_2 \cdot j\omega L_2}{R_2 + j\omega L_2} = 3.5\angle 69.2°\Omega. \tag{26.18}$$

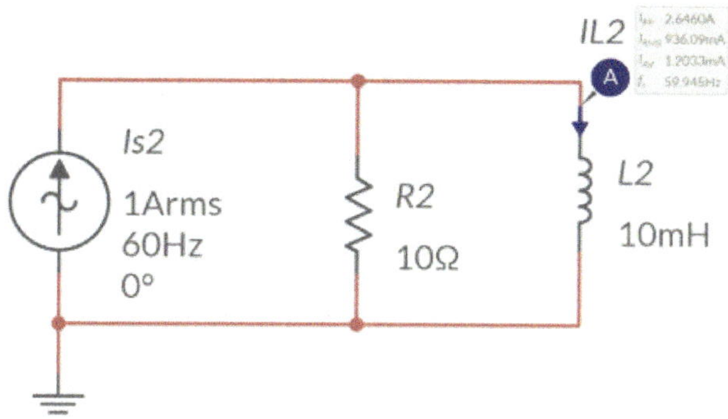

Figure 26.8 Multisim simulation confirms the calculated value for L_2.

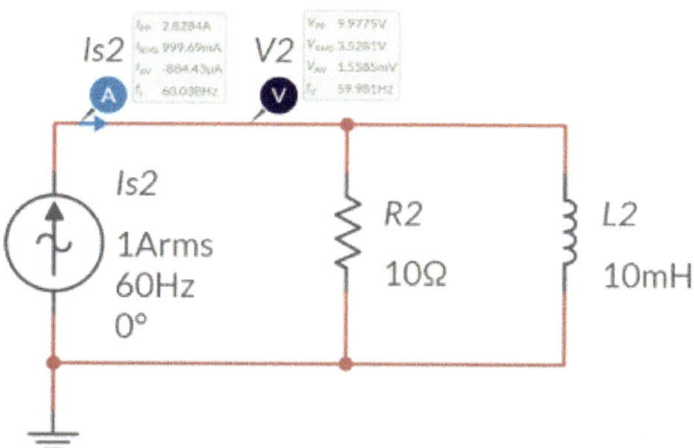

Figure 26.9 Multisim simulation confirms the calculated value of the impedance $\overline{Z}$.

To verify the calculated value of the impedance using Multisim, we can simulate the circuit and measure the *rms* values of the voltage across the parallel combination and the current circulating through it (Figure 26.9).

The magnitude of the impedance can be determined by applying Ohm's law, which involves dividing the measured *rms* values of voltage and current.

$$\left|\overline{Z}_s\right| = \frac{V_{\text{rms}}}{I_{\text{rms}}} = 3.5\ \Omega. \tag{26.19}$$

The calculated value matches the Multisim simulated value, confirming its accuracy.

3. The phasor $\overline{V}_2$ representing the voltage across the R_2-L_2 parallel combination can be calculated using the following formula:

$$\overline{V}_2 = \overline{I}_{s2} \cdot \overline{Z} = 1\angle 0^\circ \cdot 3.5\angle 69.2^\circ = 3.5\angle 69.2^\circ\text{V}. \tag{26.20}$$

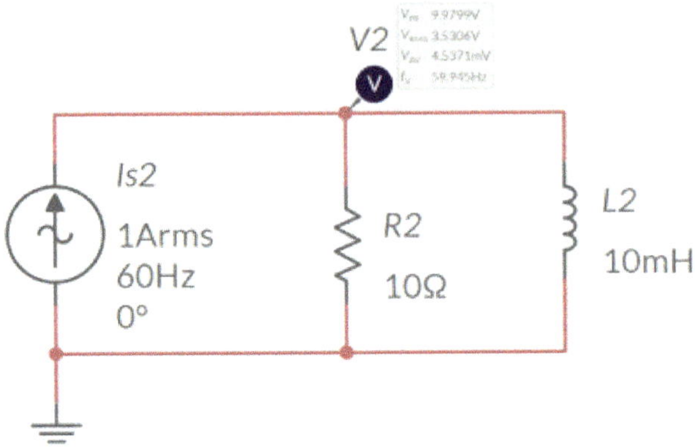

Figure 26.10 Multisim simulation confirming the *rms* calculated value of $\overline{V}_2$.

The calculated *rms* value of $\overline{V}_2$ is confirmed by the Multisim simulation (Figure 26.10).

The Multisim plot of the instantaneous values of the supply current source $i_{s2}(t)$ and of the voltage $v_2(t)$ across the parallel combination is presented in Figure 26.11.

Analyzing the plot, we can identify the time difference Δt between corresponding points on the waveforms (e.g., maxima).

The phase difference $\Delta\varphi_2$ can be estimated in degrees with the following formula:

$$\Delta\varphi_2 = \frac{360}{0.0166} \cdot 3.2 \cdot 10^{-3} = 69.4°. \quad (26.21)$$

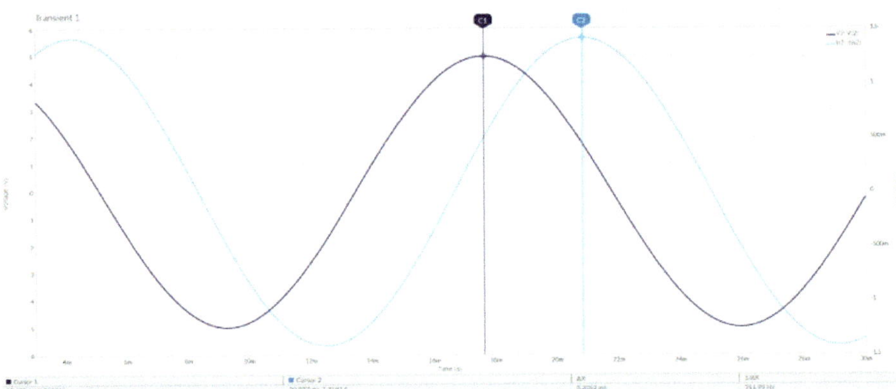

Figure 26.11 Multisim plot of the instantaneous values of the supply current source $i_{s2}(t)$ and of the voltage $v_2(t)$ across the parallel combination.

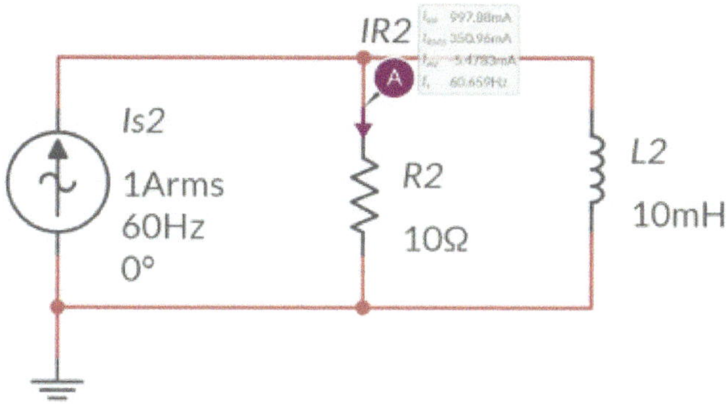

Figure 26.12 Multisim simulation confirming the calculated *rms* value of $\overline{I}_{R2}$.

The result from the Multisim plot closely matches the calculated value, confirming its accuracy.

It is evident that the voltage $\overline{V}_2$ leads the current $\overline{I}_{s2}$. This behavior is observed in ohmic-inductive circuits, where the inductance causes the current to lag behind the voltage.

4. The phasor $\overline{I}_{R2}$ representing the current flowing through the resistor R_2 can be calculated with the following formula:

$$\overline{I}_{R2} = \frac{\overline{V}_2}{R_2} = \frac{3.5\angle 69.2^\circ}{10\angle 0^\circ} = 350\angle 69.2^\circ \text{ mA}. \tag{26.22}$$

The calculated *rms* value of $\overline{I}_{R2}$ is confirmed by the Multisim simulation (Figure 26.12).

26.2.3 Problem 3

In the circuit given in Figure 26.13, the sinusoidal current source, denoted as $i_{s3}(t) = \sqrt{2} \cdot 2 \cdot \sin(6280t)$, supplies a parallel combination of impedances. Perform the following tasks:

1. Calculate the equivalent impedance $\overline{Z}$ of the parallel combination. Use Multisim simulation to verify the calculated value.
2. Determine the phasor $\overline{V}_3$ representing the voltage across the parallel combination. Compare the calculated *rms* value and phase angle of $\overline{V}_3$ with the simulation results obtained from Multisim to ensure their consistency.
3. Calculate the power absorbed by the resistor R_3.

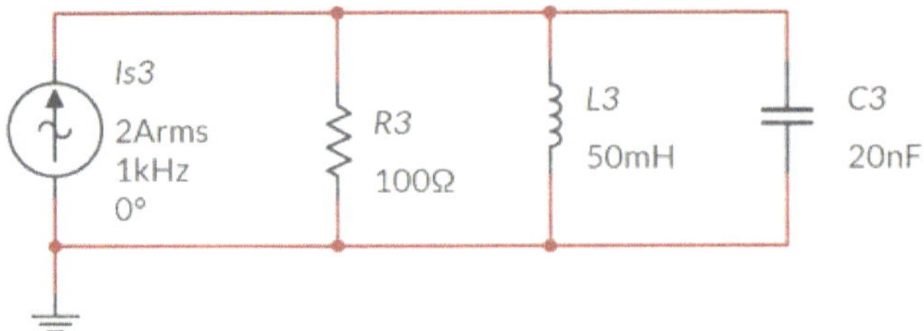

Figure 26.13 Ohmic-inductive-capacitive parallel circuit with an AC current generator.

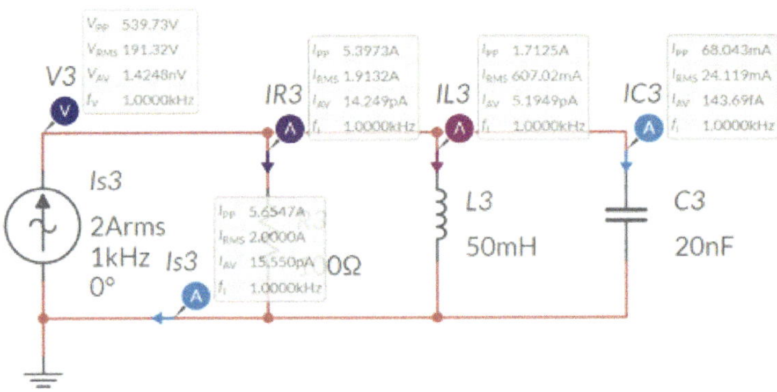

Figure 26.14 Multisim simulation providing *rms* values of the current $i_{s3}\,(t)$ and the voltage $v_3\,(t)$.

26.2.3.1 Solutions to Problem 3

1. The equivalent impedance $\overline{Z}$ of the parallel combination can be calculated with the following formula:

$$\overline{Z} = \left(\frac{1}{R} + \frac{1}{j\omega L} + j\omega C\right)^{-1} = \left(\frac{1}{100} + \frac{1}{j314.1} + j7957.7\right)^{-1}$$
$$= 91.4 + j27.9\ \Omega = 95.6\angle 17°\Omega. \tag{26.23}$$

To verify the calculated value of the impedance using Multisim, we can simulate the circuit and measure the *rms* values of the generator current $i_{s3}\,(t)$ and the voltage $v_3\,(t)$ across the parallel combination.

The magnitude of the impedance can be determined by applying Ohm's law dividing the measured *rms* values of voltage across the parallel combination and the generator current.

$$\left|\overline{Z}\right| = \frac{V_{3\ \mathrm{rms}}}{I_{s3\ \mathrm{rms}}} = 95.6\ \Omega. \tag{26.24}$$

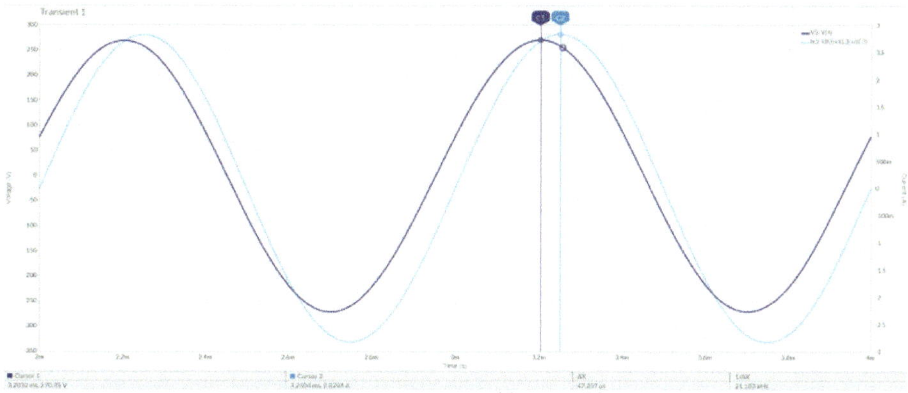

Figure 26.15 Plot of $i_{s3}(t)$ and $v_3(t)$.

The theoretical value matches the simulated result, confirming its accuracy.

To validate the calculated value of the impedance's argument, we can utilize Multisim to plot the instantaneous values of the current $i_{s3}(t)$ and the voltage $v_3(t)$ in the circuit. This simulation allows us to visually verify the phase relationship between the current and voltage (Figure 26.15).

Analyzing the plot, we can identify the time difference Δt between corresponding points on the waveforms.

The phase difference $\Delta\varphi$ can be estimated in degrees with the following formula:

$$\Delta\varphi = \frac{360}{10^{-3}} \cdot 47.2 \cdot 10^{-6} = 16.9°. \tag{26.25}$$

The result from the Multisim plot closely matches the calculated value of the argument of the total impedance, confirming its accuracy.

2. The phasor $\overline{V}_3$ representing the voltage across the parallel combination of impedances can be calculated by applying Ohm's law:

$$\overline{V}_3 = \overline{I}_{s3} \cdot \overline{Z} = 2\angle 0° \cdot 95.6\angle 17° = 191.2\angle 17° \text{ V}. \tag{26.26}$$

The Multisim simulation closely matches calculated *rms* value (Figure 26.14) and the phase angle (Figure 26.15), confirming the accuracy of calculations.

3. The power absorbed by the resistor R_3 can be calculate with the following formula:

$$P_{R3} = \frac{V_{\text{rms}}^2}{R_3} = \frac{191.2^2}{100} = 365.6 \text{ W}. \tag{26.27}$$

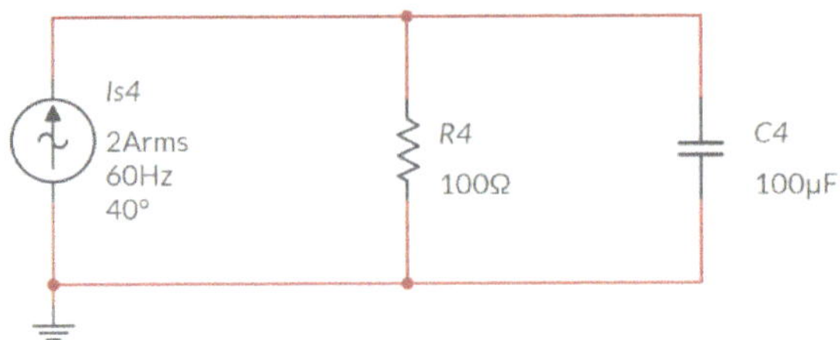

Figure 26.16 Ohmic-capacitive parallel circuit with an AC current generator.

26.3 Lab Procedure 1

In the circuit given in Figure 26.16, a sinusoidal current source, denoted as $i_{s4}(t) = \sqrt{2} \cdot 2 \cdot \sin(376.8t + 40°)$ A, supplies a parallel combination of impedances. Perform the following tasks:

1. Calculate the equivalent impedance $\overline{Z}$ of the *R*-*C* parallel combination. Use Multisim simulation to verify the calculated value. Record value in Table 26.1.
2. Determine the phasor $\overline{V}$ of the voltage across the *R*-*C* parallel combination, and its instantaneous expression $v(t)$. Simulate the circuit using Multisim to measure the *rms* value of the voltage waveform and compare it with the calculated value. Record expressions of $\overline{V}$ and $v(t)$ in Table 26.1.
3. Determine the phase difference $\Delta\varphi = \varphi_V - \varphi_I$ between the voltage $\overline{V}$ across the *R*-*C* parallel combination and the generator current $\overline{I}_{s4}$. Record value in Table 26.1.
4. With Multisim, plot the instantaneous values of the current generator $i_{s4}(t)$ and the voltage $v(t)$ across the parallel combination. By analyzing the plot, estimate the phase difference $\Delta\varphi$ in degrees between voltage and current and compare it with the calculated value to confirm its accuracy.
5. Determine the phasors $\overline{I}_{R4}$ and $\overline{I}_{C4}$ representing the currents through the resistor and the capacitor, respectively. Record values in Table 26.1. Simulate the circuit using Multisim to measure the *rms* values of the current waveforms and compare them with the calculated value.

Table 26.1 Summary of readings and calculations

$\overline{Z} = Z\angle\varphi$	$\overline{V} = V_{\text{rms}}\angle\varphi_V$	$v(t)$	$\Delta\varphi$	$\overline{I}_{R4}$	$\overline{I}_{C4}$	$\Delta\varphi_{Ic4-V}$

6. Calculate the phase difference $\Delta\varphi_{Ic4-V} = \varphi_{Ic4} - \varphi_V$ between the current $\overline{I}_{c4}$ through the capacitor and the voltage $\overline{V}$ across the *R-C* parallel. Record value in Table 26.1.
7. With Multisim, plot the instantaneous current $i_{C4}(t)$ through the capacitor and the voltage *v*(*t*) across the capacitor. By analyzing the plot, estimate the phase difference $\Delta\varphi_{IC4-V}$ in degrees between voltage and current and compare it with the calculated value.

26.4 Lab Procedure 2

In the circuit given in Figure 26.17, a sinusoidal current source, denoted as $i_{s5}(t) = \sqrt{2} \cdot 2 \cdot \sin(12,560t)$ A, supplies a parallel combination of impedances. Perform the following tasks:

1. Calculate the equivalent impedance $\overline{Z}$ of the parallel combination of impedances. Use Multisim simulation to verify the calculated value. Record value in Table 26.2.
2. Determine the phasor $\overline{V}_5$ representing the voltage across the parallel combination of impedances. Simulate the circuit using Multisim to measure the *rms* value of the voltage waveform and compare it with the calculated value. Record expression of $\overline{V}_5$ in Table 26.2.
3. With Multisim, plot the instantaneous generator current $i_{s5}(t)$ and the instantaneous voltage $v_5(t)$ across the parallel combination of impedances. By analyzing the plot, estimate the phase difference $\Delta\varphi_{I_{s5}-V_5}$ in degrees between voltage and current and compare it with the theoretical value. Record results in Table 26.2.
4. Calculate the power P_5 absorbed by the resistor R_5. Record result in Table 26.2.

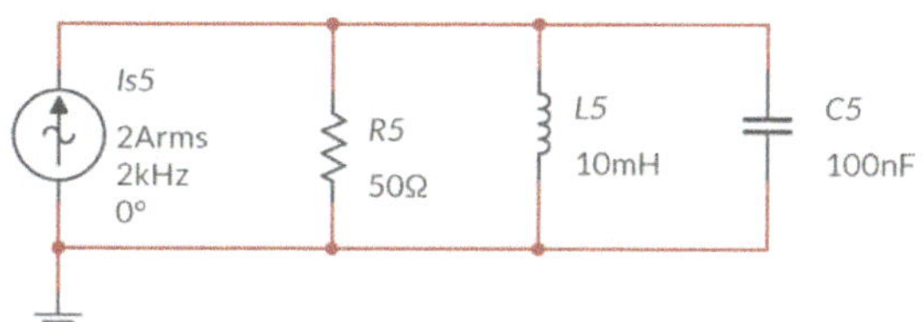

Figure 26.17 Ohmic-inductive-capacitive parallel circuit with an AC current generator.

Table 26.2 Summary of readings and calculations

$\overline{Z} = Z\angle\varphi$	$\overline{V}_5 = V_{5\ \mathrm{rms}}\angle\varphi_V$	$\Delta\varphi_{I_{s5}-V_5}$	P_5

27

Series–Parallel AC Circuits

27.1 Student Learning Outcomes

- Calculate the equivalent impedance of AC series–parallel combination circuits using theoretical principles.
- Apply knowledge of AC circuits to calculate in the phasor domain, and measure with Multisim simulations, voltage drops across and currents through resistors, inductors, and capacitors connected in AC series–parallel combination circuits.
- Apply Multisim to simulate AC series–parallel combination circuits, verify theoretical calculations, and observe the behavior of voltage drops and currents.

27.2 Prelab

AC complex circuits (Figure 27.1) with a single generator can be effectively analyzed using a systematic approach similar to that used for series and parallel circuits. By identifying series and parallel relationships within the circuit, we can simplify it and determine the equivalent impedance seen by the generator. Once the circuit is reduced to a single impedance, we can apply Ohm's law to calculate the current or voltage supplied by the generator.

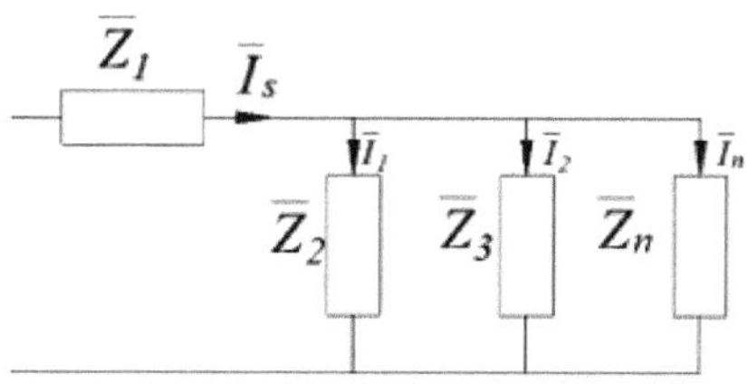

Figure 27.1 Impedances $\overline{Z}\,(j\omega)$ connected in series and in parallel in the same circuit.

Overall, the technical challenge in solving series–parallel circuits lies in effectively recognizing and simplifying the circuit topology, applying the appropriate techniques for combining series and parallel elements, working with complex impedances and phasors, and performing complex number calculations accurately and efficiently.

27.2.1 Problem 1

In the given circuit of Figure 27.2, the sinusoidal voltage source is defined as $v_s\,(t) = \sqrt{2} \cdot 24\,\sin(376.8t)$. Perform the following tasks:

1. Calculate the total impedance $\overline{Z}$ of the series–parallel combination. Use Multisim simulation to verify the calculated value.
2. Determine the phasor $\overline{V}_{12}$ of the voltage across the R_1-C_2 parallel combination. Simulate the circuit using Multisim to measure the *rms* value of the voltage waveform and compare it with the calculated value.
3. Calculate the total current $\overline{I}_s$ supplied by the generator. Simulate the circuit using Multisim to measure the *rms* value of the total current waveform and compare it with the calculated value.
4. Determine the phase difference $\Delta\varphi = \varphi_{Vs} - \varphi_{Is}$ between the phasors $\overline{V}_s$, which represents the generator's voltage, and $\overline{I}_s$, which represents the current supplied by the voltage generator.
5. With Multisim, plot the instantaneous values of the current $i_s\,(t)$ supplied by the generator and the generator voltage $v_s\,(t)$. By analyzing the plot, estimate the phase difference $\Delta\varphi$ in degrees between voltage and current waveforms and compare it with the calculated value.

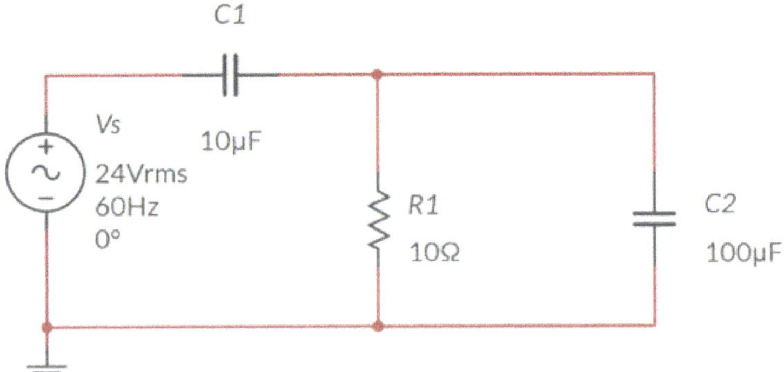

Figure 27.2 Ohmic-capacitive series–parallel combination circuit.

6. Based on the plot of the instantaneous waveforms of the current $i_s(t)$ supplied by the generator and the generator voltage $v_s(t)$, estimate the magnitude $\left|\overline{Z}\right|$ of the total impedance $\overline{Z}$.

27.2.1.1 Solutions to Problem 1

1. The equivalent impedance $\overline{Z}_{12}$ of the R_1-C_2 parallel combination can be calculated with the following formula:

$$\overline{Z}_{12} = \frac{R_1 \cdot \left(-j\frac{1}{\omega C_2}\right)}{R_1 - j\frac{1}{\omega C_2}} = \frac{10 \cdot (-j26.5)}{10 - j26.5} = 9.3\angle - 20.7°\Omega = 8.7 - \text{j}3.3\ \Omega. \tag{27.1}$$

The total impedance $\overline{Z}$ of the series–parallel combination can be calculated with the following formula:

$$\begin{aligned}\overline{Z} &= -j\frac{1}{\omega C_1} + \overline{Z}_{12} \\ &= -j265.4 + 8.7 - \text{j}3.3 = 8.7 - \text{j}268.7\ \Omega = 268.8\angle - 88.1°\Omega. \end{aligned} \tag{27.2}$$

To verify the calculated value of the impedance using Multisim, we can simulate the circuit and measure the *rms* values of voltage and current of the source (Figure 27.3).

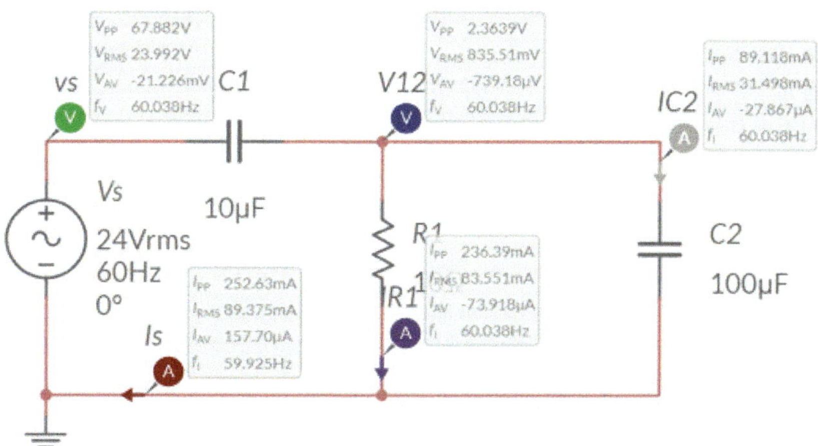

Figure 27.3 Multisim simulation providing *rms* values of voltage and current of voltage source.

The magnitude of the impedance can be determined by applying Ohm's law, which involves dividing the measured *rms* values of voltage and current of the source.

$$\left|\overline{Z}\right| = \frac{V_s}{I_s} = \frac{23.99}{89.37 \cdot 10^{-3}} = 268.4\ \Omega. \tag{27.3}$$

The theoretical value closely matches the Multisim simulated result, confirming its accuracy.

2. The phasor $\overline{V}_{12}$ representing the voltage across the R_1-C_2 parallel combination can be calculated by applying the voltage divider formula:

$$\overline{V}_{12} = 24\angle 0^\circ \frac{\overline{Z}_{12}}{-j\frac{1}{\omega C_1} + \overline{Z}_{12}} = 24\angle 0^\circ \frac{9.3\angle -20.7^\circ}{268.8\angle -88.1^\circ} = 830\ \angle 67.4^\circ \mathrm{mV}. \tag{27.4}$$

To verify the calculated value of the phasor $\overline{V}_{12}$ using Multisim, we can simulate the circuit and measure the *rms* value of the voltage across the parallel combination (Figure 27.3). The theoretical value closely matches the simulated result, confirming its accuracy.

3. The total current $\overline{I}_s$ supplied by the generator can be calculated as follows:

$$\overline{I}_s = \frac{\overline{V}_s}{\overline{Z}} = \frac{24\angle 0^\circ}{268.8\angle -88.1^\circ} = 89.3\ \angle 88.1^\circ \mathrm{mA}. \tag{27.5}$$

To confirm the magnitude of the calculated value of the phasor $\overline{I}_s$ using Multisim, we can simulate the circuit and measure the *rms* value of total current supplied by the generator (Figure 27.3). The theoretical value closely matches the simulated result, confirming its accuracy.

4. The phase difference $\Delta\varphi$ between the generator voltage $\overline{V}_s$ and the total current $\overline{I}_s$ supplied by the generator can be obtained with the following formula:

$$\Delta\varphi = \varphi_{Vs} - \varphi_{Is} = 0^\circ - 88.1^\circ = -88.1^\circ. \tag{27.6}$$

The negative value of $\Delta\varphi$ indicates that in the time domain, the current waveform leads the voltage waveform supplied by the generator. This observation aligns with the calculated value of the argument of the total impedance $\overline{Z}$, which represents the phase angle between the current and voltage in the circuit. Furthermore, the phase relationship between the current and voltage waveforms, being consistent with the expected behavior in an ohmic-capacitive circuit, reinforces the understanding of the circuit's characteristics and validates the analysis performed.

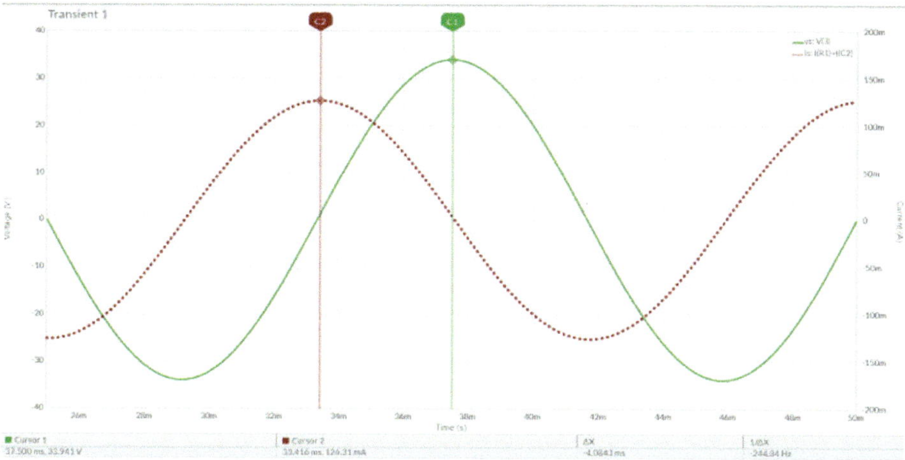

Figure 27.4 The plot of the instantaneous waveforms $i_s\left(t\right)$ and $v_s\left(t\right)$ obtained with Multisim simulation.

5. The plot of the instantaneous waveforms $i_s\left(t\right)$ and $v_s\left(t\right)$ obtained with Multisim simulation is presented in Figure 27.4.

Analyzing the plot, we can identify the time difference Δt between corresponding points on the waveforms.

The phase difference $\Delta\varphi$ can be estimated in degrees with the following formula:

$$\Delta\varphi = \frac{360}{0.0166}\left(-4.0843\right)\cdot 10^{-3} = -88.5°. \tag{27.7}$$

The result from the Multisim plot closely aligns with the calculated value.

6. Analyzing the plot of Figure 27.4, we can identify the values of the peaks of the instantaneous waveforms $i_s\left(t\right)$ and $v_s\left(t\right)$. The magnitude $\left|\overline{Z}\right|$ of the impedance can be estimated by applying Ohm's law:

$$\left|\overline{Z}\right| = \frac{V_{\mathrm{sp}}}{I_{\mathrm{sp}}} = \frac{33.9}{0.126} = 269\ \Omega. \tag{27.8}$$

The result from the Multisim plot closely matches the calculated value.

27.2.2 Problem 2

In the given circuit of Figure 27.5, the sinusoidal current source is defined as $i_{s1}\left(t\right) = \sqrt{2}\cdot 2\cdot \sin(376.8t)$. Perform the following tasks:

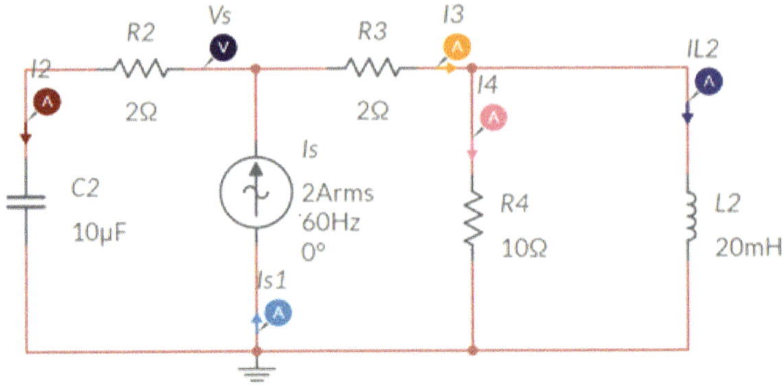

Figure 27.5 Series–parallel combination circuit with an AC current generator.

1. Determine the impedances $\overline{Z}_l$ and $\overline{Z}_r$ at the left and the right of the current generator, respectively.
2. Determine the equivalent impedance $\overline{Z}$ seen by the current generator. Use Multisim simulation to verify the calculated value.
3. Determine the phasor $\overline{I}_2$ representing the current through R_2. Confirm the calculated *rms* value with a Multisim simulation.
4. Determine the phasor $\overline{V}_s$ representing the voltage across the current generator. Confirm the calculated *rms* value with a Multisim simulation.
5. With Multisim, generate a plot illustrating the instantaneous current $i_2(t)$ through R_2 and the instantaneous voltage $v_s(t)$ across the current generator. By analyzing the plot, estimate the phase difference $\Delta\varphi$ in degrees between the two waveforms.

27.2.2.1 Solutions to Problem 2

1. The equivalent impedances $\overline{Z}_r$ and $\overline{Z}_l$ can be calculated with the following formulae:

$$\overline{Z}_r = R_3 + \frac{R_4 \cdot j\omega L_2}{R_4 + j\omega L_2} = 2 + \frac{75\angle 90^\circ}{12.5\angle 36.9^\circ} = 2 + 3.6 + j4.8\ \Omega$$
$$= 7.4\angle 40.6^\circ\ \Omega, \tag{27.9}$$

$$\overline{Z}_l = R_2 - j\frac{1}{\omega C_2} = 2 - j265.4\ \Omega = 265.4\angle{-89.6^\circ}\ \Omega. \tag{27.10}$$

2. The equivalent impedance $\overline{Z}$ seen by the current generator can be determined by calculating the parallel combination of the impedances $\overline{Z}_l$ and $\overline{Z}_r$.

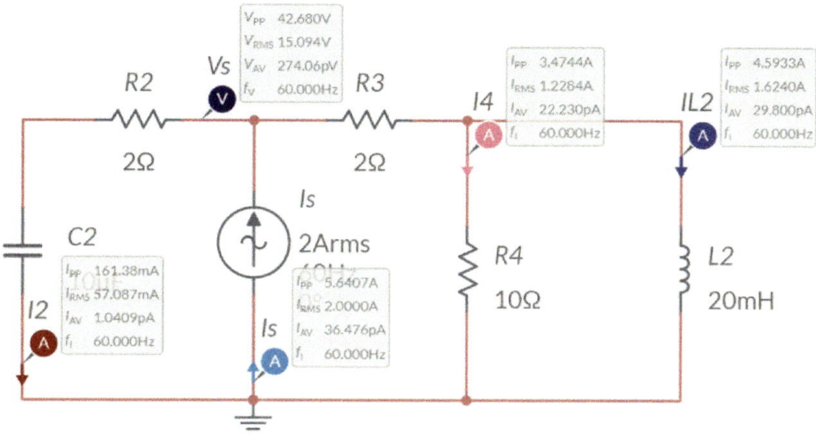

Figure 27.6 Multisim simulation providing *rms* values of $v_s\,(t)$ and $i_s\,(t)$.

$$\overline{Z} = \frac{\overline{Z}_r \cdot \overline{Z}_l}{\overline{Z}_r + \overline{Z}_l} = \frac{7.4\angle 40.6^\circ \cdot 265.4\angle{-89.6^\circ}}{5.6 + j4.8 + (2 - j265.3)}$$
$$= 7.5\angle 39.4^\circ\Omega = 5.8 + j4.8\ \Omega. \qquad (27.11)$$

To confirm the accuracy of the calculated impedance value, a simulation can be performed using Multisim. We can simulate the circuit and measure the *rms* values of the generator current $i_s\,(t)$ and of the voltage $v_s\,(t)$ across it (Figure 27.6).

The magnitude of the impedance can be determined by applying Ohm's law, which states that the magnitude of the impedance is equal to the ratio of the measured *rms* value of the voltage to the current.

$$\left|\overline{Z}\right| = \frac{V_s}{I_s} = \frac{15}{2} = 7.5\ \Omega. \qquad (27.12)$$

The simulated result matches the theoretical value, confirming its accuracy.

3. The phasor $\overline{I}_2$ representing the current through the resistor R_2 can be determined by applying the current divider formula:

$$\overline{I}_{R2} = \overline{I}_s \frac{\overline{Z}_r}{\overline{Z}_r + \overline{Z}_l} = 2\angle 0^\circ \frac{7.4\ \angle 40.6^\circ}{260.7\angle{-88.3^\circ}} = 56.7\angle 128.9^\circ\text{mA}. \qquad (27.13)$$

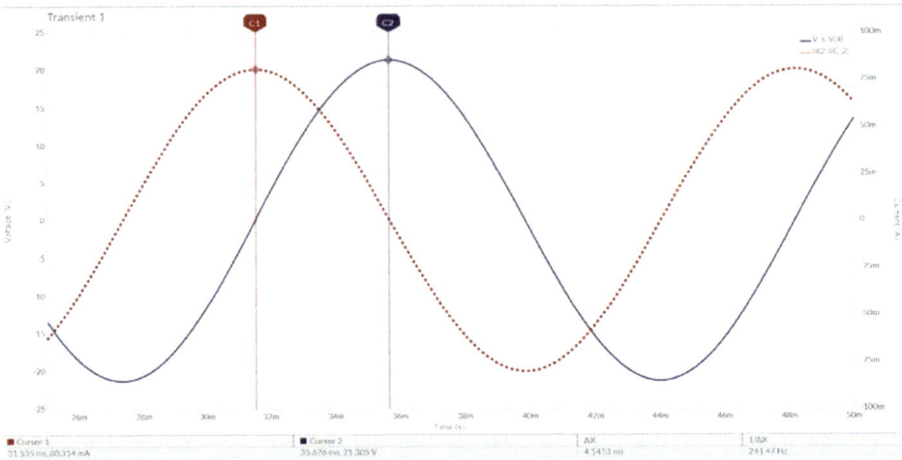

Figure 27.7 Plot of $i_2\,(t)$ and $v_s\,(t)$.

The Multisim simulation result closely matches the *rms* theoretical value, confirming its accuracy.

4. The phasor $\overline{V}_s$ representing the voltage across the current generator can be calculated by applying the Ohm's law.

$$\overline{V}_s = \overline{I}_s \cdot \overline{Z} = 2\angle 0^\circ \cdot 7.5\angle 39.4^\circ = 15\angle 39.4^\circ \text{V}. \tag{27.14}$$

The theoretical *rms* value of $\overline{V}_s$ is confirmed by the Multisim simulation shown in Figure 27.6.

5. The plot of the instantaneous current $i_2\,(t)$ and of the instantaneous voltage $v_s\,(t)$ across it, generated with Multisim simulation, is presented in Figure 27.7. Analyzing the plot, we can identify the time difference Δt between corresponding points on the waveforms.

The phase difference between $i_2\,(t)$ and $v_s\,(t)$, denoted as $\Delta\varphi$, can be estimated with the following formula:

$$\Delta\varphi = \frac{360^\circ}{0.166} \cdot 4.1413 \cdot 10^{-3} = 89.5^\circ. \tag{27.15}$$

The theoretical value of the phase difference can be calculated by subtracting the corresponding phase angles of $\overline{V}_s$ and $\overline{I}_s$:

$$\Delta\varphi = \varphi_{I_2} - \varphi_{V_s} = \;128.9^\circ - 39.4^\circ = 89.5^\circ. \tag{27.16}$$

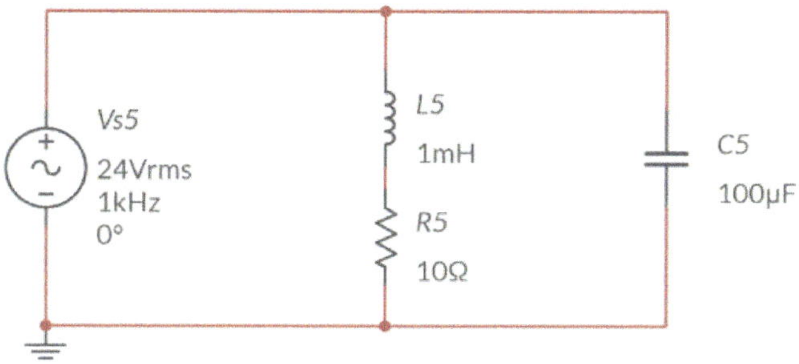

Figure 27.8 Ohmic-inductive-capacitive parallel circuit with an AC voltage generator.

The theoretical value of the phase difference $\Delta\varphi$ is confirmed by the Multisim simulation. The agreement between the theoretical calculation and the simulation results reinforces the accuracy and reliability of the analysis conducted in this circuit.

The observed phase relationship clearly indicates that the voltage $\overline{V}_s$ lags behind the current $\overline{I}_2$, which is consistent with the behavior of an ohmic-capacitive circuit.

27.2.3 Problem 3

In the circuit given in Figure 27.8, the sinusoidal voltage source $v_{s5}(t) = \sqrt{2} \cdot 24 \cdot \sin(6280t)$ supplies an ohmic-inductive-capacitive parallel combination circuit.

1. Calculate the impedance $\overline{Z}$ seen by the voltage generator. Verify the calculated value using Multisim simulation.
2. Find the phasors $\overline{V}_{R_5}$ and $\overline{V}_{L_5}$ representing the voltages across the resistor R_5 and the inductor L_5, respectively. Verify your derived rms values' accuracy by contrasting them with the outcomes of a Multisim simulation.
3. Find the phase difference $\Delta\varphi = \varphi_{V_{L_5}} - \varphi_{V_{R_5}}$ between the phasors $\overline{V}_{L5}$, which represents the voltage across the inductor L_5, and $\overline{V}_{R_5}$, which represents the voltage across the resistor R_5.
4. Plot the instantaneous voltages $v_{L_5}(t)$ and $v_{R_5}(t)$ using Multisim. Estimate the phase difference $\Delta\varphi$ between the voltage and current waveforms in degrees by analyzing the plot, and then compare your estimate to the theoretical value.
5. Find the phasor $\overline{I}_{C_5}$ representing the current through the capacitor C_5. Verify the calculated value using Multisim simulation.

27.2.3.1 Solutions to Problem 3

1. The impedance $\overline{Z}$ seen by the voltage generator can be calculated by taking the parallel combination of the series impedance $\overline{Z}_{L5-R5}$ and the capacitor impedance $\overline{Z}_{C5}$.

$$\overline{Z}_{L5-R5} = R_5 + j\omega L_5 = 10 + j6.28\ \Omega = 11.8\ \angle 32.1°\Omega, \quad (27.17)$$

$$\overline{Z}_{C5} = -j\frac{1}{\omega C_5} = -j1.59. \quad (27.18)$$

The impedance seen by the voltage generator can be expressed as follows:

$$\overline{Z} = \frac{\overline{Z}_{L5-R5} \cdot \overline{Z}_{C5}}{\overline{Z}_{L5-R5} + \overline{Z}_{C5}} = 0.2 - j1.7\ \Omega = 1.7\angle{-83}°\Omega. \quad (27.19)$$

To confirm the calculated value of the impedance, we can simulate the circuit using Multisim and measure the *rms* values of the generator voltage and the current supplied by it (Figure 27.9). The comparison between the theoretical and simulated values provides validation for the calculated impedance value.

We can apply Ohm's law, which includes dividing the recorded *rms* values of voltage and current to determine the magnitude of the impedance.

$$\left|\overline{Z}\right| = \frac{V_{s5}}{I_{s5}} = 1.7\ \Omega. \quad (27.20)$$

The correctness of the theoretical value is supported by how closely the Multisim simulated result fits it.

2. The phasors $\overline{V}_{R_5}$ and $\overline{V}_{L_5}$ can be determined by using the voltage divider rule:

$$\overline{V}_{R_5} = \overline{V}_{S_5} \cdot \frac{R_5}{R_5 + j\omega L_5} = 24\angle 0° \cdot \frac{10\angle 0°}{10 + j6.28} = 20.3\angle{-32.1}°\ \text{V}, \quad (27.21)$$

$$\overline{V}_{L_5} = \overline{V}_{S_5} \cdot \frac{j\omega L_5}{R_5 + j\omega L_5} = 24\angle 0° \cdot \frac{j6.28}{10 + j6.28} = 12.8\angle 57.9°\ \text{V}. \quad (27.22)$$

The computed *rms* values are supported by the Multisim simulation (Figure 27.10).

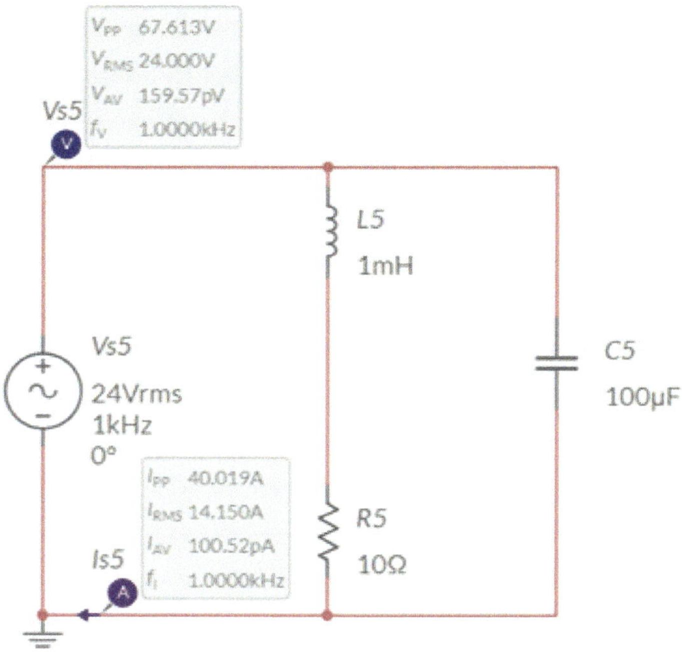

Figure 27.9 Multisim simulation providing *rms* values of supply voltage and current.

3. The phase difference between the phasors can be calculated as follows: The formula for calculating the phase difference between the phasors $\overline{V}_{R_5}$ and $\overline{V}_{L_5}$ is

$$\Delta\varphi = \varphi_{V_{L_5}} - \varphi_{V_{R_5}} = 57.9^\circ + 32.1^\circ = 90^\circ. \tag{27.23}$$

The same current flows through resistor R_5 and inductor L_5; the voltage phasor $\overline{V}_{R_5}$ across R_5 is always in phase with the current. On the other hand, the voltage phasor $\overline{V}_{L_5}$ across L_5 always leads the current by 90° due to the inductive nature of the component. Therefore, regardless of any other parameter values in the circuit, the phase difference between $\overline{V}_{R_5}$ and $\overline{V}_{L_5}$ is always 90°. This relationship is consistent and does not change.

4. The plot of the instantaneous waveforms $v_{L_5}(t)$ and $v_{R_5}(t)$ obtained with Multisim simulation is presented in Figure 27.11.

The following formula can be used to estimate the phase difference $\Delta\varphi$ in degrees:

$$\Delta\varphi = \frac{360^\circ}{T} \cdot \Delta t = \frac{360^\circ}{0.001} \cdot 251.67 \cdot 10^{-6} = 90.6^\circ. \tag{27.24}$$

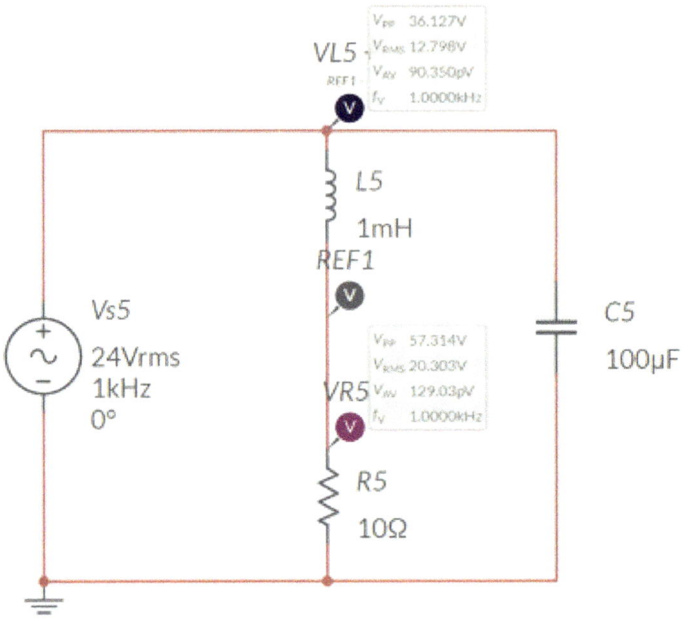

Figure 27.10 Multisim simulation showing *rms* values of $\overline{V}_{R_5}$ and $\overline{V}_{L_5}$.

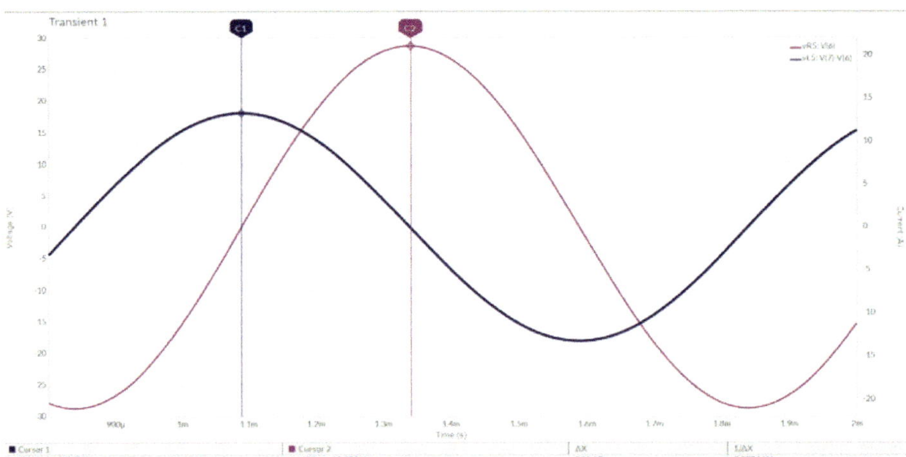

Figure 27.11 Plot of the instantaneous voltages $v_{L_5}(t)$ and $v_{R_5}(t)$.

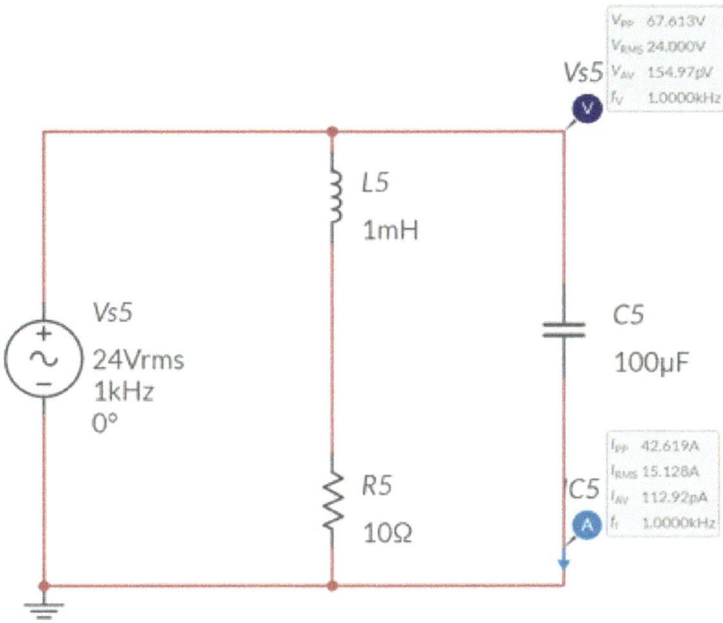

Figure 27.12 Multisim simulation showing *rms* values of $\overline{I}_{C_5}$.

The result from the Multisim plot closely aligns with the calculated value, confirming the accuracy and consistency of our analysis.

5. The phasor $\overline{I}_{C_5}$representing the current through the capacitor C_5 can be calculated by applying Ohm's law as the ratio of the generator voltage to the capacitor impedance.

$$\overline{I}_{C_5} = \frac{\overline{V}_{S_5}}{-j\frac{1}{\omega C_5}} = \frac{24\ \angle 0^{\circ}}{1.6\ \angle{-90^{\circ}}} = 15\ \angle 90^{\circ}\ \text{A}. \tag{27.25}$$

To confirm the calculated value of the current, we can simulate the circuit using Multisim and measure the *rms* values of the current (Figure 27.12).

27.3 Lab Procedure 1

In the circuit given in Figure 27.13, which consists of a series–parallel configuration, the sinusoidal current source $i_{s2}\,(t) = \sqrt{2} \cdot 5 \cdot \sin\,(376.8t)$ A is applied. Perform the following tasks:

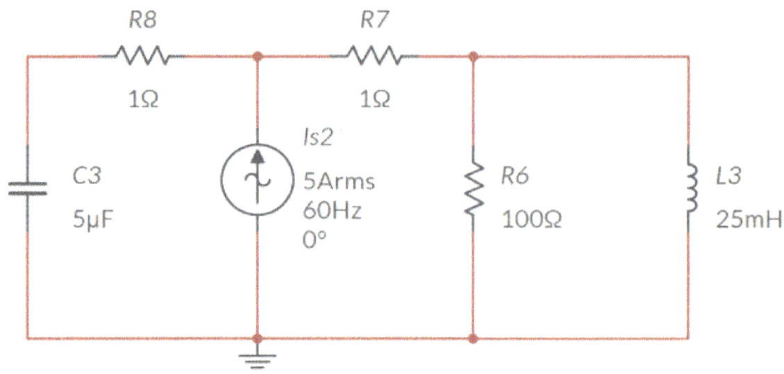

Figure 27.13 Ohmic-capacitive parallel combination circuit with an AC current generator.

Table 27.1 Summary of readings and calculations

$\overline{Z} =$ $Z\angle\varphi$	$\overline{I}_{R_8} =$ $I_{R_8} \angle\varphi_{I_8}$	$\overline{V}_{S_2} =$ $V_{S_2} \angle\varphi_{S_2}$	$\Delta\varphi$

1. Determine the equivalent impedance $\overline{Z}$ seen by the current generator. Use Multisim simulation to verify the calculated value. Record results in Table 27.1.
2. Determine the phasor $\overline{I}_{R_8}$ representing the current through R_8. Use Multisim simulation to verify the calculated *rms* value. Record value in Table 27.1.
3. Determine the phasor $\overline{V}_{s2}$ representing the voltage across the current generator. Use Multisim simulation to verify the calculated *rms* value. Record value in Table 27.1.
4. With Multisim, plot the instantaneous current $i_{R8}(t)$ through R_8 and the instantaneous voltage $v_{s2}(t)$ across the current generator. Estimate their phase difference $\Delta\varphi$ in degrees.

27.4 Lab Procedure 2

In the circuit given in Figure 27.14, the sinusoidal voltage source $v_{s6}(t) = \sqrt{2} \cdot 12 \cdot \sin(6280t)$ supplies a series–parallel configuration. Perform the following tasks:

1. Calculate the impedance $\overline{Z}$ seen by the voltage generator. Use Multisim simulation to verify the calculated value. Record values in Table 27.2.

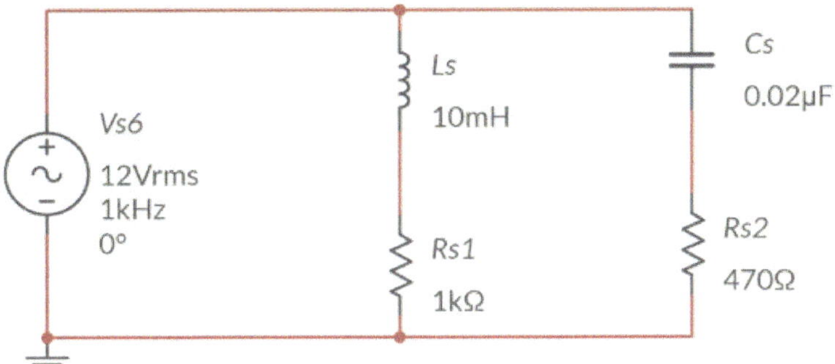

Figure 27.14 Ohmic-inductive-capacitive parallel circuit with an AC voltage generator.

Table 27.2 Summary of readings and calculations

$\overline{Z} = Z\angle\varphi$	$\overline{I}_{s6} = I_{s6}\angle\varphi I_{s6}$	$\overline{I}_{Ls} = I_{Ls}\angle\varphi I_{Ls}$	$\overline{I}_{Cs} = I_{Cs\,\mathrm{rms}}\angle\varphi I_{Cs}$	$\Delta\varphi_L$	$\Delta\varphi_C$

2. Determine the phasor $\overline{I}_{s6}$ representing the current supplied by the voltage generator. Use Multisim simulation to verify the calculated value. Record values in Table 27.2.
3. Determine the phasors $\overline{I}_{Ls}$ and $\overline{I}_{Cs}$ representing the currents flowing through the inductance L_S and the capacitor C_S, respectively. Use Multisim simulation to verify the calculated value. Record results in Table 27.2.
4. With Multisim, plot the waveform $v_{s6}(t)$ of the source voltage and the waveform $i_{Ls}(t)$ of the current through the indictor L_s. By analyzing the plot, estimate the phase difference $\Delta\varphi_L$ in degrees between the two sinusoids and compare it with the calculated value. Record results in Table 27.2.
5. With Multisim, plot the waveform $v_{s6}(t)$ of the source voltage and the waveform $i_{Cs}(t)$ of the current through the capacitor C_s. By analyzing the plot, estimate the phase difference $\Delta\varphi_C$ in degrees between the two sinusoids and compare it with the calculated value. Record results in Table 27.2.

28

Superposition Principle in AC circuits

28.1 Student Learning Outcomes

- Recall and explain the superposition principle.
- Apply the superposition principle in circuits containing both AC and DC sources.
- Understand and recognize the limitations of the superposition principle, verifying that it cannot be applied to electric power calculations.

28.2 Prelab

In AC circuits, the *superposition* principle plays a crucial role in the analysis and calculation of currents and voltages. By utilizing complex numbers, the superposition principle allows us to apply the same approach as in DC circuits, with slight modifications. In this chapter, we will explore its application in scenarios involving both DC and AC sources.

When applying the superposition principle in AC circuits, we calculate the phasors representing currents and voltages by considering each generator individually while de-energizing the other sources. For a linear circuit with n generators, we analyze n different circuits, each with one source activated while the remaining n-1 sources are deactivated. To deactivate ideal AC voltage sources, we replace them with *short circuits*, while ideal AC current sources are substituted with *open circuits*. This selective de-energizing allows us to focus on the contribution of each individual source.

To determine the current or voltage in a specific branch of the circuit, we algebraically sum the phasors of the responses to the single generator that is active. By combining (i.e., superimposing) the individual responses, we can obtain the overall response of the circuit.

Importantly, the superposition principle can also be effectively applied in scenarios where both DC and AC sources coexist. By treating the DC and AC

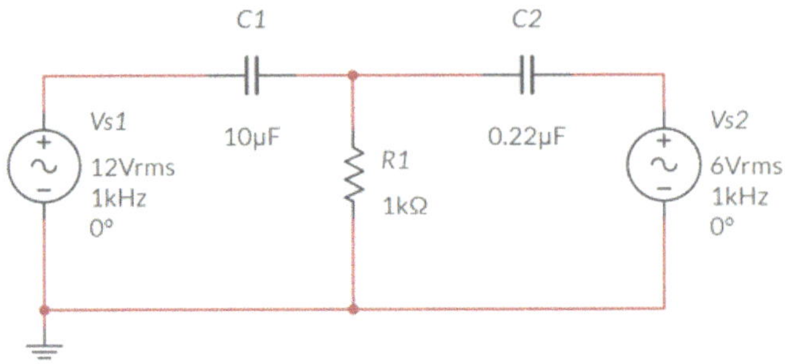

Figure 28.1 Circuit with two voltage generators.

sources separately, we can analyze their individual effects on the circuit and then combine the results to obtain the complete response.

Through the rigorous application of the superposition principle in AC circuits, taking into account the complex number representation and the selective de-energizing of sources, we can effectively analyze and calculate the currents and voltages in a wide range of linear circuit configurations.

28.2.1 Problem 1

In the given circuit depicted in Figure 28.1, we have two sinusoidal voltage sources, $v_{s1}(t) = \sqrt{2} \cdot 12\sin(6280t)$ and $v_{s2}(t) = \sqrt{2} \cdot 6\sin(6280t)$, operating at the same frequency f = 1 kHz.

To analyze the circuit using the superposition principle, perform the following tasks:

1. Determine the phasor $\overline{I}'_{R1}$ representing the current circulating through the resistor R_1 when the voltage generator $v_{s2}(t)$ is short-circuited. Confirm the correctness of the calculated *rms* value of $\overline{I}'_{R1}$ by performing a Multisim simulation.
2. Determine the phasor $\overline{I}''_{R1}$ representing the current circulating through the resistor R_1 when the voltage generator $v_{s1}(t)$ is short-circuited. Confirm the correctness of the calculated *rms* value of $\overline{I}''_{R1}$ with a Multisim simulation.
3. Determine the phasor $\overline{I}_{R1}$ representing the total current flowing through the resistor R_1 by considering the combined effect of both voltage sources. Confirm the correctness of the calculated *rms* value of $\overline{I}_{R1}$ with a Multisim simulation.

4. Calculate the electric powers P_1', P_1'', and P_1, associated with the currents $\overline{I}'_{R1}$, $\overline{I}''_{R1}$, and $\overline{I}_{R1}$, respectively. Verify that the superposition principle is not applicable to electric power, as the powers associated with individual sources cannot be added or subtracted to obtain the total power.

28.2.1.1 Solutions to Problem 1

1. We can start by calculating the impedances of the two capacitors C_1 and C_2 as follows:

$$\overline{Z}_{C1} = -j\frac{1}{\omega C_1} = -j15.9\ \Omega = 15.9\ \angle -90°\Omega, \tag{28.1}$$

$$\overline{Z}_{C2} = -j\frac{1}{\omega C_2} = -j723.8\ \Omega = 723.8\ \angle -90°\Omega. \tag{28.2}$$

To apply the superposition principle, we start by short-circuiting the voltage generator $v_{s2}(t)$, resulting in a parallel combination of the resistor R_1 and the capacitor C_2. The equivalent impedance $\overline{Z}_{12}$ of this parallel combination can be calculated using the following formula:

$$\overline{Z}_{12} = \frac{R_1 \cdot \left(-j\frac{1}{\omega C_2}\right)}{R_1 - j\frac{1}{\omega C_2}} = \frac{1000\angle 0° \cdot 723.8\angle -90°}{1000 - j723.8} = 586.3\angle -54.1°\Omega$$
$$= 343.8 - \mathrm{j}474.9\ \Omega. \tag{28.3}$$

To calculate the phasor $\overline{V}'_{R1}$ representing the voltage across the resistor R_1, we can apply the voltage divider rule.

$$\overline{V}'_{R1} = \overline{V}_{s1} \cdot \frac{\overline{Z}_{12}}{\overline{Z}_{12} + \overline{Z}_{C1}} = 12\angle 0° \frac{586.3\angle -54.1°}{343.8 - \mathrm{j}474.9 - \mathrm{j}15.9} = 11.7\angle 0°\mathrm{V}. \tag{28.4}$$

By applying Ohm's law, we can determine the phasor $\overline{I}'_{R1}$ representing the current circulating through resistor R_1 when the voltage generator $v_{s2}(t)$ is short-circuited. Ohm's law states that the current flowing through a resistor is equal to the voltage across the resistor divided by its resistance.

$$\overline{I}'_{R1} = \frac{\overline{V}'_{R1}}{R_1} = \frac{11.7\angle 0°}{1,000} = 11.7\angle 0°\ \mathrm{mA}. \tag{28.5}$$

To verify with Multisim the calculated *rms* value of $\overline{I}'_{R1}$, we can simulate the circuit configured with the short-circuit condition for $v_{s2}(t)$ (Figure 28.2).

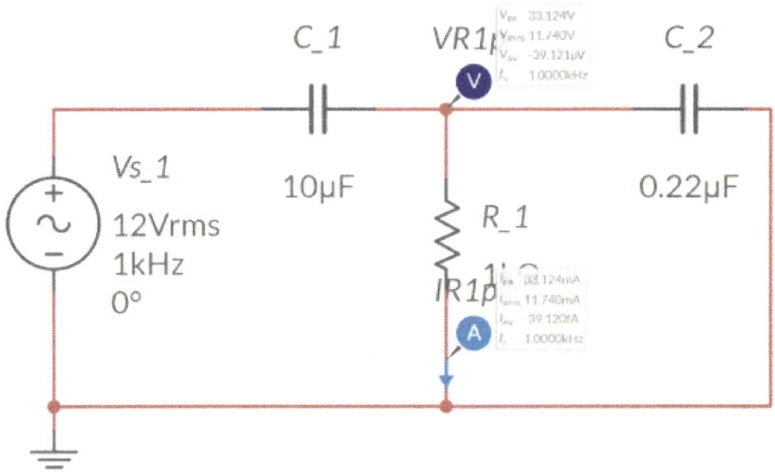

Figure 28.2 Multisim simulation showing the *rms* value of $\overline{I}'_{R1}$.

The agreement between the calculated values and the simulated result obtained from Multisim serves to validate the accuracy of the calculations. This positive verification enhances our confidence in the application of the superposition principle to analyze and understand the behavior of complex circuits.

2. To continue the application of the superposition principle to determine the phasor $\overline{I}''_{R1}$, we short-circuit the voltage generator $v_{s1}(t)$, resulting in a parallel combination of the resistor R_1 and the capacitor C_1. The equivalent impedance $\overline{Z}_{11}$ of this parallel combination can be calculated using the following formula:

$$\begin{aligned}\overline{Z}_{11} &= \frac{R_1 \cdot \overline{Z}_{C1}}{R_1 + \overline{Z}_{C1}} \\ &= \frac{1000\,\angle 0° \cdot 15.9\,\angle -90°}{1000 - j15.9} = 15.9\,\angle -89.1°\Omega = 0.2 - \text{j}15.9\,\Omega.\end{aligned} \tag{28.6}$$

Thus, by applying the voltage divider rule, the voltage drop $\overline{V}''_{R1}$ across the resistor R_1 is given by

$$\overline{V}''_{R1} = \overline{V}_{s2} \cdot \frac{\overline{Z}_{11}}{\overline{Z}_{11} + \overline{Z}_{C2}} = 6\angle 0° \frac{15.9\angle -89.1°}{0.2 - j15.9 - j723.8\,\Omega} = 128\,\angle 1°\text{mV}. \tag{28.7}$$

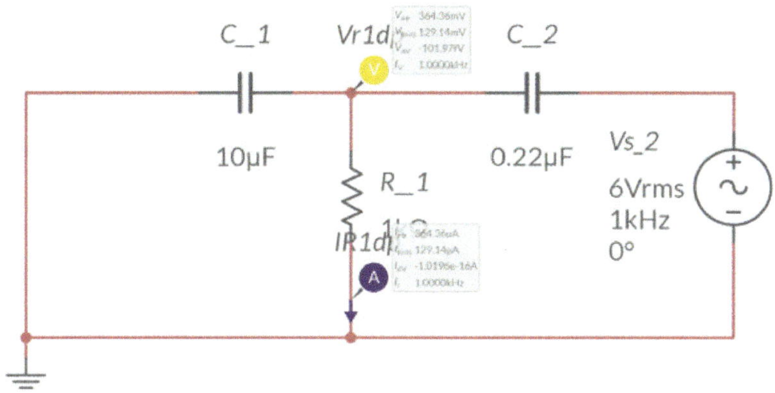

Figure 28.3 Multisim simulation showing the *rms* value of $\overline{I}''_{R1}$.

By applying Ohm's law, we can determine the phasor $\overline{I}''_{R1}$ representing the current circulating through resistor R_1 when the voltage generator $v_{s1}(t)$ is short-circuited:

$$\overline{I}''_{R1} = \frac{\overline{V}''_{R1}}{R_1} = \frac{128\ \angle 1^\circ}{1,000} = 0.128\angle 1^\circ \text{mA} = 0.128 + \text{j}0.002\ \text{mA}. \quad (28.8)$$

The *rms* calculated value of $\overline{I}''_{R1}$ is confirmed by a Multisim simulation.

To verify with Multisim the calculated *rms* value of $\overline{I}''_{R1}$, we can simulate the circuit configured with the short-circuit condition for $v_{s1}(t)$ (Figure 28.3).

The agreement between the calculated value and the simulated result obtained from Multisim confirms the accuracy of the calculations.

3. The phasor $\overline{I}_{R1}$ representing the current flowing through the resistor R_1 with both sources energized can be calculated by adding the phasors of the single responses $\overline{I}'_{R1}$and $\overline{I}''_{R1}$:

$$\overline{I}_{R1} = \overline{I}'_{R1} + \overline{I}''_{R1} = (11.7 + 0.128 + \text{j}0.002)\ \text{mA} = 11.8\ \angle 0^\circ \text{mA}. \quad (28.9)$$

To verify with Multisim the calculated *rms* value of $\overline{I}_{R1}$, we can simulate the circuit configured with both voltage generators active (Figure 28.4).

The calculated *rms* value of $\overline{I}_{R1}$ is confirmed by the Multisim simulation.

4. The electric powers P'_1, P''_1, and P_1 can be calculated with the following formulae:

$$P'_1 = I'^2_{R1} \cdot R_1 = 136.9\ \text{mW}, \quad (28.10)$$

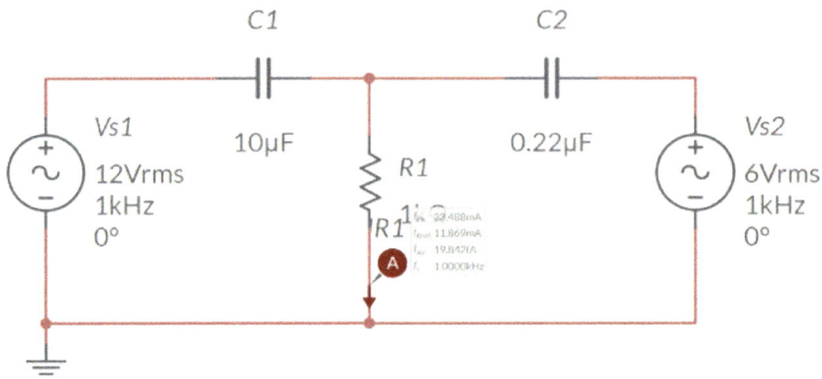

Figure 28.4 Multisim simulation showing the *rms* value of $\overline{I}_{R1}$.

$$P_1'' = I_{R1}''^2 \cdot R_1 = 0.0163 \text{ mW}, \tag{28.11}$$

$$P_1 = I_{R1}^2 \cdot R_1 = 139.2 \text{ mW}. \tag{28.12}$$

It is evident that the superposition principle, which is successfully applied to linear quantities such as currents and voltages, cannot be extended to nonlinear quantities such as electric power. Therefore, the powers associated with individual voltage sources, represented by P_1' and P_1'', cannot be simply added or subtracted to determine the total power in the circuit.

28.2.2 Problem 2

In the given circuit depicted in Figure 28.5, two sources are present: a sinusoidal current source $i_s(t) = \sqrt{2} \cdot 4\sin(376.8t)$ A and a DC voltage source V_s = 12 V. By applying the superposition principle perform the following tasks:

1. Determine the phasor $\overline{I}'_{R3}$, which represents the current flowing through the resistor R_3 when the sinusoidal current generator $i_s(t)$ is open-circuited. Confirm the calculated *rms* value of $\overline{I}'_{R3}$ by conducting a Multisim simulation to compare the results.
2. Determine the phasor $\overline{I}''_{R3}$ representing the current circulating through the resistor R_3 when the DC voltage generator V_s is short-circuited. Confirm the calculated *rms* value of $\overline{I}''_3$ by conducting a Multisim simulation.

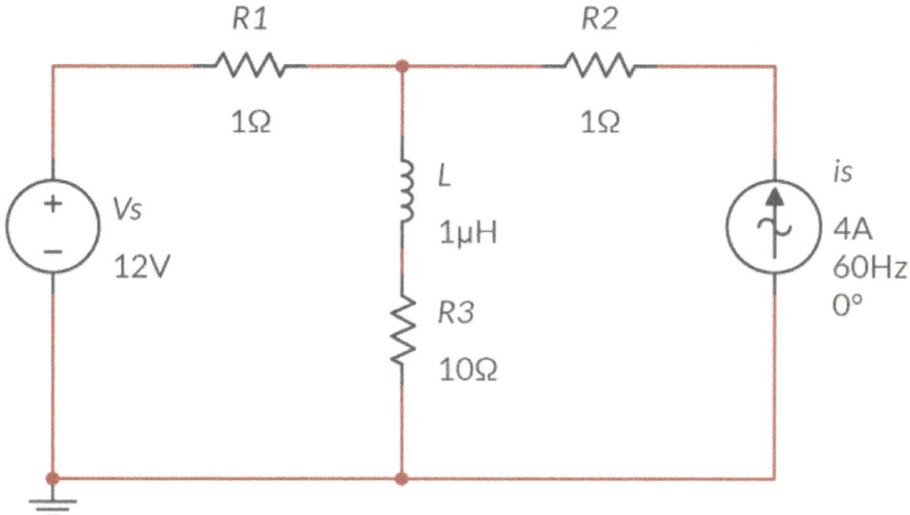

Figure 28.5 Circuit with a DC voltage generator and an AC current generator.

3. Determine the expression of the instantaneous current $i_{R3}\,(t)$ and, by employing Multisim simulation, plot the waveform and gain insights into its characteristics.

28.2.2.1 Solutions to Problem 2

1. By open-circuiting the sinusoidal current generator $i_s\,(t)$, R_1 and R_3 are connected in series, while the inductor L is a short circuit. Consequently, the equivalent resistance R_{13} is obtained by summing the individual resistances:

$$R_{13} = R_1 + R_3 = 11\ \Omega. \tag{28.13}$$

Applying Ohm's law, the DC current $\overline{I}'_{R3}$ flowing through resistor R_3 is determined as the ratio of the applied voltage V_s to the resistance R_{13}:

$$\overline{I}'_{R3} = \frac{12\ \text{V}}{11\ \Omega} = 1.09\ \angle 0°\text{A}. \tag{28.14}$$

To verify with Multisim the calculated *rms* value of $\overline{I}'_{R3}$, we can simulate the circuit configured with the open-circuit condition for the sinusoidal current source $i_s\,(t)$ (Figure 28.6).

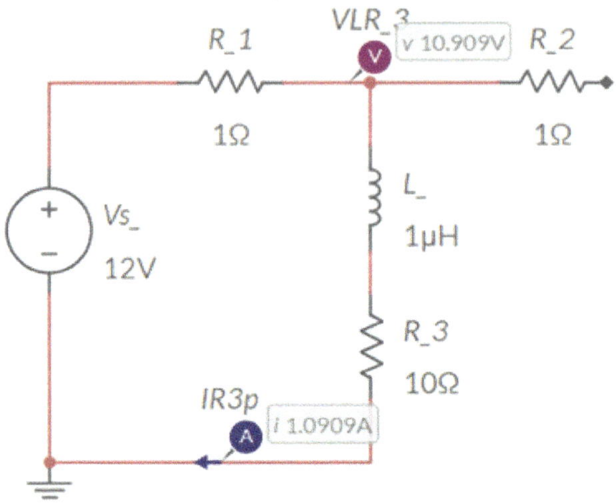

Figure 28.6 Multisim simulation showing the value of I'_{R3}.

The calculated *rms* value of $\overline{I}'_{R3}$ is confirmed by the Multisim simulation.

2. The phasor $\overline{I}''_{R3}$ representing the current circulating through resistor R_3 when the DC voltage generator V_s is short-circuited can be determined applying the current divider formula:

$$\overline{I}''_{R3} = 4\angle 0^{\circ}\frac{1\angle 0^{\circ}}{1+10+j376.8\cdot 10^{-6}} = 363\angle 0^{\circ}\text{mA}. \qquad (28.15)$$

To verify with Multisim the calculated *rms* value of the phasor $\overline{I}''_{R3}$, we can simulate the circuit configured with the short-circuit condition for the DC voltage source V_s (Figure 28.7).

The calculated *rms* value of $\overline{I}''_{3}$ is confirmed by the Multisim simulation.

3. The expression of the instantaneous total current $i_{R3}(t)$ flowing through resistor R_3 when all the generators are energized is given by the sum of the DC current component I'_{R3} and the time-varying component $i''_{R3}(t)$:

$$i_{R3}(t) = I'_{R3} + i''_{R3}(t) = 1.09\text{ A} + \sqrt{2}\cdot 0.363\sin(376.8t)\text{ A}. \qquad (28.16)$$

The Multisim simulation results in Figure 28.8 illustrate the plot of $i_{R3}(t)$. The plot showcases the time-dependent behavior of $i_{R3}(t)$, providing a visual representation of the current waveform.

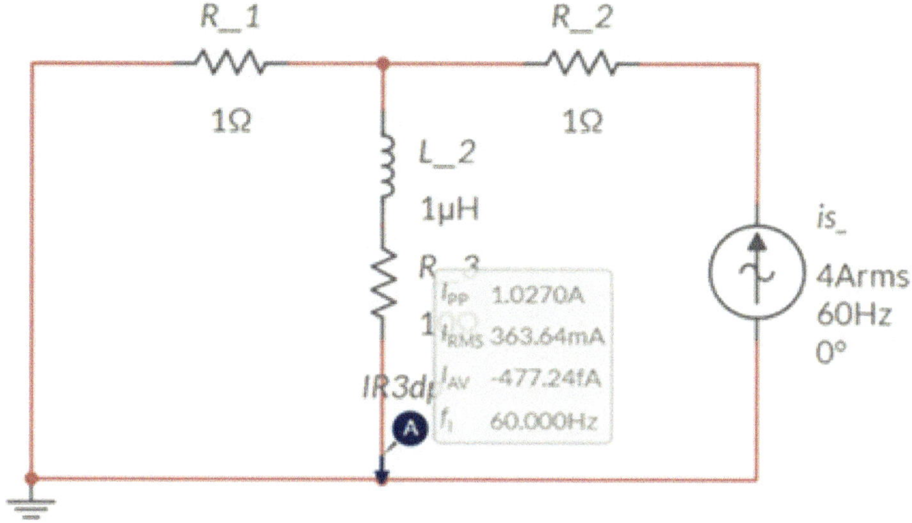

Figure 28.7 Multisim simulation showing the *rms* value of $\overline{I}''_{R3}$.

The plot in Figure 28.8 clearly illustrates that the total current $i_{R3}(t)$ consists of a sinusoidal component superimposed on a DC level of 1.09 A. This observation highlights the presence of both time-varying and constant components in the current waveform.

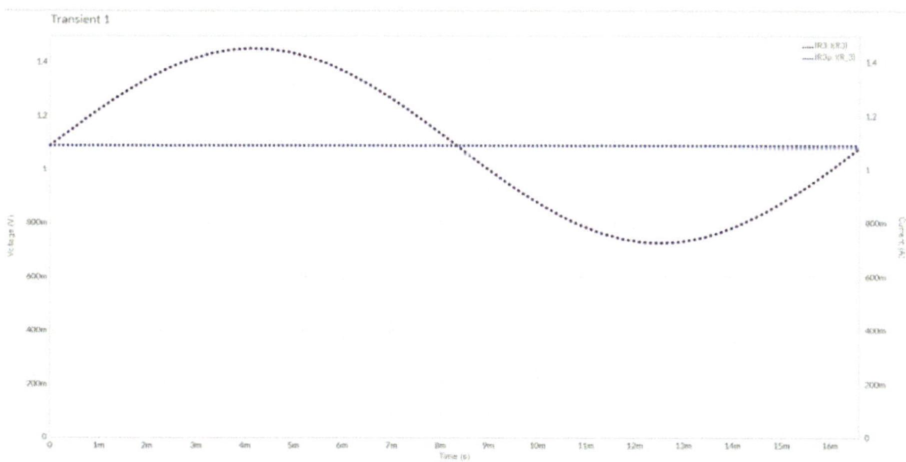

Figure 28.8 Plot of $i_{R3}(t)$ obtained with a Multisim simulation.

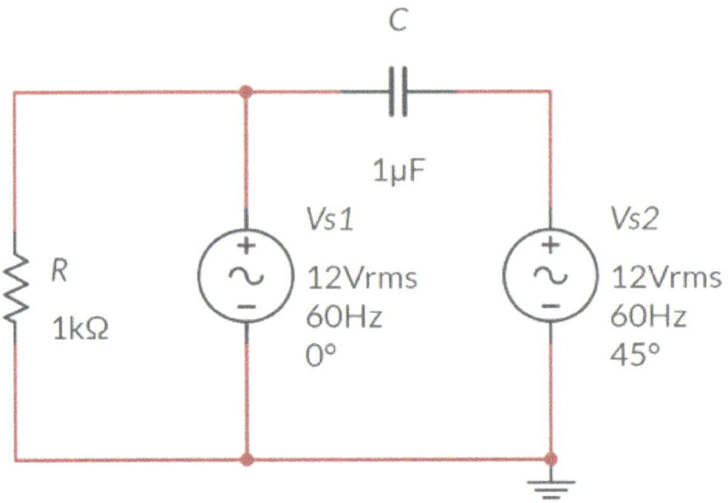

Figure 28.9 Circuit with two sinusoidal voltage generators.

28.2.3 Problem 3

In the given circuit depicted in Figure 28.9, two sinusoidal voltage sources of the same frequency f = 60 Hz are present: $v_{s1}(t) = \sqrt{2} \cdot 12 \sin(6280t)$ and $v_{s2}(t) = \sqrt{2} \cdot 6 \sin(6280t + 45°)$. By applying the superposition principle perform the following tasks:

1. Determine the phasor $\overline{I}'_R$ representing the current circulating through the resistor R when the voltage generator $v_{s2}(t)$ is short-circuited. Confirm the calculated *rms* value of $\overline{I}'_R$ by conducting a Multisim simulation to compare the results.
2. Determine the phasor $\overline{I}''_R$ representing the current circulating through the resistor R when the voltage generator $v_{s1}(t)$ is short-circuited. Confirm the calculated *rms* value of $\overline{I}''_R$ by conducting a Multisim simulation.
3. Determine the phasor $\overline{I}_R$ representing the total current flowing through the resistor R by considering the combined effect of both voltage sources. Confirm the accuracy of the calculated *rms* value of $\overline{I}_R$ with a Multisim simulation.

28.2.3.1 Solutions to problem 3

1. By short-circuiting the voltage generator $v_{s2}(t)$, the resistor R and the capacitor C are connected in parallel. The phasor $\overline{I}'_R$ representing the

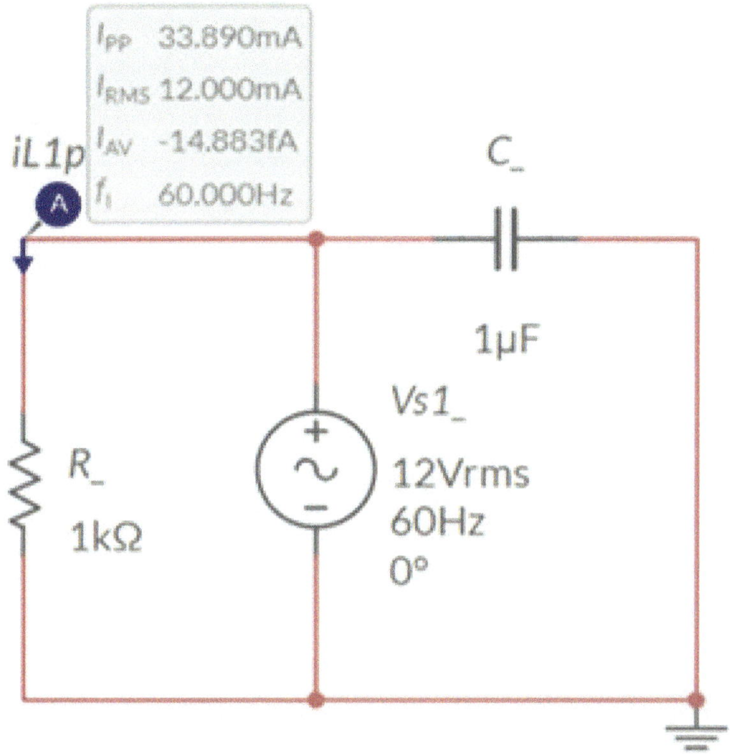

Figure 28.10 Multisim simulation showing the *rms* value of $\overline{I}'_R$.

current circulating through the resistor *R* can be calculated by applying Ohm's law:

$$\overline{I}'_R = \frac{\overline{V}_{s1}}{R} = \frac{12\angle 0^\circ}{1,000} = 12\angle 0^\circ \text{mA}. \tag{28.17}$$

To verify with Multisim the calculated *rms* value of $\overline{I}'_R$, we can simulate the circuit configured with the short-circuit condition for the sinusoidal voltage source $v_{s2}(t)$ (Figure 28.10).

The calculated *rms* value of $\overline{I}'_R$ is confirmed by Multisim simulation.

2. The phasor $\overline{I}''_R$ representing the current circulating through the resistor *R* when the voltage generator $v_{s1}(t)$ is short-circuited is zero. This is due to the nature of the ideal short-circuit, which provides a zero-resistance path that prevents the current flow through *R* from the active

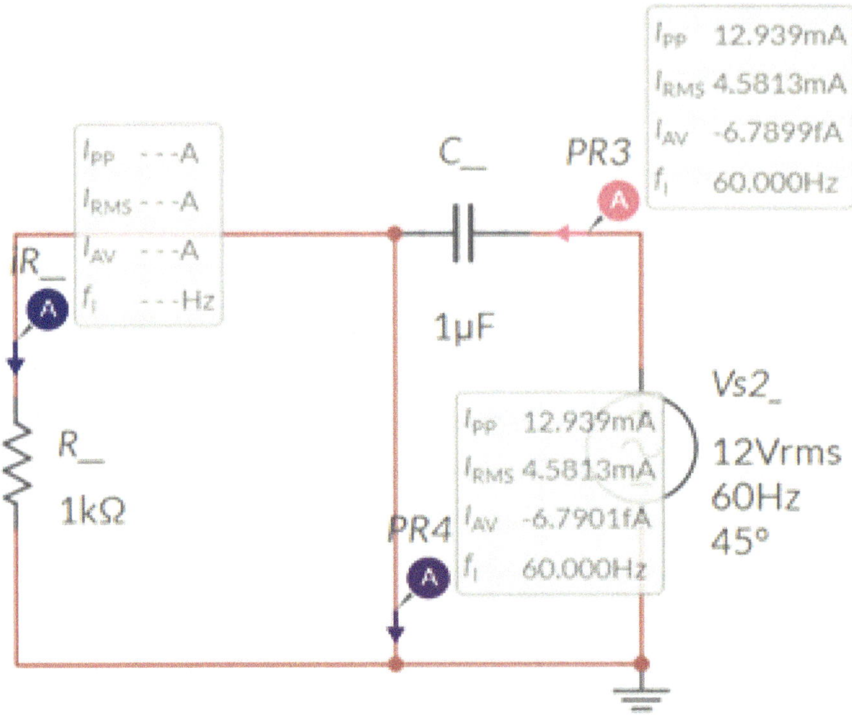

Figure 28.11 Multisim simulation showing the *rms* value of $\overline{I}''_R$.

generator $v_{s2}(t)$. We can simulate with Multisim the circuit configured with the short-circuit condition for the sinusoidal voltage source $v_{s1}(t)$ (Figure 28.11).

Multisim confirms that the current through the resistor *R* is zero.

3. The phasor $\overline{I}_R$ of the total current flowing through the resistor R_1 is given by: The phasor $\overline{I}_R$ representing the current flowing through the resistor *R* when both sources are energized can be calculated by adding the phasors of the single responses $\overline{I}'_R$and $\overline{I}''_R$:

$$\overline{I}_R = \overline{I}'_R + \overline{I}''_R = 12\ \angle 0^\circ \text{mA} \tag{28.18}$$

The Multisim simulation confirms the calculated *rms* value (Figure 28.12).

The agreement between the theoretical *rms* value and the simulated result obtained from Multisim serves to validate the accuracy of the calculations.

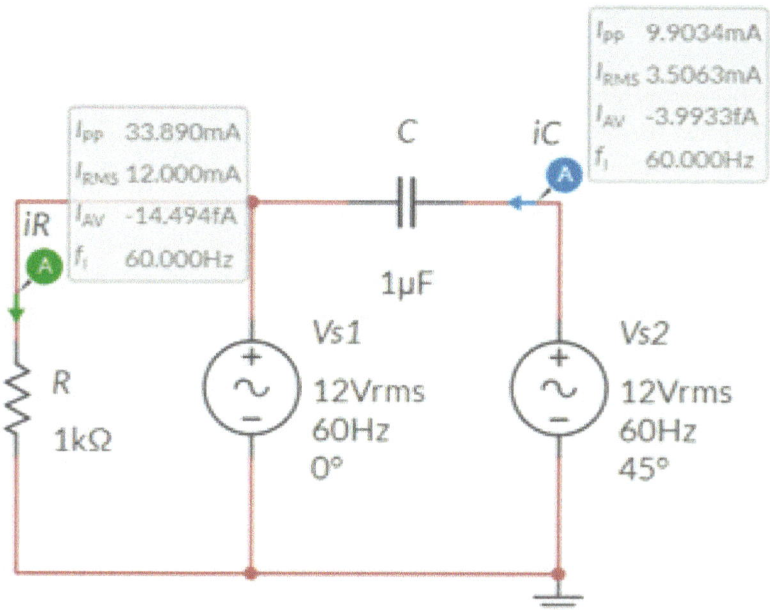

Figure 28.12 Multisim Simulation showing *rms* value of $\overline{I}_R$.

28.3 Lab Procedure 1

In the given circuit depicted in Figure 28.13, we have two sinusoidal voltage sources, $v_{s3}\,(t) = \sqrt{2}\cdot 10\,\sin(6280t + 30°)$ and $v_{s4}\,(t) = \sqrt{2}\cdot 6\,\sin(6280t)$, operating at the same frequency f = 1 kHz. To analyze the circuit using the superposition principle, perform the following tasks:

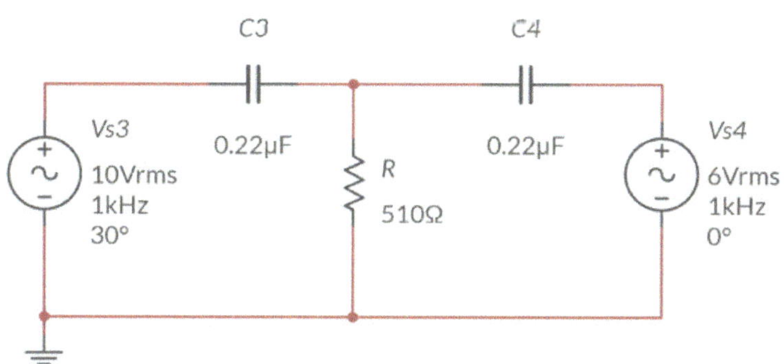

Figure 28.13 Ohmic-capacitive circuit with two AC voltage generators.

Table 28.1 Summary of readings and calculations

$\overline{I}'_R = I'_R\angle\varphi'$	$\overline{I}''_R = I''_R\angle\varphi''$	$\overline{I}_R = I_R\angle\varphi$	P'	P''	P

1. Determine the phasor $\overline{I}'_R$ representing the current circulating through the resistor *R* when the voltage generator $v_{s4}(t)$ is de-energized. Confirm the correctness of the calculated *rms* value of $\overline{I}'_R$ by performing a Multisim simulation. Record result in Table 28.1.
2. Determine the phasor $\overline{I}''_R$ representing the current circulating through the resistor *R* when the voltage generator $v_{s3}(t)$ is de-energized. Confirm the correctness of the calculated *rms* value of $\overline{I}''_R$ by performing a Multisim simulation. Record result in Table 28.1.
3. Determine the phasor $\overline{I}_R$ representing the total current flowing through the resistor *R* by considering the combined effect of both voltage sources. Confirm the correctness of the calculated *rms* value of $\overline{I}_R$ with a Multisim simulation. Record result in Table 28.1.
4. Calculate the electric powers P', P'', and *P*, associated with the currents $\overline{I}'_R$, $\overline{I}''_R$, and $\overline{I}_R$, respectively. Verify that the superposition principle is not applicable to electric power. Record results in Table 28.1.

28.4 Lab Procedure 2

In the given circuit depicted in Figure 28.14, two sources are present: a sinusoidal current source $i_{s1}(t) = \sqrt{2} \cdot 4\sin(376.8t)$ A and a DC voltage source V_{s1} = 12 V. By applying the superposition principle perform the following analysis:

1. Determine the phasor $\overline{I}'_{R3}$, which represents the current flowing through the resistor R_3 when the sinusoidal current generator $i_{s1}(t)$ is de-energized. Confirm the calculated *rms* value of $\overline{I}'_{R3}$ by conducting a Multisim simulation to compare the results. Record result in Table 28.2.
2. Determine the phasor $\overline{I}''_{R3}$, which represents the current circulating through the resistor R_3 when the DC voltage generator V_{s1} is de-energized. Confirm the calculated *rms* value of $\overline{I}''_3$ by conducting a Multisim simulation. Record result in Table 28.2.

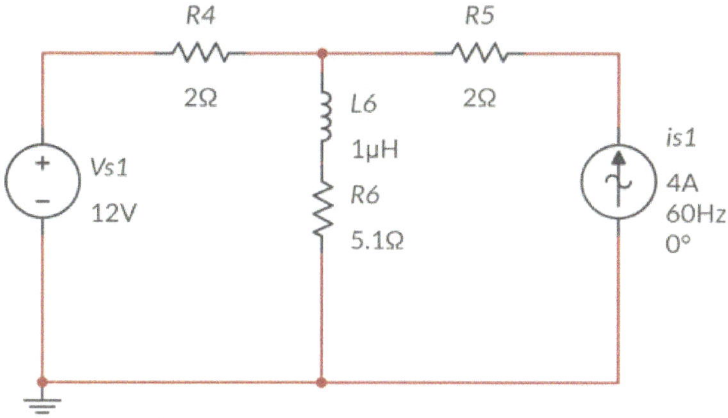

Figure 28.14 Circuit with a DC voltage generator and an AC current generator.

Table 28.2 Summary of readings and calculations

$\overline{I}'_{R3} = I'_{R3}\angle\varphi'$	$\overline{I}''_{R3} = I''_{R3}\angle\varphi''$	$i_{R3}(t)$

3. Determine the expression of the instantaneous current $i_{R3}(t)$ and, by employing Multisim simulation, plot the waveform and gain insights into its characteristics. Record result in Table 28.2.

29

Thevenin's Theorem and Maximum Power Transfer in AC

29.1 Student Learning Outcomes

- Apply Thevenin's theorem to simplify complex AC linear circuits.
- Verify Thevenin's theorem through simulations using simulation software Multisim.
- Determine the optimal load impedance that results in maximum power transfer from an AC source.

29.2 Prelab

In AC circuits, the Thevenin's theorem provides a method to simplify complex linear circuits, which consist of resistors, inductors, and capacitors. This theorem allows us to replace such circuits with an equivalent circuit comprising a single voltage source $\overline{V}_{\mathrm{th}}$ in series with an impedance $\overline{Z}_{\mathrm{th}}$. This replacement can be made at any two terminals A and B of the original circuit when connected to a load impedance $\overline{Z}_L$ (Figure 29.1).

The phasor $\overline{V}_{\mathrm{th}}$ represents the voltage across the terminals A and B once the load impedance $\overline{Z}_L$ is disconnected. On the other hand, the impedance $\overline{Z}_{\mathrm{th}} = R_{\mathrm{th}}+jX_{\mathrm{th}}$ represents the impedance observed from terminals A and B

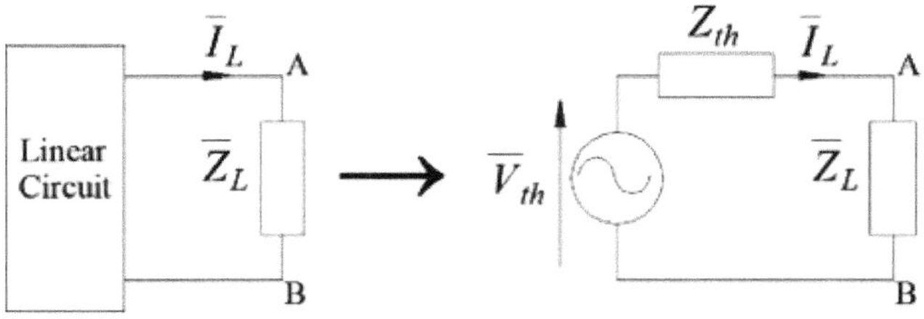

Figure 29.1 Thevenin's theorem.

when all voltage and current sources are de-energized, replaced respectively by short-circuits and open circuits. It is worth noting that R_{th} and X_{th} may vary depending on the frequency of the AC source.

To determine the phasor $\overline{I}_L$ representing the current circulating through the load, we can use the following expression:

$$\overline{I}_L = \frac{\overline{V}_{\text{th}}}{\overline{Z}_{\text{th}} + \overline{Z}_L}. \tag{29.1}$$

The magnitude of the phasor $\overline{I}_L$can be calculated as follows:

$$|I_L| = \frac{|V_{\text{th}}|}{\sqrt{(R_L + R_{\text{th}})^2 + (X_L + X_{\text{th}})^2}}. \tag{29.2}$$

In AC circuits, the Thevenin's theorem can also be employed to maximize the real power transferred, denoted as P_L, from the electric network, which is replaced by its Thevenin equivalent, to a load impedance $\overline{Z}_L = R_L + jX_L$. The power P_L can be calculated using the following expression:

$$P_L = \frac{V_L^2}{R_L} = I_L^2 \, R_L. \tag{29.3}$$

Consequently, the expression for power is

$$P_L = \frac{|V_{\text{th}}|^2}{(R_L + R_{\text{th}})^2 + (X_L + X_{\text{th}})^2} R_{L.} \tag{29.4}$$

To achieve maximum power transfer, we aim to maximize P_L with respect to R_L and X_L. The term $X_L + X_{\text{th}}$ does not absorb any power and if it had a zero value, would increase P_L. Therefore, by selecting $X_L = -X_{\text{th}}$, we can zero this term. Furthermore, to determine the value of R_L that maximizes P_L, we differentiate the expression with respect to R_L and equate the derivative to zero. This condition leads to $R_L = R_{\text{th}}$.

Hence, the maximum power transfer occurs when the load impedance $\overline{Z}_L$ is the *complex conjugate* of the Thevenin equivalent impedance, i.e., $\overline{Z}_L = R_{\text{th}} - jX_{\text{th}} = \overline{Z}_{\text{th}}^*$.

29.2.1 Problem 1

In the given linear circuit depicted in Figure 29.2, we have a sinusoidal voltage source, $v_s(t) = \sqrt{2} \cdot 12\sin(6280t)$, and a load R_L connected to the terminals A and B. To analyze the circuit using the Thevenin's theorem, perform the following tasks:

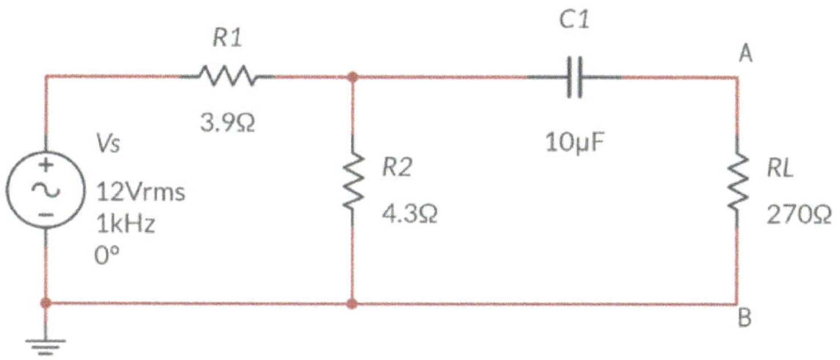

Figure 29.2 A linear circuit with a sinusoidal voltage generator.

1. By utilizing Multisim simulation, measure the *rms* values of the phasor $\overline{V}_{R_L}$ representing the voltage across the load R_L and the phasor representing the current $\overline{I}_{R_L}$ through the load R_L. Remove the load R_L and measure the *rms* value of the Thevenin equivalent voltage $\overline{V}_{\text{th}}$ across the open-circuited terminals A and B.
2. Calculate the Thevenin equivalent phasor $\overline{V}_{\text{th}}$ representing the voltage at the open-circuited terminals A and B. Confirm the correctness of the calculated *rms* value of $\overline{V}_{\text{th}}$ with that obtained with Multisim simulation.
3. Calculate the Thevenin equivalent impedance $\overline{Z}_{\text{th}}$ observed at the open-circuited terminals A and B.
4. Using the calculated values of $\overline{V}_{\text{th}}$ and $\overline{Z}_{\text{th}}$, and the load R_L, conduct a Multisim simulation of the Thevenin equivalent circuit. Measure the *rms* values for the current $\overline{I}_{R_L}$ and the voltage $\overline{V}_{R_L}$ across the load R_L. Compare these simulation readings with the corresponding values measured in the original circuit.
5. Using Multisim simulation, evaluate magnitude and argument of $\overline{Z}_{\text{th}}$ in the original circuit. By analyzing the simulation results, compare these values with the calculated value of $\overline{Z}_{\text{th}}$.

29.2.1.1 Solutions to Problem 1

1. To measure with Multisim the *rms* values of the phasor $\overline{V}_{R_L}$ representing the voltage across the load R_L and of the phasor $\overline{I}_{R_L}$ representing the current through the load R_L, we can conduct a simulation of the circuit (Figure 29.3).

The *rms* value of the voltage $\overline{V}_{\text{th}}$ across the open-circuited terminals A and B is shown in Figure 29.4.

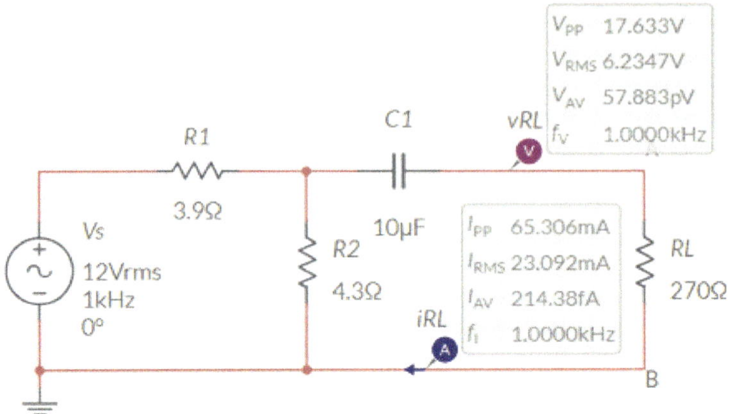

Figure 29.3 Multisim simulation providing the *rms* values of $\overline{V}_{R_L}$ and $\overline{I}_{R_L}$.

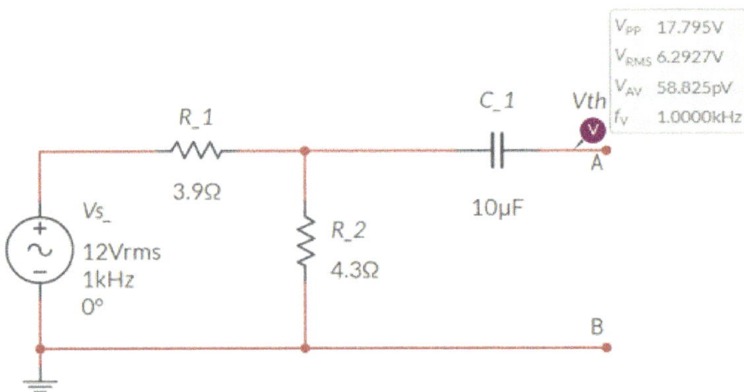

Figure 29.4 Multisim simulation providing the *rms* value of $\overline{V}_{\text{th}}$ across the open-circuited terminals A and B.

When the load resistor R_L is removed, no current is flowing through the circuit beyond the terminals A and B. In this case, the capacitor C_1 will not experience any current circulation and, consequently, no voltage drop will occur across its terminals.

2. The phasor $\overline{V}_{\text{th}}$ representing the Thevenin equivalent voltage across the open-circuited terminals A and B can be calculated by applying the voltage divider formula:

$$\overline{V}_{\text{th}} = \overline{V}_s \frac{R_2}{R_1 + R_2} = 6.29\ \angle 0°\text{V}. \tag{29.5}$$

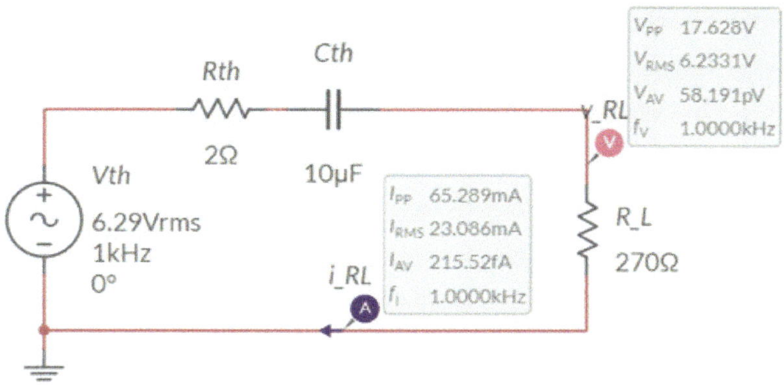

Figure 29.5 Multisim simulation of the Thevenin equivalent circuit providing the *rms* values of $\overline{I}_{R_L}$ and $\overline{V}_{R_L}$.

The theoretical *rms* value of $\overline{V}_{\text{th}}$ matches the result of the Multisim simulation.

3. The Thevenin equivalent impedance $\overline{Z}_{\text{th}}$ observed at the terminals A and B can be calculated by short-circuiting the voltage generator $v_s(t)$ and removing R_L. In this configuration, we can simplify the circuit by combining the two resistors R_1 and R_2 in parallel. This parallel combination is then in series with the impedance of the capacitor C_1.

$$\overline{Z}_{\text{th}} = \frac{R_1 \cdot R_2}{R_1 + R_2} - j\frac{1}{\omega C_1} = 2 - j15.9\ \Omega = 16\angle - 82.8°\Omega. \tag{29.6}$$

The reactive part of the impedance $\overline{Z}_{\text{th}}$ is exclusively defined by the capacitor C_1.

4. The Multisim simulation of the Thevenin equivalent circuit is provided in Figure 29.5.

The agreement between the *rms* values for the current $\overline{I}_{R_L}$ and the voltage $\overline{V}_{R_L}$ and the simulated results of the original circuit (Figure 29.2) serves to validate the accuracy of the Thevenin equivalent circuit.

5. The magnitude of the Thevenin equivalent impedance $\overline{Z}_{\text{th}}$ can be determined by replacing the original voltage source with a short-circuit and applying a test source $v_{\text{test}}(t)$ of same frequency and arbitrary *rms* value at the load terminals A and B (Figure 29.6).

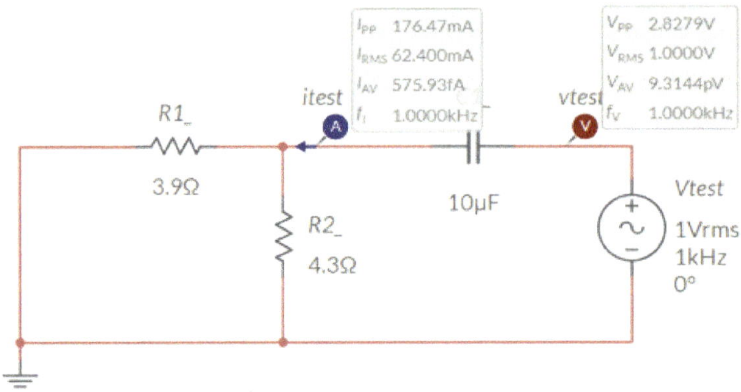

Figure 29.6 Multisim simulation setup for the determination of $\overline{Z}_{\text{th}}$.

The magnitude of $\overline{Z}_{\text{th}}$ can be determined by applying Ohm's law:

$$\left|\overline{Z}_{\text{th}}\right| = \frac{V_{\text{test}}}{I_{\text{test}}} = \frac{1}{0.0624} = 16\ \Omega. \tag{29.7}$$

The value of the magnitude of $\overline{Z}_{\text{th}}$ based on Multisim simulation matches the calculated value.

6. The argument of $\overline{Z}_{\text{th}}$ can be determined by estimating the phase difference $\Delta\varphi$ between the waveforms $v_{\text{test}}(t)$ and $i_{\text{test}}(t)$. The plot of the waveforms $v_{\text{test}}(t)$ and $i_{\text{test}}(t)$, obtained with Multisim simulation, is presented in Figure 29.7.

Analyzing the plot, we can identify the time difference Δt between corresponding points on the waveforms.

The phase difference $\Delta\varphi$ can be estimated in degrees with the following formula:

$$\Delta\varphi = \frac{360}{T} \cdot \Delta t = \frac{360}{0.001} \cdot \left(-230.24 \cdot 10^{-6}\right) = -82.9^{\circ}. \tag{29.8}$$

The result from the Multisim plot closely aligns with the calculated value of the argument of $\overline{Z}_{\text{th}}$.

29.2.2 Problem 2

In the given linear circuit depicted in Figure 29.8, we have a sinusoidal current generator $i_s(t) = \sqrt{2}{\cdot}2\sin(376.8t)$ and a load R_L connected to the terminals A and B. To analyze the circuit using the Thevenin's theorem, perform the following tasks:

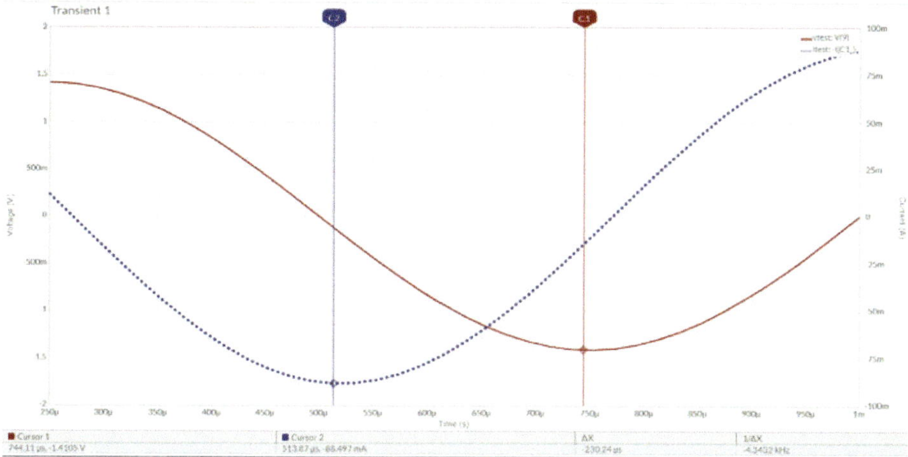

Figure 29.7 Plot of $v_{\text{test}}(t)$ and $i_{\text{test}}(t)$ for the determination of the argument of the Thevenin impedance $\overline{Z}_{\text{th}}$.

1. By utilizing Multisim simulation, measure the *rms* values of the phasor $\overline{V}_{R_L}$ representing the voltage across the load R_L and the phasor representing the current $\overline{I}_{R_L}$through the load R_L. Remove the load R_L and measure the *rms* value of the Thevenin equivalent voltage $\overline{V}_{\text{th}}$ across the open-circuited terminals A and B.
2. Calculate the Thevenin equivalent phasor $\overline{V}_{\text{th}}$ representing the voltage at the open-circuited terminals A and B. Confirm the correctness of the calculated *rms* value of $\overline{V}_{\text{th}}$ with that obtained with Multisim simulation.

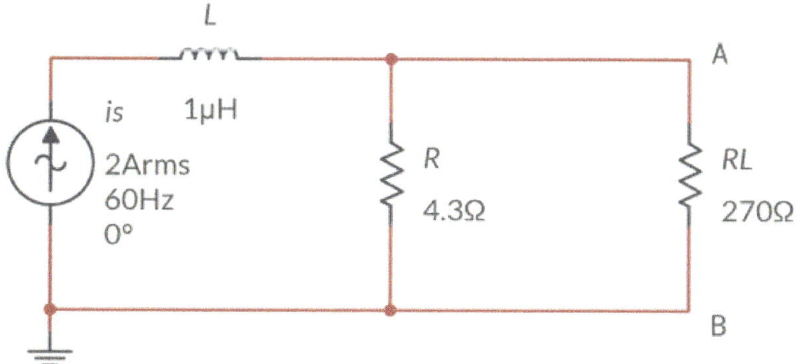

Figure 29.8 A linear ohmic-inductive circuit with a sinusoidal current generator.

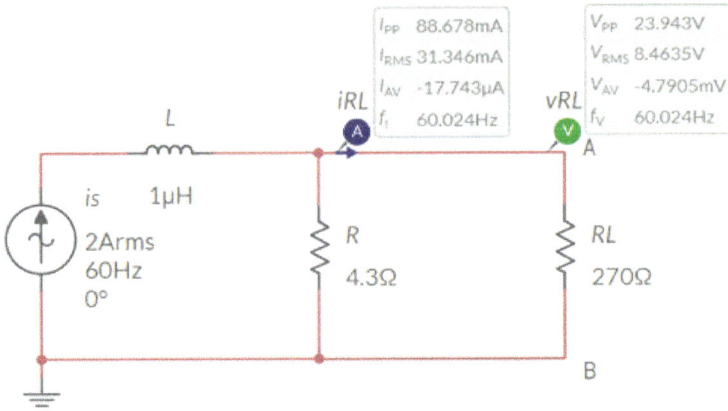

Figure 29.9 Multisim simulation providing the *rms* values of $\overline{V}_{R_L}$ and $\overline{I}_{R_L}$.

3. Calculate the Thevenin equivalent impedance $\overline{Z}_{\text{th}}$ observed at the terminals A and B.
4. Using the calculated values of $\overline{V}_{\text{th}}$ and $\overline{Z}_{\text{th}}$, and the load R_L, conduct a Multisim simulation of the Thevenin equivalent circuit. Measure the *rms* values for the current $\overline{I}_{R_L}$ and the voltage $\overline{V}_{R_L}$ across the load R_L. Compare these simulation readings with the corresponding values measured in the original circuit.
5. Using Multisim simulation, evaluate magnitude and argument of $\overline{Z}_{\text{th}}$ in the original circuit. By analyzing the simulation results, compare these values with the calculated value of $\overline{Z}_{\text{th}}$.

29.2.2.1 Solutions to Problem 2

1. To measure with Multisim the *rms* values of the phasor $\overline{V}_{R_L}$ representing the voltage across the load R_L and of the phasor $\overline{I}_{R_L}$ representing the current through the load R_L, we can conduct a simulation of the circuit (Figure 29.9).

The *rms* value of the voltage $\overline{V}_{\text{th}}$ across the open-circuited terminals A and B is shown in Figure 29.10.

2. The phasor $\overline{V}_{\text{th}}$ representing the Thevenin equivalent voltage across the open-circuited terminals A and B can be calculated by applying Ohm's law:

$$\overline{V}_{\text{th}} = \overline{I}_s R = 2\angle 0° \cdot 4.3 = 8.6\angle 0° \text{V}. \tag{29.9}$$

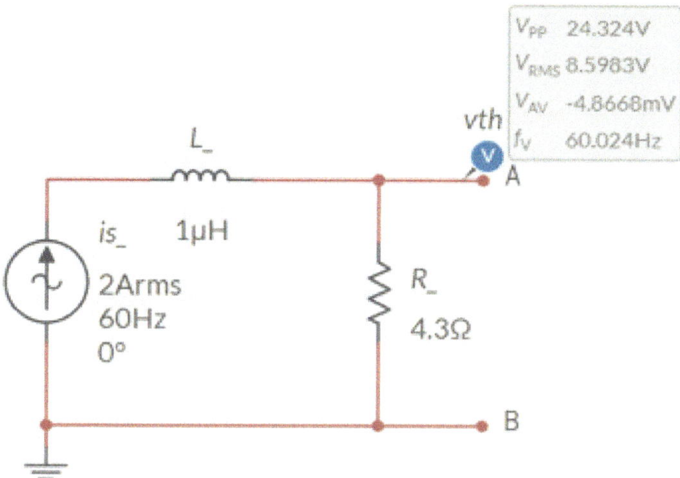

Figure 29.10 Multisim simulation providing the *rms* value of $\overline{V}_{\text{th}}$ across the open-circuited terminals A and B.

The above calculated *rms* value matches that provided by the simulation (Figure 29.10). The theoretical *rms* value of $\overline{V}_{\text{th}}$ matches the result of the Multisim simulation.

3. The Thevenin equivalent impedance $\overline{Z}_{\text{th}}$ observed at the terminals A and B can be calculated by open-circuiting the current generator $i_s(t)$ and removing R_L. In this configuration, the Thevenin equivalent impedance is determined to be the resistor *R*.

$$\overline{Z}_{\text{th}} = R = 4.3\angle 0^{\circ}\ \Omega. \tag{29.10}$$

4. The Multisim simulation of the Thevenin equivalent circuit is provided in Figure 29.11.

In the simulation of Figure 29.11, the *rms* values for the current $\overline{I}_{R_L}$ and the voltage $\overline{V}_{R_L}$ match those obtained in the simulation of the original circuit (Figure 29.8). The agreement between the *rms* values for the current $\overline{I}_{R_L}$ and the voltage $\overline{V}_{R_L}$ and the simulated results of the original circuit serves to validate the accuracy of the Thevenin equivalent circuit.

5. The magnitude of the Thevenin equivalent impedance $\overline{Z}_{\text{th}}$ can be determined by de-energizing the original current source and applying a test source $v_{\text{test}}(t)$ of same frequency and arbitrary *rms* value at the load terminals A and B (Figure 29.12).

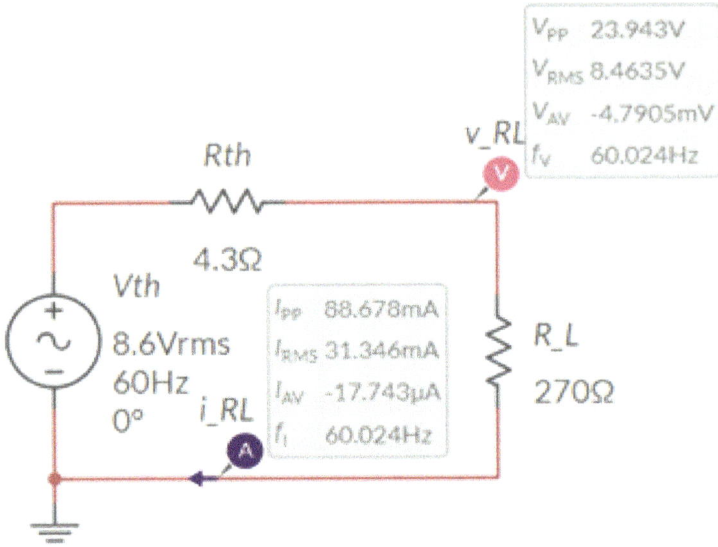

Figure 29.11 Multisim simulation providing the *rms* values of $\overline{I}_{R_L}$ and $\overline{V}_{R_L}$.

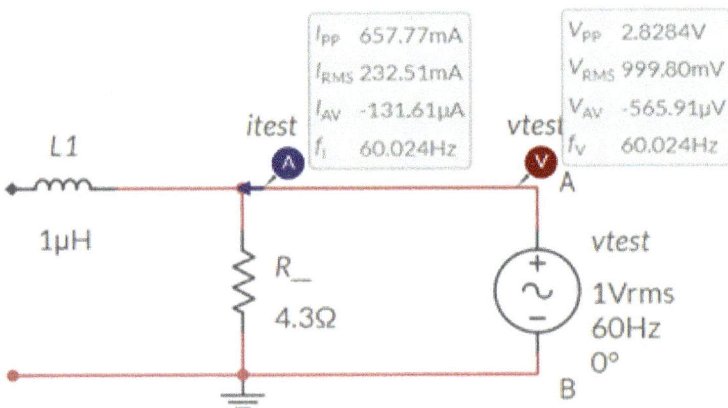

Figure 29.12 Multisim simulation setup for the determination of $\overline{Z}_{\text{th}}$.

The magnitude of $\overline{Z}_{\text{th}}$ can be determined by applying Ohm's law:

$$\left|\overline{Z}_{\text{th}}\right| = \frac{V_{\text{test rms}}}{I_{\text{test rms}}} = \frac{1}{0.232} = 4.3\ \Omega. \tag{29.11}$$

The magnitude of the Thevenin equivalent impedance $\overline{Z}_{\text{th}}$ obtained from Multisim simulation matches the theoretical value.

The argument of $\overline{Z}_{\text{th}}$ can be determined by estimating the phase difference between $v_{\text{test}}(t)$ and $i_{\text{test}}(t)$. Since the circuit is purely resistive, the waveforms of voltage and current are in phase (i.e., $\varphi = 0$), which aligns with the theoretical argument of $\overline{Z}_{\text{th}}$ previously calculated in eqn (29.10).

29.2.3 Problem 3

In the given linear circuit depicted in Figure 29.13, the Thevenin equivalent parameters at terminals A and B are $\overline{V}_{\text{th}} = \sqrt{2} \cdot 10 \angle 45° \text{ V}$ and $\overline{Z}_{\text{th}} = 10.8 \angle 68° \Omega$. The operating frequency of the circuit is $f = 100$ Hz. Perform the following tasks:

1. Determine the optimum impedance $\overline{Z}_L$ of the load that allows the maximum power transfer.
2. Under maximum power transfer, determine the phasors $\overline{I}_L$ and $\overline{V}_L$, respectively, representing the current through and the voltage across the optimum impedance $\overline{Z}_L$. Conducting a Multisim simulation, confirm the calculated peak values of the voltage $\overline{V}_L$ and the current $\overline{I}_L$ and their phase difference $\Delta\varphi$ in degrees.
3. With a Multisim simulation, plot the current $i_L(t)$ through the optimum impedance $\overline{Z}_L$ and the Thevenin voltage $v_{\text{th}}(t)$. Evaluate their phase difference $\Delta\varphi$ in degrees and compare with the theoretical value.
4. Calculate the power P_L absorbed by $\overline{Z}_L$ and the power P_{th} supplied by the Thevenin's equivalent source $\overline{V}_{\text{th}}$.

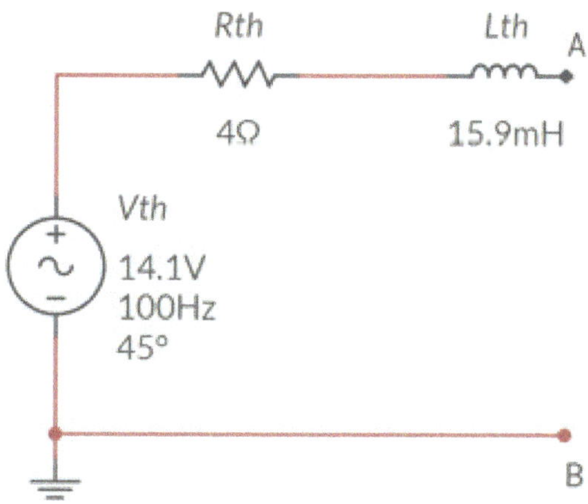

Figure 29.13 Thevenin equivalent circuit.

29.2.3.1 Solutions to Problem 3

1. To determine the impedance $\overline{Z}_L$ of the load that allows the maximum power transfer, we need to make $\overline{Z}_L$ equal to the complex conjugate of the Thevenin impedance $\overline{Z}_{\text{th}}$. Therefore, the optimum load impedance can be calculated as follows:

$$\overline{Z}_L = \overline{Z}_{\text{th}}^* = 10.8\angle{-68°}\Omega = 4 - \text{j}10\,\Omega. \tag{29.12}$$

The capacitance C_{th} associated with the optimum load impedance $\overline{Z}_L$ can be calculated as follows:

$$C_{\text{th}} = \frac{1}{\omega X_C} = \frac{1}{628 \cdot 10} = 159\,\mu\text{F}. \tag{29.13}$$

2. In the condition of maximum power transfer, the reactive components cancel each other out, resulting in a purely resistive circuit. Thus, the phasor $\overline{I}_L$ can be calculated as follows:

$$\overline{I}_L = \frac{\overline{V}_{\text{th}}}{2R_{\text{th}}} = \frac{\sqrt{2} \cdot 10\,\angle 45°}{8} = 1.76\,\angle 45°\,\text{A} = 1.2 + \text{j}1.2\,\Omega. \tag{29.14}$$

The phasor $\overline{V}_L$ representing the voltage across the optimum load can be calculated by applying Ohm's law:

$$\overline{V}_L = \overline{I}_L \cdot \overline{Z}_L = 1.76\angle 45° \cdot 10.8\angle{-68°} = 19\,\angle{-23°}\text{ V}. \tag{29.15}$$

The peak values of $\overline{I}_L$ and $\overline{V}_L$ are confirmed with a Multisim simulation (Figure 29.14).

To verify the calculated peak values of the phasors $\overline{I}_L$ and $\overline{V}_L$, we can use Multisim simulation to simulate the Thevenin equivalent circuit with the optimum load connected, as shown in Figure 29.14.

The theoretical values of the peaks are confirmed by the Multisim simulation. The agreement between the theoretical calculation and the simulation results reinforces the accuracy and reliability of the analysis conducted in this circuit.

To verify the phase difference between the phasors $\overline{I}_L$ and $\overline{V}_L$, with Multisim, we can generate a plot of the instantaneous current $i_L(t)$ and of the instantaneous voltage $v_L(t)$ across the optimum load impedance $\overline{Z}_L$ (Figure 29.15).

Analyzing the plot, we can identify the time difference Δt between corresponding points on the waveforms. The phase difference between $v_L(t)$

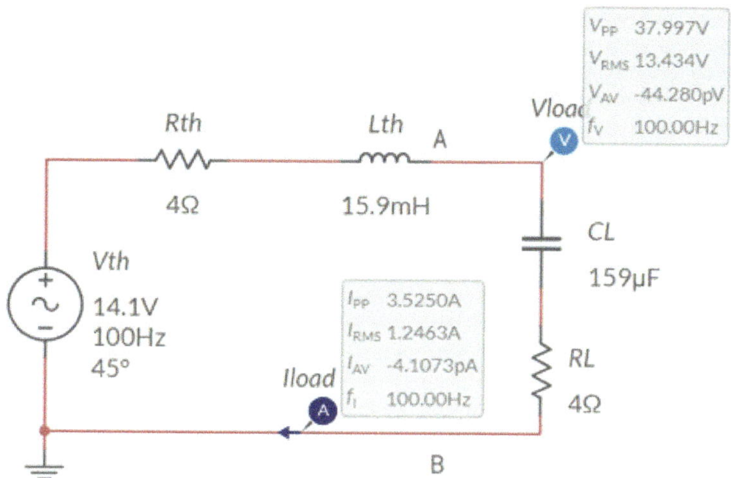

Figure 29.14 Multisim simulation showing the peak values of $\overline{I}_L$ and $\overline{V}_L$.

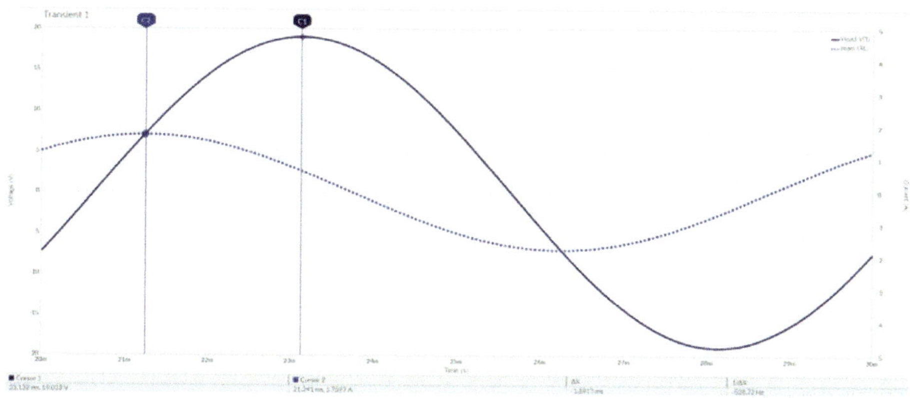

Figure 29.15 Multisim plot of the instantaneous current $i_L(t)$ and instantaneous voltage $v_L(t)$ across the impedance $\overline{Z}_L$.

and $i_L(t)$, denoted as $\Delta\varphi$, can be estimated with the following formula:

$$\Delta\varphi = \frac{360}{T} \cdot \Delta t = \frac{360}{0.01} \cdot \left(-1.8913 \cdot 10^{-3}\right) = -68°. \tag{29.16}$$

The theoretical value of the phase difference $\Delta\varphi = -23° - 45° = -68°$ is confirmed by the Multisim simulation.

3. To verify the phase difference between the waveforms $i_L(t)$ and $v_{\text{th}}(t)$, we can generate a plot with Multisim (Figure 29.16).

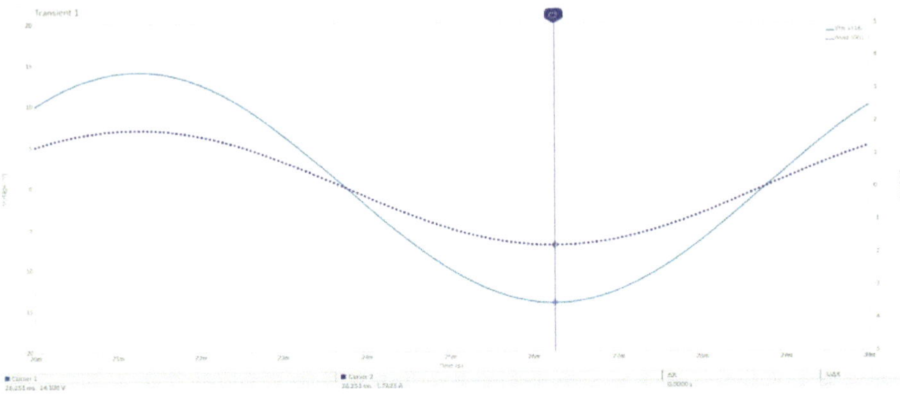

Figure 29.16 Multisim plot of $i_L\,(t)$ and $v_{\text{th}}(t)$.

The phase difference between $i_L\,(t)$ and $v_{\text{th}}\,(t)$, denoted as $\Delta\varphi$, can be estimated with the following formula:

$$\Delta\varphi = \frac{360}{T} \cdot \Delta t = \frac{360}{0.01} \cdot 0 = 0^{\circ}. \tag{29.17}$$

In maximum power transfer conditions, the circuit is purely resistive, which means that the current waveform $i_L(t)$ and the Thevenin voltage waveform $v_{\text{th}}(t)$ are in phase, resulting in $\Delta\varphi = 0$. The Multisim plot confirms this result by showing that there is no phase difference between the two waveforms.

4. The power P_L absorbed by $\overline{Z}_L$ in maximum power transfer conditions can be calculated as follows:

$$P_L = \frac{1}{2} I_L^2 R_{\text{th}} = \frac{1}{2} \cdot 1.76^2 \cdot 4 = 6.2\ \text{W}. \tag{29.18}$$

The power supplied by the Thevenin's equivalent source $\overline{V}_{\text{th}}$ can be calculated as follows:

$$P_{\text{th}} = \frac{1}{2} V_{\text{th}} I_{\text{th}} = 12.4\ \text{W}. \tag{29.19}$$

The result obtained is consistent with the theory. The power P_{th} supplied by the Thevenin's equivalent source is twice the value of the power P_L absorbed by the optimum load. This difference in power is due to the power dissipation across the Thevenin resistance R_{th} of the source itself. Therefore, the power delivered by the source is higher than the power consumed by the load, with the excess power being dissipated within the Thevenin equivalent circuit.

29.3 Lab Procedure 1

In the given linear circuit depicted in Figure 29.17, we have a sinusoidal voltage source, $v_s(t) = \sqrt{2} \cdot 120\sin(6280t)$, and a load R_3 connected to the terminals A and B. To analyze the circuit using the Thevenin's theorem, perform the following tasks:

1. By utilizing Multisim simulation, measure the *rms* values of the phasor $\overline{V}_{R_3}$ representing the voltage across the load R_3 and the phasor representing the current $\overline{I}_{R_3}$ through the load R_3. Remove the load R_3 and measure the *rms* value of the Thevenin equivalent voltage $\overline{V}_{\text{th}}$ across the open-circuited terminals A and B. Record values in Table 29.1.
2. Calculate the Thevenin equivalent phasor $\overline{V}_{\text{th}}$ representing the voltage at the open-circuited terminals A and B. Confirm the correctness of the calculated *rms* value of $\overline{V}_{\text{th}}$ with that obtained with Multisim simulation. Record results in Table 29.1.
3. Calculate the Thevenin equivalent impedance $\overline{Z}_{\text{th}}$ observed at the terminals A and B. Record values in Table 29.1.
4. Using the calculated values of $\overline{V}_{\text{th}}$ and $\overline{Z}_{\text{th}}$, and the load R_3, conduct a Multisim simulation of the Thevenin equivalent circuit. Measure the *rms* values for the current $\overline{I}_{R_3}$ and the voltage $\overline{V}_{R_3}$ across the load R_3. Compare these simulation readings with the corresponding values measured in the original circuit.

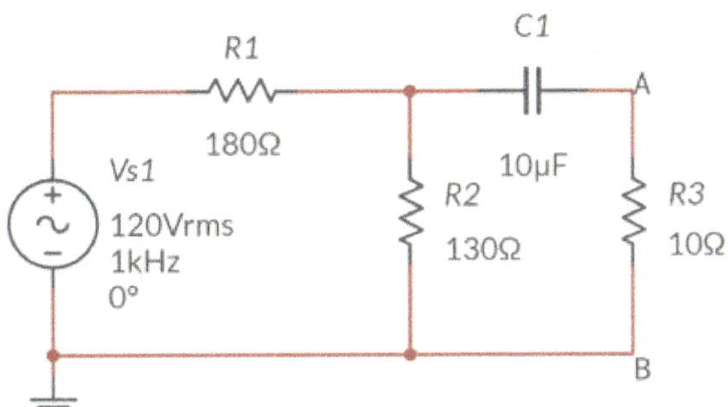

Figure 29.17 A linear ohmic-capacitive circuit with a sinusoidal voltage generator.

Table 29.1 Summary of readings and calculations

V_{R3}	I_{R3}	$\overline{V}_{\text{th}} = V_{\text{th}}\angle\varphi$	$\overline{Z}_{\text{th}} = Z_{\text{th}}\angle\varphi_Z$

5. Using Multisim simulation, evaluate magnitude and argument of the impedance $\overline{Z}_{\text{th}}$ in the original circuit. By analyzing the simulation results, compare these values with the calculated value of $\overline{Z}_{\text{th}}$.

29.4 Lab Procedure 2

The Thevenin equivalent parameters at terminals A and B of a linear circuit are $\overline{V}_{\text{th}} = \sqrt{2} \cdot 120\angle 30°$ V and $\overline{Z}_{\text{th}} = \overline{Z}_{\text{th}} = 100\angle{-45}°\Omega$. The operating frequency of the circuit is f = 100 Hz. Perform the following tasks:

1. Determine the optimum impedance $\overline{Z}_L$ of the load that allows the maximum power transfer. Record values in Table 29.2.
2. Under maximum power transfer, determine the phasors $\overline{I}_L$ and $\overline{V}_L$, respectively, representing the current through and the voltage across the optimum impedance $\overline{Z}_L$. Conducting a Multisim simulation, confirm the calculated peak values of the voltage $\overline{V}_L$ and the current $\overline{I}_L$ and their phase difference $\Delta\varphi$ in degrees. Record values in Table 29.2.
3. With a Multisim simulation, plot the current $i_L(t)$ through the optimum impedance $\overline{Z}_L$ and the Thevenin voltage $v_{\text{th}}(t)$. Evaluate their phase difference $\Delta\varphi$ in degrees and compare with the theoretical value.
4. Calculate the power P_L absorbed by $\overline{Z}_L$ and the power P_{th} supplied by the Thevenin's equivalent source $\overline{V}_{\text{th}}$.

Table 29.2 Summary of readings and calculations

$\overline{Z}_L = Z_L\angle\varphi_{Z_L}$	$\overline{I}_L = I_L\angle\varphi_{I_L}$	$\overline{V}_L = V_L\angle\varphi_{V_L}$	$\Delta\varphi$	P_L	P_{th}

30

Series Resonant Circuits

30.1 Student Learning Outcomes

- Given a series *R-L-C* circuit, calculate resonant frequency and bandwidth.
- Given a series *R-L-C* circuit, calculate electrical quantities in the circuit (e.g., impedance, voltage, and current).
- Use Multisim simulation to verify magnitude of parameters of series *R-L-C* circuits as a function of the frequency.

30.2 Prelab

Series resonance is a phenomenon that occurs in a series *R-L-C* circuit at a specific frequency, denoted as f_r, when the capacitive reactance X_C and the inductive reactance X_L are equal in magnitude. As a result, the imaginary components of the impedance cancel each other out, and the impedance $\overline{Z}(\omega) = R + j\,(X_L - X_C)$ becomes purely resistive (Figure 30.1).

The resonant frequency f_r at which $X_L = X_C$ can be determined using the following equation:

$$2\pi f_r L = \frac{1}{2\pi f_r C} \Rightarrow f_r = \frac{1}{2\pi\sqrt{LC}}. \tag{30.1}$$

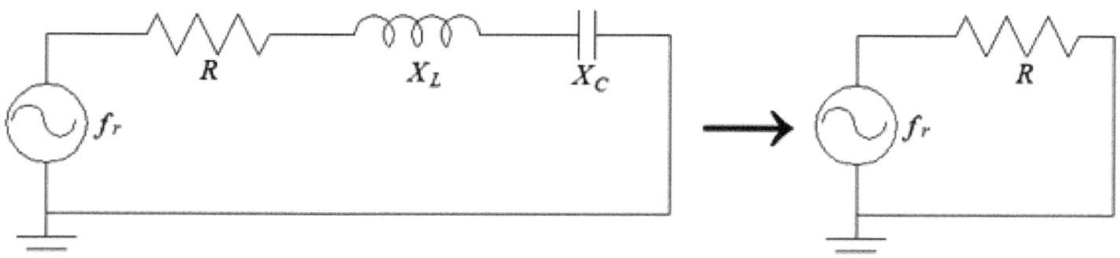

Figure 30.1 Resonant condition.

At resonance, the impedance is equivalent to the resistance and reaches its minimum value. The current is then at its maximum and in phase with the applied voltage. As the frequency deviates from f_r, the current decreases because the reactances no longer cancel each other out. In resonance, the voltage drop across the resistor *R* is also maximum and equal to the generator voltage. For frequencies above f_r, the term $X_L - X_C$ is positive, resulting in an ohmic-inductive circuit, and vice versa, for frequencies below f_r, the circuit is ohmic-capacitive.

Based on this frequency-dependent behavior, the resonant series circuit, also known as an *acceptor*, exhibits frequency selectivity by *accepting* currents within a specific range known as the *bandwidth*, while rejecting others.

To quantify the reduction in current, we define the *half-power* frequencies f_1 and f_2 as the frequencies at which the power absorbed in the circuit is half the power dissipated at resonance, or alternatively, when the magnitude of the current is $\frac{1}{\sqrt{2}}$ times the magnitude of the current at resonance. These frequencies are also referred to as the *low cutoff* and *high cutoff* frequencies. The cutoff frequencies can be further defined as the frequencies at which the current decreases by 3 dB compared to its maximum resonant value.

The analytical expressions of f_1 and f_2 can be calculated by setting the magnitude of $\overline{Z}$ equal to $\sqrt{2}R$ and writing:

$$\sqrt{R^2 + \left(2\pi f L - \frac{1}{2\pi f C}\right)^2} = \sqrt{2}R. \tag{30.2}$$

Solving for *f*, we obtain

$$f_1 = \frac{-R + \sqrt{R^2 + \frac{4L}{C}}}{4\pi L}, \tag{30.3}$$

$$f_2 = \frac{R + \sqrt{R^2 + \frac{4L}{C}}}{4\pi L}. \tag{30.4}$$

The *bandwidth* (*BW*) of the circuit is defined as follows:

$$\text{BW} = f_2 - f_1 = \frac{R}{2\pi L}. \tag{30.5}$$

A smaller bandwidth indicates a more frequency-selective circuit. The selectivity of the circuit in resonance can be measured by the *quality factor*

(Q), defined as follows:

$$Q = \frac{2\pi f_r \cdot L}{R} = \frac{1}{2\pi f_r \cdot C \cdot R} = \frac{1}{R}\sqrt{\frac{L}{C}} = \frac{f_r}{\mathrm{BW}}. \tag{30.6}$$

A circuit with a smaller bandwidth has a higher quality factor, indicating greater selectivity, and vice versa.

30.2.1 Problem 1

In the circuit of Figure 30.2, a series *R-L-C* circuit is connected to a sinusoidal voltage generator $v_s(t) = \sqrt{2} \cdot 3.5\sin(\omega t)$. Perform the following tasks:

1. Calculate the resonance frequency f_r, the cutoff frequencies f_1 and f_2, and the bandwidth *BW*.
2. At resonance, determine the peak-to-peak values of the current $i(t)$ flowing through the circuit and the voltage $v_R(t)$ across the resistor *R*. Compare these results with a Multisim simulation and verify that the current is in phase with the applied voltage.
3. Using a Multisim simulation (*AC Sweep*[1]), plot the magnitude of the current $i(t)$ as function of frequency and confirm the occurrence of its maximum at resonance. Additionally, confirm the calculated values of the cutoff frequencies.

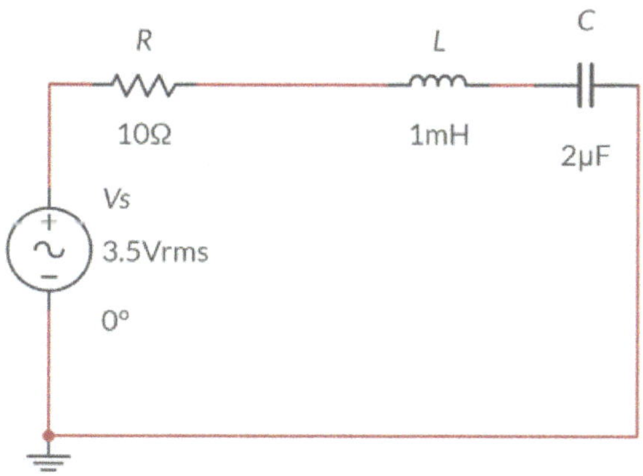

Figure 30.2 Series *R-L-C* circuit with a sinusoidal voltage generator.

[1]In Multisim, the *AC* Sweep simulation calculates the frequency response of linear circuits.

4. Verify with a Multisim simulation that at resonance, the *rms* values of the voltages $v_C(t)$ across the capacitor and the voltage $v_L(t)$ across the inductor are equal.

30.2.1.1 Solutions to Problem 1

1. The resonance frequency can be calculated as follows:

$$f_r = \frac{1}{2\pi\sqrt{LC}} = 3.6 \text{ kHz}. \tag{30.7}$$

The cutoff frequencies are calculated as follows:

$$f_1 = \frac{-R + \sqrt{R^2 + \frac{4L}{C}}}{4\pi L} = 2.85 \text{ kHz}, \tag{30.8}$$

$$f_2 = \frac{R + \sqrt{R^2 + \frac{4L}{C}}}{4\pi L} = 4.44 \text{ kHz}. \tag{30.9}$$

The bandwidth can be calculated as follows:

$$\text{BW} = f_2 - f_1 = 1.59 \text{ kHz}. \tag{30.10}$$

2. At resonance, the circuit is purely resistive. The peak value of the current can be calculated as follows:

$$\overline{I} = \frac{\overline{V}_s}{R} = \frac{\sqrt{2} \cdot 3.5\angle 0^\circ}{10} = 0.49\ \angle 0^\circ \text{A}. \tag{30.11}$$

The associated peak-to-peak value can be calculated as follows:

$$I_{pp} = 0.98 \text{ A}. \tag{30.12}$$

The peak-to-peak value of the voltage $\overline{V}_R$ across the resistor *R* can be calculated by using Ohm's law:

$$V_{R\,pp} = I_{pp} \cdot R = 9.8 \text{ V}. \tag{30.13}$$

To measure with Multisim the *peak-to-peak* values of the current $i(t)$ and the voltage $v_R(t)$, we can conduct a simulation of the circuit (Figure 30.3).

The theoretical *peak-to-peak* values of the current $i(t)$ and the voltage $v_R(t)$ match the result of the Multisim simulation.

The plot of the waveforms $i(t)$ and the voltage $v_s(t)$ at resonance, obtained with Multisim simulation, is presented in Figure 30.4.

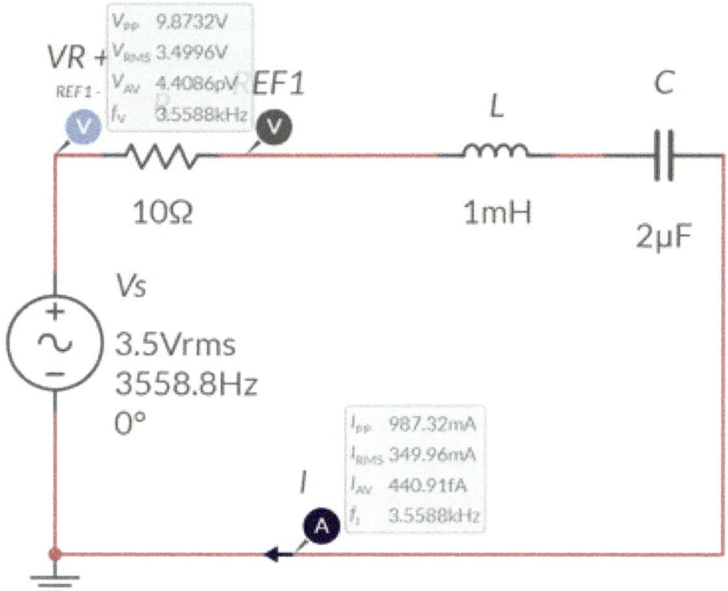

Figure 30.3 Multisim simulation showing peak-to-peak values of $i(t)$ and $v_R(t)$.

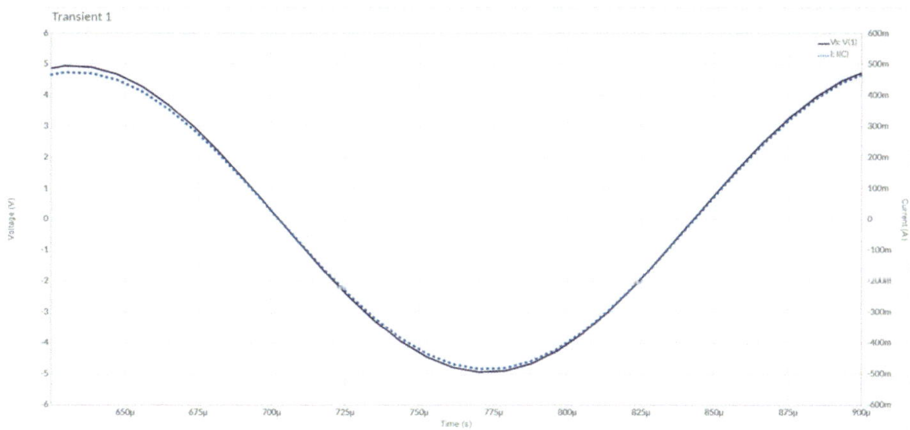

Figure 30.4 Multisim simulation confirming that at resonance the current $i(t)$ is in phase with the supply voltage $v_s(t)$.

The plot confirms that at the frequency f_r = 3558.8 Hz the current $i(t)$ and the supply voltage $v_s(t)$ are in phase. This alignment of phase indicates that the circuit is operating at resonance conditions.

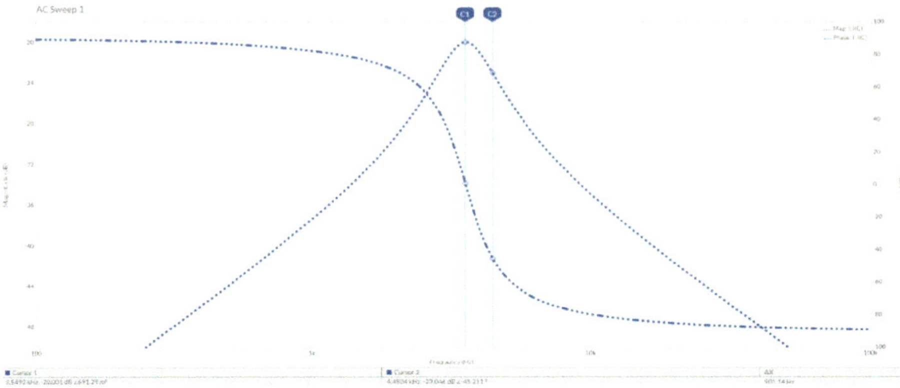

Figure 30.5 Plot of magnitude and phase angle of the current $i(t)$ as function of the frequency.

3. The plot of the magnitude and phase angle of the current $i(t)$ as function of frequency, obtained with Multisim AC Sweep is given in Figure 30.5.

The plot confirms that the current $i(t)$ reaches its maximum magnitude at the resonant frequency, as expected in a resonant series *R-L-C* circuit. Additionally, the phase angle of $i(t)$ is zero at the resonant frequency, which indicates that the current is in phase with the supply voltage. This observation aligns with the theory, which states that at resonance, the circuit becomes purely resistive, causing the current and voltage to be in phase.

The Multisim plot of Figure 30.6 confirms the analytical results by showing the cutoff frequencies at which the current magnitude decreases by 3 dB compared to its maximum resonant value.

Furthermore, the close match between the simulated and theoretical values for the bandwidth confirms the accuracy of the calculations.

4. The Multisim simulation of the circuit at resonance is provided in Figure 30.7.

The simulation shows that the *rms* values of $v_C\,(t)$ and $v_L(t)$ closely match at the resonant frequency. This confirms that the voltages across the capacitor and inductor are identical. This observation is consistent with the expected behavior of a series *R-L-C* circuit at resonance. The corresponding phasors $\overline{V}_L$ and $\overline{V}_C$ are in opposition, therefore they cancel out in resonance.

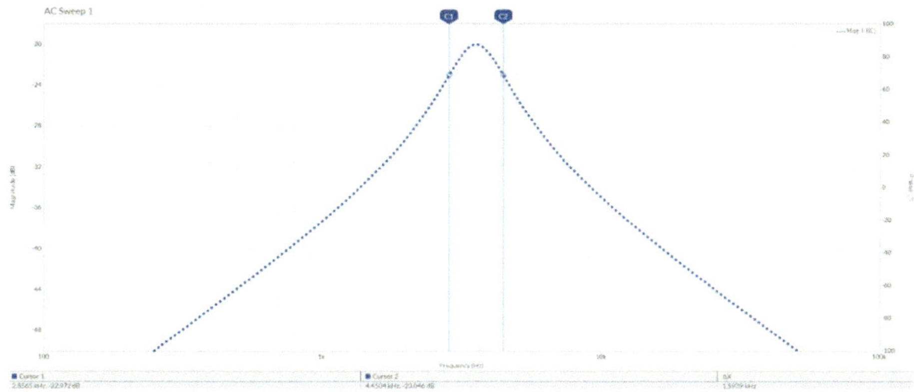

Figure 30.6 Plot of magnitude of the current $i(t)$ as a function of the frequency showing the cutoff frequencies.

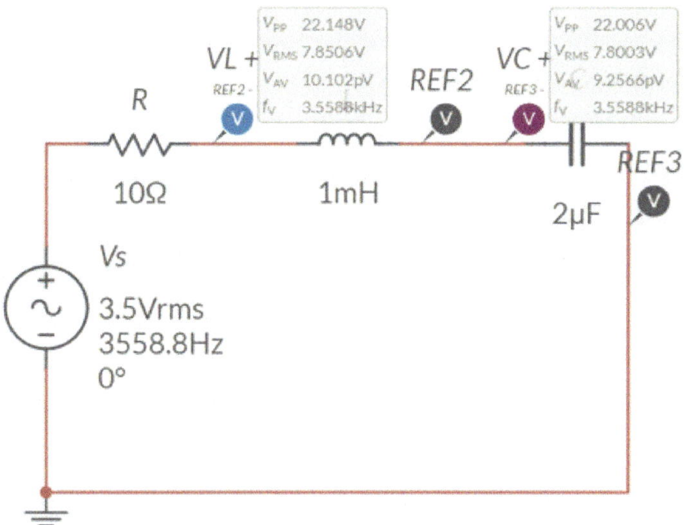

Figure 30.7 Multisim simulation providing the *rms* values of $v_C(t)$ and $v_L(t)$.

30.2.2 Problem 2

In the circuit depicted in Figure 30.8, a series *R-L-C* circuit is connected to a sinusoidal voltage generator $v_s(t) = \sqrt{2} \cdot 12\sin(\omega t)$. Perform the following tasks:

1. Calculate the resonance frequency f_r, the cutoff frequencies f_1 and f_2 and the bandwidth BW.

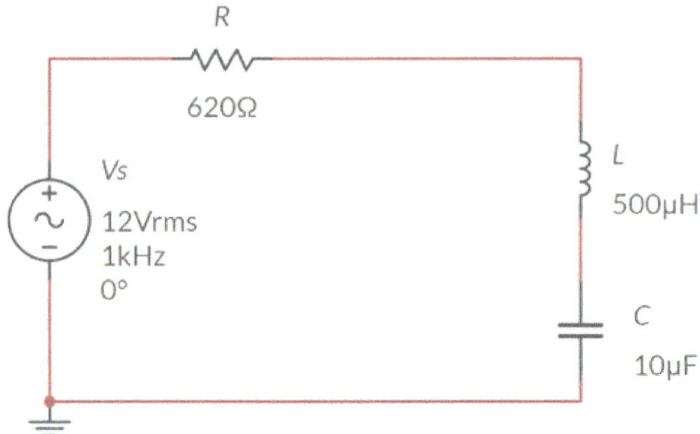

Figure 30.8 Series *R-L-C* circuit with a sinusoidal voltage generator.

2. Determine the impedance $\overline{Z}(\omega)$ of the circuit at resonance and at the cutoff frequencies.
3. Calculate the phasor $\overline{I}(\omega)$ representing the current circulating through the circuit at resonance and at the cutoff frequencies. Compare your results with those obtained from a Multisim simulation.
4. Calculate the phasor representing the voltage $\overline{V}_{LC}(\omega)$ across the series combination of the inductor and the capacitor at resonance and cutoff frequencies. Compare your results with those obtained from a Multisim simulation.

30.2.2.1 Solutions to Problem 2

1. The resonance frequency can be calculated as follows:

$$f_r = \frac{1}{2\pi\sqrt{LC}} = 2250.8 \text{ Hz}. \tag{30.14}$$

The cutoff frequencies are calculated as follows:

$$f_1 = \frac{-R + \sqrt{R^2 + \frac{4L}{C}}}{4\pi L} = 25.7 \text{ Hz}, \tag{30.15}$$

$$f_2 = \frac{R + \sqrt{R^2 + \frac{4L}{C}}}{4\pi L} = 197.5 \text{ kHz}. \tag{30.16}$$

The bandwidth can be calculated as follows:

$$\text{BW} = f_2 - f_1 = 197.4 \text{ kHz}. \tag{30.17}$$

Table 30.1 Summary of calculations

Frequency	R [Ω]	$jX_L = j2\pi fL$ [Ω]	$-jX_C = -j\frac{1}{2\pi f}$ [Ω]	$\overline{Z} = R + j(X_L - X_C)$ [Ω]
f_r	620	$j7.1$	$-j7.1$	620
f_1	620	$j0.1$	$-j620.1$	$620 - j620$
f_2	620	$j620.1$	$-j0.1$	$620 + j620$

Table 30.2 Summary of calculations

Frequency	$\overline{I} = \frac{\overline{V}_s}{\overline{Z}}$ [mA]
f_r	$19.3\angle 0°$
f_1	$13.6\angle 45°$
f_2	$13.6\angle -45°$

2. The impedance $\overline{Z}(\omega)$ of the circuit at resonance and at the cutoff frequencies is tabulated in Table 30.1.

3. The phasor $\overline{I}(\omega)$ representing the current circulating through the circuit in resonance and at the cutoff frequencies can be determined by applying the Ohm's law $\overline{I} = \frac{\overline{V}_s}{\overline{Z}}$. The calculated values are presented in Table 30.2.

To measure with Multisim the *rms* value of the phasor $\overline{I}(\omega)$ at a specific frequency, we can conduct a simulation of the circuit (Figure 30.9).

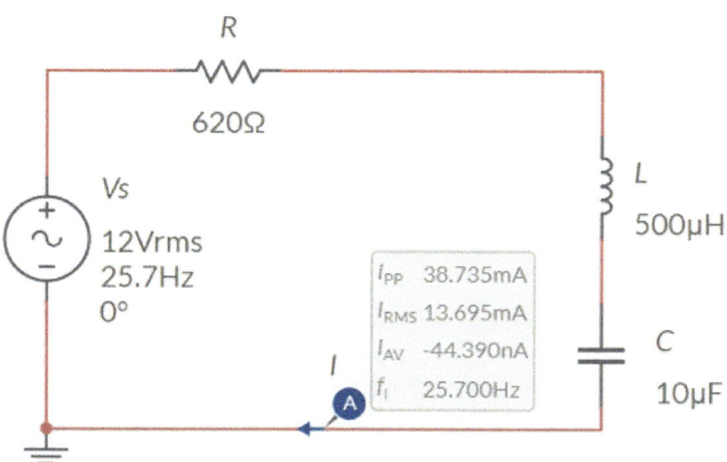

Figure 30.9 Simulation showing the *rms* value of the current $\overline{I}$ for $f = f_1$.

Table 30.3 Summary of calculations

Frequency	$jX_L = j2\pi fL$ [Ω]	$-jX_C = -j\frac{1}{2\pi f}$ [Ω]	$\overline{Z}_{LC} = j(X_L - X_C)$ [Ω]	$\overline{V}_{LC}(\omega)$ [V]
f_r	$j7$	$-j7$	0	0
f_1	$j0.1$	$-j620$	$-j619.9$	$8.4\angle-45°$
f_2	$j620$	$-j0.1$	$j619.9$	$8.4\angle45°$

Figure 30.9 illustrates the simulation results for the cutoff frequency f_1 = 25.7 Hz. The Multisim simulation results are consistent with the analytical calculations for the current phasor $\overline{I}$.

4. The phasor $\overline{V}_{LC}(\omega)$ representing the voltage across the series combination of the inductor and the capacitor can be calculated at the resonance and cutoff frequencies by applying the Ohm's law:

$$\overline{V}_{LC}(\omega) = \overline{I}(\omega) \cdot \overline{Z}_{LC}(\omega), \tag{30.18}$$

where $\overline{Z}_{LC}(\omega)$ is the impedance of the series combination of the inductor and the capacitor.

The calculated values of $\overline{Z}_{LC}$ and $\overline{V}_{LC}(\omega)$ at resonance and cutoff frequencies are shown in Table 30.3.

To measure with Multisim the *rms* value of the phasor $\overline{V}_{LC}(\omega)$ at a specific frequency, we can conduct a simulation of the circuit (Figure 30.10).

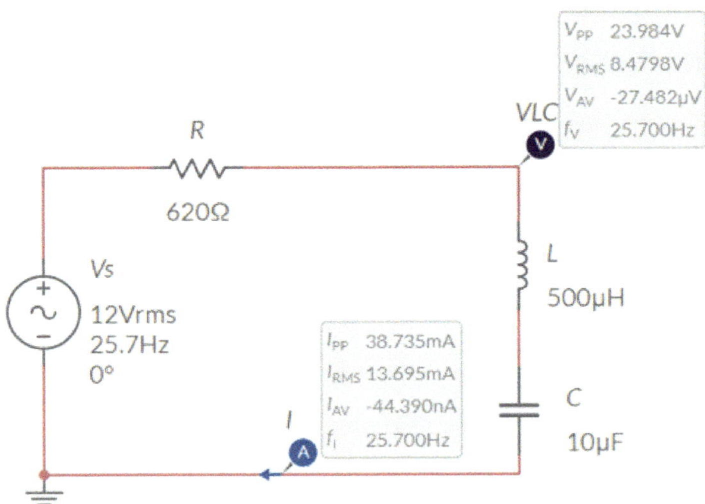

Figure 30.10 Multisim simulation showing the *rms* value of $\overline{V}_{LC}$ for the cutoff frequency f_1.

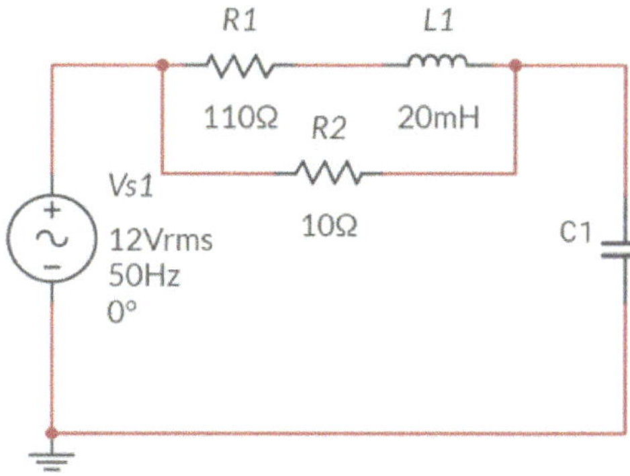

Figure 30.11 Resonant circuit with a sinusoidal voltage generator.

Figure 30.10 illustrates the simulation results for the cutoff frequency f_1 = 25.7 Hz. The Multisim simulation results are consistent with the analytical calculations for the voltage phasor $\overline{V}_{LC}$.

30.2.3 Problem 3

Given the circuit shown in Figure 30.11, where a sinusoidal voltage generator $v_s\,(t) = \sqrt{2} \cdot 12\sin\,(314t)$ supplies a series–parallel combination load, we need to perform the following tasks:

1. Calculate the value of the capacitance C_1 at which the circuit resonates for $f_r = 50$ Hz. Confirm the results using Multisim simulation.
2. Determine the phasor $\overline{I}_t$ representing the total current circulating in the series circuit at resonance. Confirm the result using Multisim simulation and verify that $\overline{I}_t$ is in phase with the generator voltage $\overline{V}_s$.
3. Calculate the cutoff frequencies f_1 and f_2 and the bandwidth *BW*.
4. Increase the frequency of the generator $v_s\,(t)$ to 60 kHz and calculate the phasor $\overline{I}_t$ representing the total current circulating in the circuit. Confirm the calculation with a Multisim simulation.

30.2.3.1 Solutions to Problem 3

1. To determine the capacitance C_1 that allows resonance in the circuit at f_r = 50 Hz, we first need to find the partial impedance $\overline{Z}_P$ of components.

$$\overline{Z}_P = (R_1 + jX_{L1}) \parallel R_2 = \frac{(R_1 + jX_{L1}) \cdot R_2}{(R_1 + jX_{L1}) + R_2} = 9.2\ \angle 0.3°\Omega$$
$$= 9.2 + j0.05\ \Omega. \tag{30.19}$$

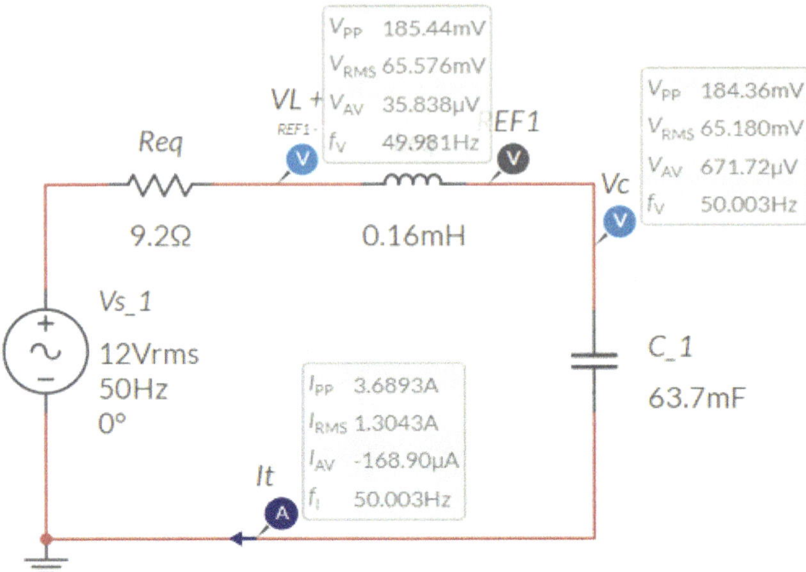

Figure 30.12 Multisim simulation of the resonant circuit.

The inductance L_{eq} associated with $\overline{Z}_p$ is given by

$$L_{\text{eq}} = \frac{X_L}{\omega} = \frac{0.05}{6.28 \cdot 50} = 0.16 \text{ mH}. \tag{30.20}$$

We determine the capacitance C_1 by imposing that at resonance $X_L = X_C$. Hence,

$$C_1 = \frac{1}{\omega X_L} = 63.7 \text{ mF}. \tag{30.21}$$

To verify with Multisim that calculated value C_1 of the capacitance is correct, we can conduct a simulation of the circuit (Figure 30.12).

The simulation shows that the *rms* values of $v_{C1}(t)$ and $v_L(t)$ closely match, indicating that the circuit has achieved resonance. This provides validation for the calculated value C_1 for the capacitance.

2. The phasor $\overline{I}_t$ representing the total current circulating in the series circuit at resonance can be calculated by applying Ohm's law:

$$\overline{I}_t = \frac{\overline{V}_{s1}}{R_{\text{eq}}} = \frac{12\angle 0^\circ}{9.2} = 1.3\angle 0^\circ \text{A}. \tag{30.22}$$

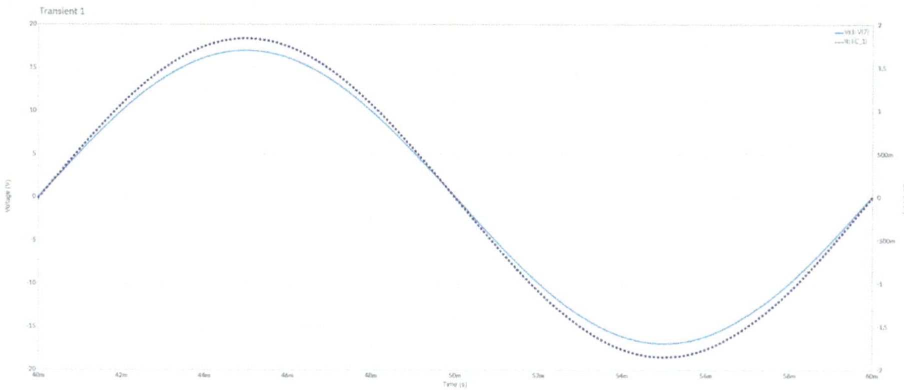

Figure 30.13 Multisim simulation showing that at resonance $i_t(t)$ and $v_{s1}(t)$ are in phase.

The Multisim simulation of Figure 30.12 confirms the correctness of the calculated *rms* value for $\overline{I}_t$.

The Multisim simulation further demonstrates that at resonance the current $i_t(t)$ is in phase with the generator voltage $v_{s1}(t)$ (Figure 30.13).

This observation confirms the theoretical prediction that at resonance, the circuit behaves purely resistive, resulting in the current being in phase with the applied voltage. The simulation results provide evidence supporting the analysis conducted on the circuit.

3. The cutoff frequencies can be calculated as follows:

$$f_1 = \frac{-R + \sqrt{R^2 + \frac{4L}{C}}}{4\pi L} = 0.27 \text{ Hz}, \tag{30.23}$$

$$f_2 = \frac{R + \sqrt{R^2 + \frac{4L}{C}}}{4\pi L} = 9.15 \text{ kHz}. \tag{30.24}$$

The bandwidth is given by

$$\text{BW} = f_2 - f_1 = 9.14 \text{ kHz}. \tag{30.25}$$

The plot of the magnitude and phase angle of the total current $i(t)$ as function of frequency, obtained with Multisim, is provided in Figure 30.14.

The Multisim plot confirms that the current $i(t)$ reaches its maximum magnitude at the resonant frequency, as expected in a resonant series *R-L-C* circuit. Additionally, the phase angle of $i(t)$ is zero at the resonant frequency,

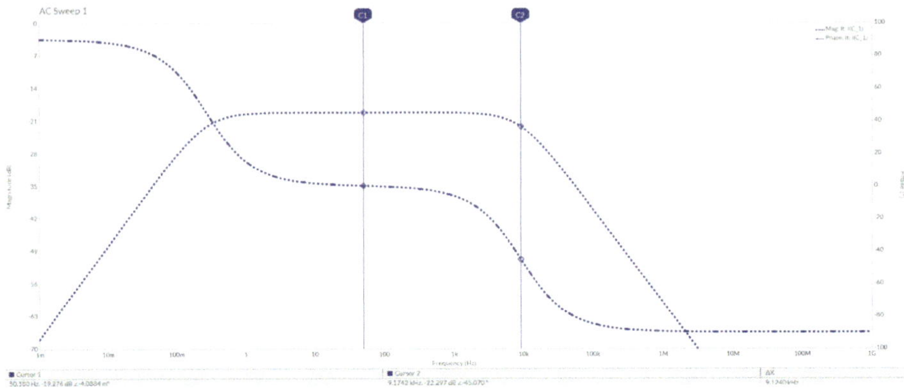

Figure 30.14 Plot of magnitude and phase angle of the total current $i(t)$ as function of the frequency.

which indicates that the current is in phase with the supply voltage. This observation aligns with the theory.

The Multisim plot also confirms the calculated value of the cutoff frequencies, at which the current magnitude decreases by 3 dB compared to its maximum resonant value.

4. With the increased frequency of the generator $v_s\,(t)$ to 60 kHz the phasor $\overline{I}$ representing the total current circulating in the circuit can be calculated by applying Ohm's law:

$$\overline{I} = \frac{\overline{V}_{s1}}{\overline{Z}_{\text{eq}}} = \frac{12\angle 0^\circ}{9.2 + j(60.29 - 4.17 \cdot 10^{-5})} = 197\ \angle{-81.3^\circ}\,\mathrm{mA}. \qquad (30.26)$$

The calculation is confirmed by Multisim simulation, which validates the accuracy of the calculated *rms* value for $i_t(t)$ at the increased frequency of 60 kHz (Figure 30.15).

For frequencies greater than the resonant frequency, the inductive reactance becomes dominant in the circuit and the current $i(t)$ lags the voltage $\overline{V}_{s1}$. The phase difference $\Delta\varphi$ between $\overline{I}_t$ and $\overline{V}_{s1}$ is confirmed by a Multisim plot (Figure 30.16).

For frequencies greater than the resonant frequency (i.e., f_r = 50 Hz, in our case), the inductive reactance becomes dominant in the circuit, causing the current $i(t)$ to lag the supply voltage $v_{s1}(t)$. The phase difference $\Delta\varphi$ between the current and voltage is confirmed by a Multisim plot shown in Figure 30.16.

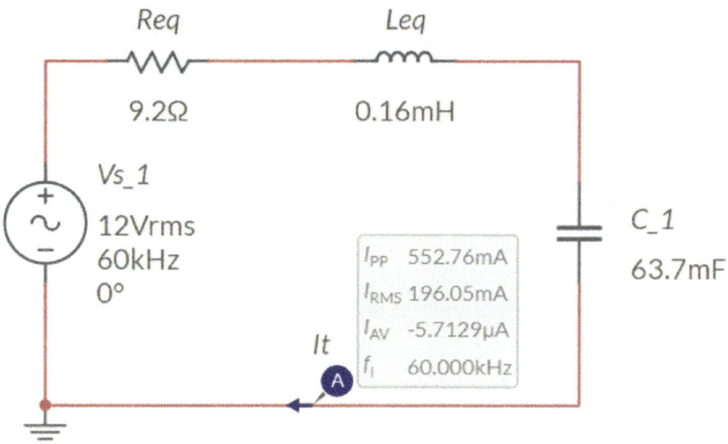

Figure 30.15 Multisim simulation showing the *rms* value of $i_t(t)$.

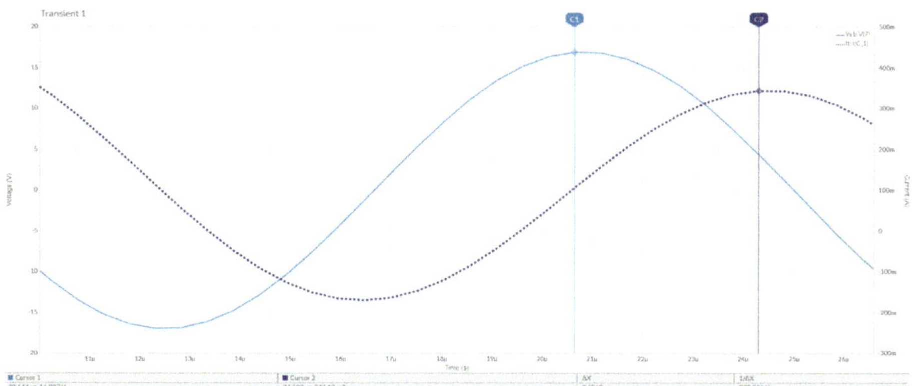

Figure 30.16 Multisim simulation showing the phase difference $\Delta\varphi$ between $v_{s1}(t)$ and $i_\iota(t)$.

The plot clearly illustrates the lagging nature of the current with respect to the voltage, further validating the expected behavior of the circuit at frequencies above the resonance. To estimate the phase difference $\Delta\varphi$ in degrees between the waveforms, we can calculate the time difference Δt between corresponding points on the plots. The formula to convert the time difference to phase difference in degrees is given by

$$\Delta\varphi = \frac{360^\circ}{T} \cdot \Delta t = \frac{360^\circ}{1.66 \cdot 10^{-5}} \cdot 3.65 \cdot 10^{-6} {=} 79.1^\circ. \tag{30.27}$$

The Multisim simulation closely matches the theoretical value of the phase difference $\Delta\varphi$. It is important to note that the Multisim plot approximates $\Delta\varphi$ due to limitations in the program's ability to accurately measure the exact difference between any two adjacent time points.

30.3 Lab Procedure 1

In the circuit depicted in Figure 30.17, a series combination *R-L-C* circuit is connected to a sinusoidal voltage generator $v_{s3}(t) = \sqrt{2} \cdot 10\sin(\omega t)$. Perform the following tasks:

1. Calculate the resonance frequency f_r, the cutoff frequencies f_1 and f_2 and the bandwidth BW. Record values in Table 30.4.
2. At resonance, determine the peak-to-peak values of the current $i(t)$ flowing through the circuit and the voltage $v_{R3}(t)$ across the resistor R_3. Compare these results with a Multisim simulation and verify that $i(t)$ in phase with $v_{R3}(t)$. Record values in Table 30.4.
3. Using a Multisim simulation (*AC Sweep*), plot the magnitude of the current $i(t)$ as function of frequency and confirm the occurrence of its maximum at resonance. Additionally, confirm the calculated values of the cutoff frequencies.
4. Verify with a Multisim simulation that at resonance, the *rms* values of the voltages $v_C(t)$ across the capacitor and the voltage $v_L(t)$ across the inductor are equal. Record values in Table 30.4.

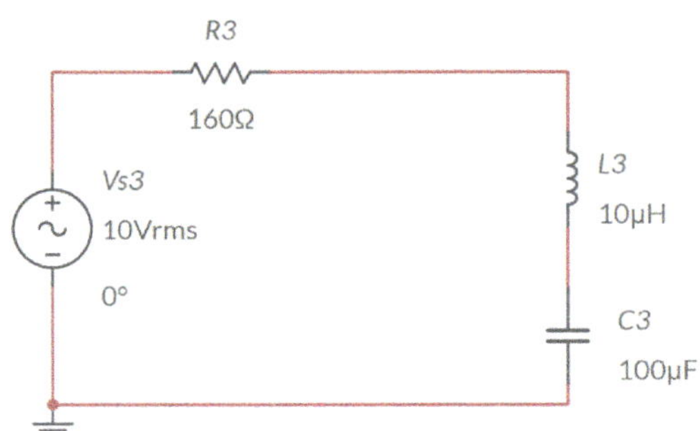

Figure 30.17 Series *R-L-C* circuit with a sinusoidal voltage generator.

Table 30.4 Summary of readings and calculations

f_r	f_1	f_2	BW	I_{pp}	$V_{R3\ pp}$	$V_{C3\ \text{rms}}$	$V_{L3\ \text{rms}}$

30.4 Lab Procedure 2

In the circuit depicted in Figure 30.18, a series combination *R-L-C* circuit is connected to a sinusoidal voltage generator $v_s(t) = \sqrt{2}\cdot 24\sin(\omega t)$. Perform the following tasks:

1. Calculate the resonance frequency f_r, the cutoff frequencies f_1 and f_2 and the bandwidth BW. Record values in Table 30.5.
2. Calculate the impedance $\overline{Z}(\omega)$ of the circuit at resonance and at the cutoff frequencies. Record values in Table 30.5.
3. Calculate the phasor $\overline{I}(\omega)$ representing the current circulating through the circuit at resonance and at the cutoff frequencies. Compare your results with those obtained from a Multisim simulation. Record values in Table 30.6.
4. Calculate the phasor representing the voltage $\overline{V}_{LC}(\omega)$ across the series combination of the inductor and the capacitor at resonance and cutoff frequencies. Compare your results with those obtained from a Multisim simulation. Record values in Table 30.6.

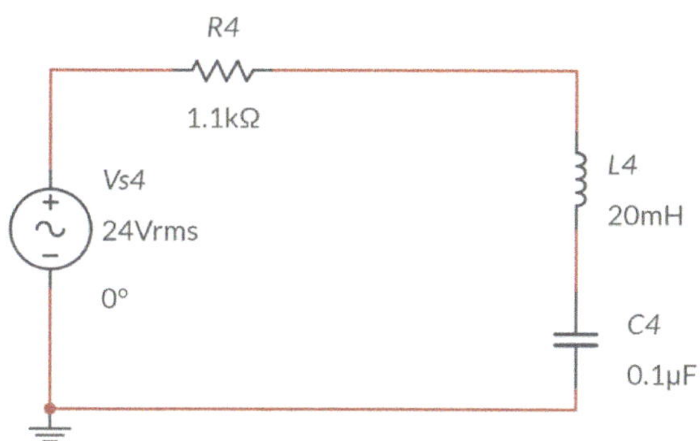

Figure 30.18 Series *R-L-C* circuit with a sinusoidal voltage generator.

Table 30.5 Summary of calculations

f_r	f_1	f_2	BW	$\overline{Z}(f_r) = Z\angle\varphi_r$	$\overline{Z}(f_1) = Z\angle\varphi_1$	$\overline{Z}(f_2) = Z\angle\varphi_2$

Table 30.6 Summary of calculations

$\overline{I}(f_r) = I\angle\varphi_r$	$\overline{I}(f_1) = I\angle\varphi_1$	$\overline{I}(f_2) = I\angle\varphi_2$	$\overline{V}_{LC}(f_r) = V_{LC}\angle\varphi_r$	$\overline{V}_{LC}(f_1) = V_{LC}\angle\varphi_1$	$\overline{V}_{LC}(f_2) = V_{LC}\angle\varphi_2$

31

Parallel Resonant Circuits

31.1 Student Learning Outcomes

- Calculate the resonant frequency, bandwidth, and quality factor of parallel *RLC* circuits.
- Calculate the magnitude of parameters in parallel *RLC* circuits and analyze the effect of frequency.
- Determine the parameters of the equivalent parallel circuit of a practical inductor.
- Validate calculated values of parameters in parallel *RLC* circuits using Multisim simulation.

31.2 Prelab

In a resonant parallel circuit, it is common to have a current source driving the circuit (Figure 31.1), but resonance can also occur with voltage sources. In parallel circuits, the equivalent admittance $\overline{Y}_{\text{eq}}$ at the generator terminals is determined by the summation of the conductance $\overline{G}$, the capacitive susceptance $\overline{B}_C$, and the inductive susceptance $\overline{B}_L$:

$$\overline{G} = \frac{1}{R\angle 0^\circ} = G\angle 0^\circ, \tag{31.1}$$

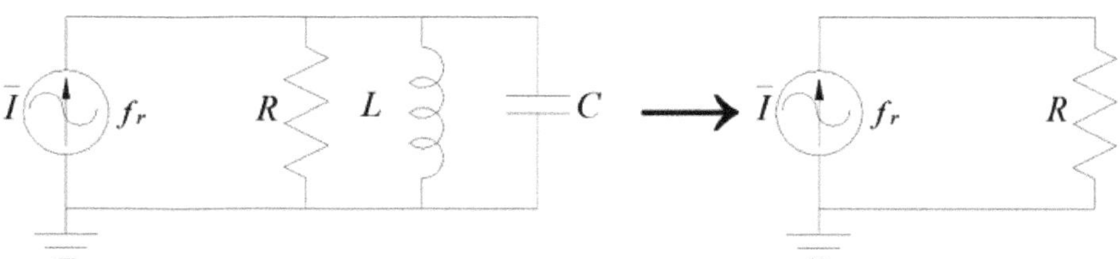

Figure 31.1 Parallel resonant condition.

$$\overline{B}_C = \frac{1}{-jX_C} = j\omega C = jB_C, \tag{31.2}$$

$$\overline{B}_L = \frac{1}{jX_L} = -j\frac{1}{\omega L} = -jB_L, \tag{31.3}$$

$$\overline{Y}_{\text{eq}} = \frac{1}{R} + j\left(\omega C - \frac{1}{\omega L}\right). \tag{31.4}$$

The *parallel resonance* is a condition that can occur in a parallel *RLC* circuit at a specific frequency f_r, where the phasor $\overline{V}$ of the voltage across the parallel and the phasor $\overline{I}$ of the current of the generator are in phase, indicating unity power factor. Therefore, the equivalent admittance $\overline{Y}_{\text{eq}}$ is a real quantity, and the coefficient of its imaginary part is zero. The parallel LC combination acts like an open circuit, and the entire current flows through *R* (Figure 31.1).

The resonant frequency can be determined by solving the following equation:

$$2\pi f_r C = \frac{1}{2\pi f_r L}, \tag{31.5}$$

which yields

$$f_r = \frac{1}{2\pi\sqrt{LC}}. \tag{31.6}$$

We obtained the same result as we did for series resonance.

At resonance, the equivalent impedance of the parallel circuit is purely resistive and reaches its maximum value. Consequently, the voltage across each element in the circuit is also at its maximum. This behavior is a characteristic feature of resonant parallel circuits. At frequencies above and below f_r, the voltage decreases because the susceptances no longer cancel each other out.

Due to this frequency-dependent performance, the resonant parallel circuit is also known as a *rejecter* circuit, as it rejects currents with frequencies equal to its resonant frequency. Additionally, the circuit behaves as inductive below resonance and capacitive above resonance. At frequencies $f < f_r$, the capacitor acts as an open circuit, making the circuit inductive. Conversely, at frequencies $f > f_r$, the inductor acts as an open circuit, making the circuit capacitive.

To quantify the rejection of current, we define the *half-power* frequencies f_1 and f_2. These frequencies correspond to the points where voltage and impedance are $\frac{1}{\sqrt{2}}$ times their values at resonance. They are also referred to as the *low cutoff* and *high cutoff frequencies*, respectively. Alternatively, the cutoff frequencies can be defined as the frequencies at which the voltage decreases by 3 dB compared to its maximum resonant value.

The analytical expressions for f_1 and f_2 are

$$f_1 = \frac{-\frac{1}{R} + \sqrt{\frac{1}{R^2} + \frac{4C}{L}}}{4\pi C}, \tag{31.7}$$

$$f_2 = \frac{\frac{1}{R} + \sqrt{\frac{1}{R^2} + \frac{4C}{L}}}{4\pi C}. \tag{31.8}$$

We define the *bandwidth* BW of the circuit as: The bandwidth (BW) of the circuit is defined as the difference between f_2 and f_1 and can be calculated as follows:

$$\text{BW} = f_2 - f_1 = \frac{1}{2\pi RC}. \tag{31.9}$$

The *quality factor* (Q) of a resonant parallel circuit is a measure of its selectivity or sharpness at resonance. It is defined as the ratio of the reactive power to the real power in the circuit.

$$Q = \frac{V^2/X}{V^2/R} = \frac{R}{2\pi f_r L} = 2\pi f_r RC = \frac{f_r}{\text{BW}}. \tag{31.10}$$

Greater values for the resistance *R* do increase the quality factor, since the bandwidth decreases.

The *quality factor* is directly proportional to the resistance, therefore, greater values of resistance in the circuit result in a narrower bandwidth. This means that the circuit becomes more selective, allowing a narrower range of frequencies to pass through the circuit at maximum amplitude.

31.2.1 Practical inductor

In a parallel resonant circuit, the internal series coil resistance (R_L) of a practical inductor needs to be considered, as it significantly affects the resonance condition. Figure 31.2 illustrates a parallel resonant circuit that includes the internal coil resistance R_L.

To find the equivalent parallel network for the impedance $R_L + j\omega L$ of the practical inductor, we need to determine the equivalent parallel components R_{Lp} and jX_{Lp} (Figure 31.2). The transformation is done by equating the

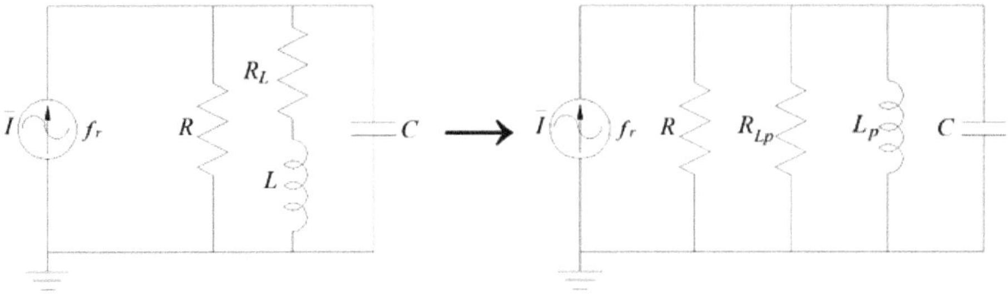

Figure 31.2 Parallel resonant circuit including the internal coil resistance R_L of the inductor.

practical inductor admittance $\overline{Y}_L$ to the admittance of the equivalent parallel combination:

$$\overline{Y}_L = \frac{1}{R_L + j\omega L} = \frac{1}{R_{Lp}} + \frac{1}{jX_{Lp}}. \tag{31.11}$$

To isolate the coefficients of the real and imaginary parts, we multiply both sides of the equation by the complex conjugate of the denominator. This leads to

$$\overline{Y}_L = \frac{1}{R_L + j\omega L} \cdot \frac{R_L - j\omega L}{R_L - j\omega L} = \frac{R_L - j\omega L}{R_L^2 + (\omega L)^2} = \frac{R_L}{R_L^2 + (\omega L)^2} - j\frac{\omega L}{R_L^2 + (\omega L)^2}. \tag{31.12}$$

Substituting eqn (31.12) into eqn (31.11), and equating real parts with real parts and imaginary parts with imaginary parts, we obtain

$$R_{Lp} = \frac{R_L^2 + (\omega L)^2}{R_L}, \tag{31.13}$$

$$X_{Lp} = \frac{R_L^2 + (\omega L)^2}{\omega L} = \omega L_p. \tag{31.14}$$

This equation yields the parallel inductance L_p:

$$L_p = \frac{R_L^2 + (\omega L)^2}{\omega^2 L}. \tag{31.15}$$

In Figure 31.2, *R* and *C* represent the values from the original circuit, while R_{Lp} and X_{Lp} are the parallel transformed values derived from the original practical inductor. It is important to note that R_{Lp} depends on the frequency. In this equivalent circuit, R_{Lp} and *R* are in parallel and can be combined into a single resistor R_p. The equivalent admittance $\overline{Y}_{\text{eq}}$ is given by

$$\overline{Y}_{\text{eq}} = \frac{1}{R_p} + j\left(\frac{1}{X_c} - \frac{1}{X_{Lp}}\right). \tag{31.16}$$

At resonance, the phasors $\overline{V}$ representing the voltage across the parallel combination, and the phasor $\overline{I}$ representing the current impressed by the generator, are in phase (i.e., unity power factor). This implies that the reactive component of the admittance $\overline{Y}_{\text{eq}}$, which is the coefficient of the imaginary part, is zero:

$$\frac{1}{X_c} = \frac{1}{X_{Lp}}, \tag{31.17}$$

$$\frac{1}{\omega_r C} = \frac{R_L^2 + (\omega_r L)^2}{\omega_r L}. \tag{31.18}$$

We can therefore determine the resonant frequency f_r as follows:

$$f_r = \frac{1}{2\pi L}\sqrt{\frac{L}{C} - R_L^2}. \tag{31.19}$$

The quality factor Q_r remains the ratio of the reactive power to the real power:

$$Q_r = \frac{V^2/X_{Lp}}{V^2/R_p} = \frac{R||R_{Lp}}{X_{Lp}} = \frac{R||R_{Lp}}{X_C}. \tag{31.20}$$

Here, R_{Lp}, X_{Lp}, and X_C are calculated at the resonant frequency.

The cutoff frequencies f_1 and f_2 can be determined using the following formulae:

$$f_1 = \frac{f_r}{2}\left(-\frac{1}{Q_r} + \sqrt{\frac{1}{Q_r^2} + 4}\right), \tag{31.21}$$

$$f_2 = \frac{f_r}{2}\left(\frac{1}{Q_r} + \sqrt{\frac{1}{Q_r^2} + 4}\right). \tag{31.22}$$

31.2.2 Problem 1

In the given circuit depicted in Figure 31.3, a sinusoidal current generator $i(t) = \sqrt{2} \cdot \sin(\omega t)$ A is connected to a parallel *RLC* circuit. To analyze the circuit, perform the following tasks:

1. Calculate the resonant frequency f_r, the cutoff frequencies f_1 and f_2, the bandwidth *BW* and the quality factor *Q*.
2. At resonance, determine the *rms* value of the voltage $v_{R1}(t)$ across the resistor R_1. Compare these results with a Multisim simulation and verify that the generator current is in phase with $v_{R1}(t)$.

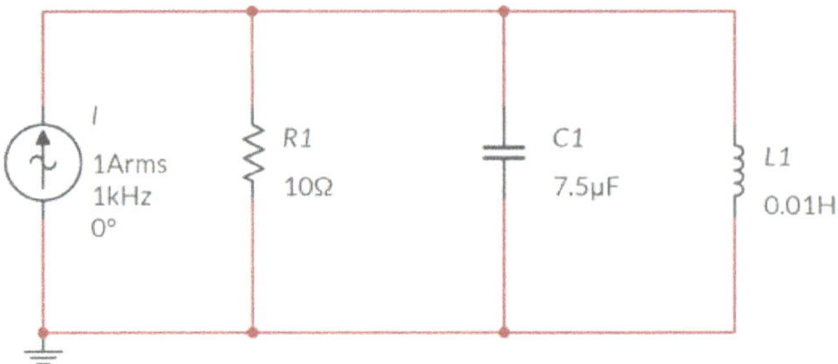

Figure 31.3 A parallel *RLC* circuit with a current generator.

3. Using a Multisim simulation (*AC Sweep*[1]), plot the magnitude of the voltage $v_{R1}(t)$ as function of frequency and confirm the occurrence of its maximum at resonance. Additionally, confirm the calculated values of the cutoff frequencies.
4. Verify with a Multisim simulation that at resonance the current $i_{C1}(t)$ through the capacitor and the current $i_{L1}(t)$ through the inductor are equal.

31.2.2.1 Solutions to Problem 1

1. The resonance frequency can be calculated as follows:

$$f_r = \frac{1}{2\pi\sqrt{LC}} = 581.2 \text{ Hz}. \tag{31.23}$$

The cutoff frequencies are calculated as follows:

$$f_1 = \frac{-\frac{1}{R} + \sqrt{\frac{1}{R^2} + \frac{4C}{L}}}{4\pi C} = 148.8 \text{ Hz}, \tag{31.24}$$

$$f_2 = \frac{\frac{1}{R} + \sqrt{\frac{1}{R^2} + \frac{4C}{L}}}{4\pi C} = 2.27 \text{ kHz}. \tag{31.25}$$

The bandwidth is given by

$$\text{BW} = f_2 - f_1 = 2.12 \text{ kHz}, \tag{31.26}$$

$$Q = \frac{f_r}{\text{BW}} = 0.27. \tag{31.27}$$

[1] In Multisim, the AC sweep simulation calculates the frequency response of linear circuits.

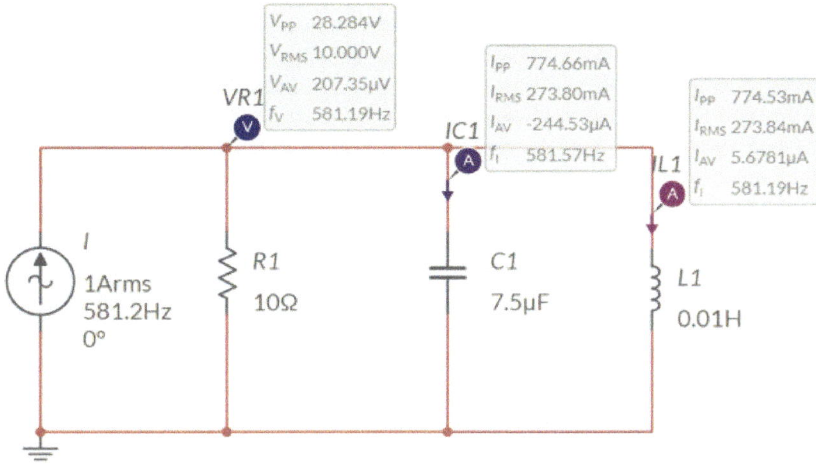

Figure 31.4 Multisim simulation confirming the theoretical *rms* value for $v_{R1}(t)$.

2. At resonance, the circuit is purely resistive. The *rms* value of the voltage $v_R(t)$ can be calculated as follows:

$$\overline{V}_R = \overline{I}R = 10 \angle 0^\circ \text{ V}. \tag{31.28}$$

To measure in Multisim the *rms* value of the voltage $v_{R1}(t)$, we can conduct a simulation of the circuit (Figure 31.4).

The theoretical *rms* value of the voltage $v_{R1}(t)$ matches the result of the Multisim simulation.

The plot of the waveforms $i(t)$ and the voltage $v_{R1}(t)$ at resonance, obtained with Multisim simulation, is presented in Figure 31.5. The plot confirms that at the frequency f_r = 581.2 Hz current and voltage are in phase. This alignment of phase indicates that the circuit is operating at resonance conditions.

3. The plot of the magnitude of the voltage $v_{R1}(t)$ as function of frequency, obtained with Multisim, is given in Figure 31.6.

The plot confirms that the magnitude of the voltage $v_{R1}(t)$ reaches its maximum at a frequency of 583.8 Hz, which closely matches the analytical result for the resonant frequency.

The plot of Figure 31.7 shows the values of the cutoff frequencies, at which the magnitude of $v_{R1}(t)$ is approximately $\frac{1}{\sqrt{2}}$ times the value at resonance.

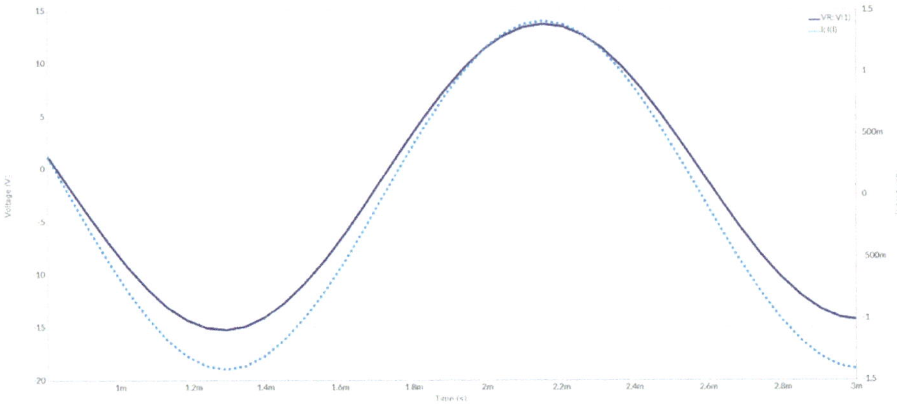

Figure 31.5 Multisim simulation confirms that at resonance the current $i(t)$ is in phase with the voltage $v_{R1}(t)$.

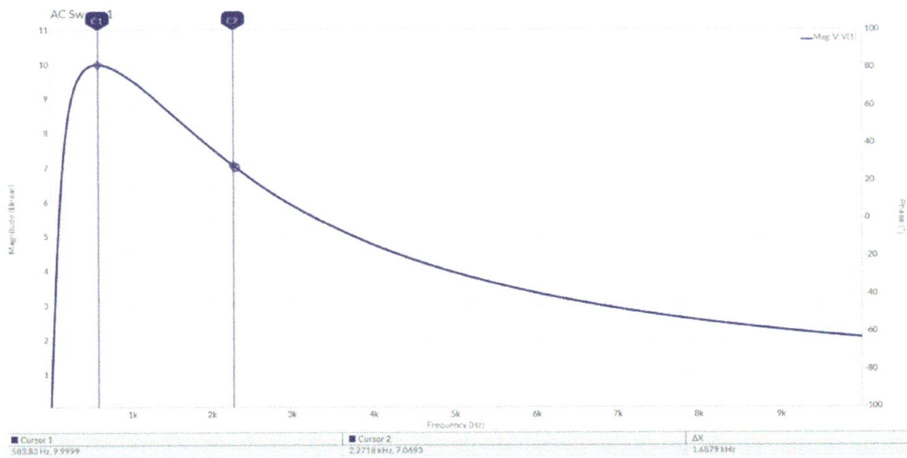

Figure 31.6 Plot of the magnitude of the voltage $v_{R1}(t)$ as a function of the frequency.

The values f_1 = 152.39 Hz and f_2 = 2.27 kHz closely match the calculated values of the cutoff frequencies. The plot also shows that the bandwidth is equal to 2.12 kHz, which is in tight agreement with the calculation.

4. The Multisim simulation of Figure 31.4 shows that the *rms* values of the currents $i_{C1}(t)$ and $i_{L1}(t)$ closely match at the resonant frequency. This confirms that the currents through the capacitor and the inductor are equal. This observation is consistent with the expected behavior of a parallel *RLC* circuit at resonance.

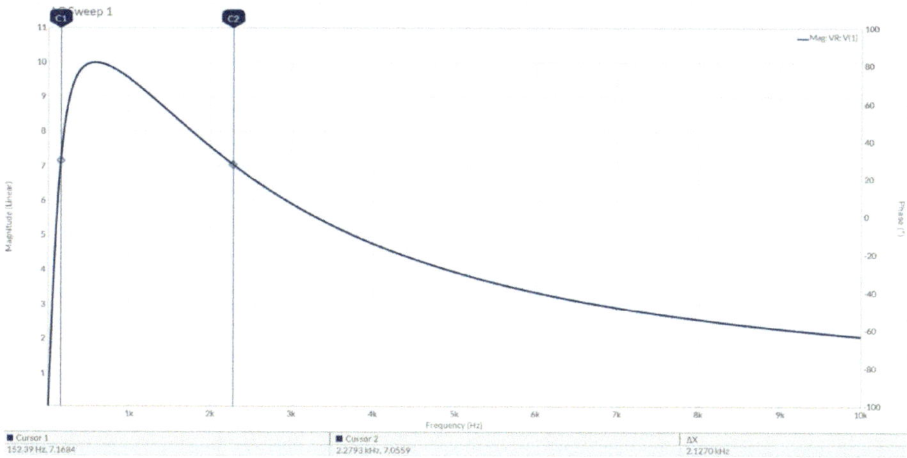

Figure 31.7 Plot of the magnitude of the voltage $v_{R1}(t)$ as a function of the frequency showing the cutoff frequencies and the bandwidth.

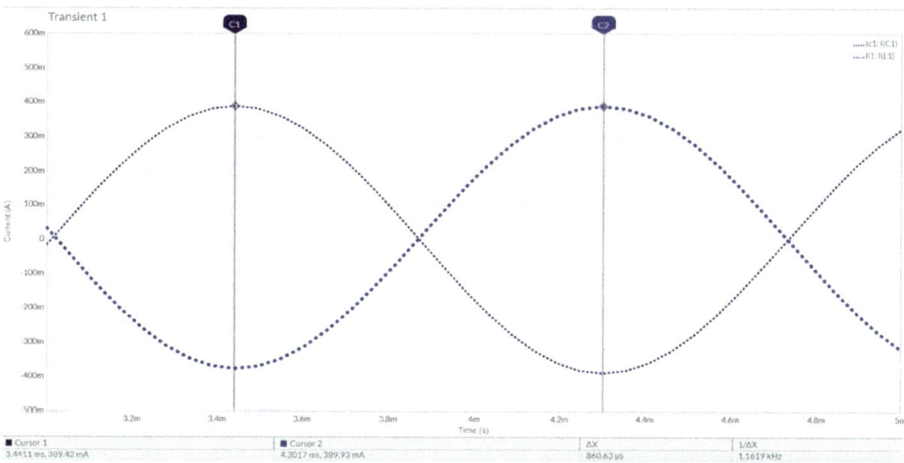

Figure 31.8 Plot of the waveforms $i_{L1}(t)$ and $i_{C1}(t)$ in opposition at resonance.

At resonance, the waveforms $i_{C1}(t)$ and $i_{L1}(t)$ are in opposition (i.e., they have a phase difference of 180°) and therefore they cancel each other out. This can be observed in the Multisim plot shown in Figure 31.8.

To estimate the phase difference $\Delta\varphi$ in degrees between the waveforms, we can calculate the time difference Δt between corresponding points on the plots. The formula to convert the time difference to phase difference in

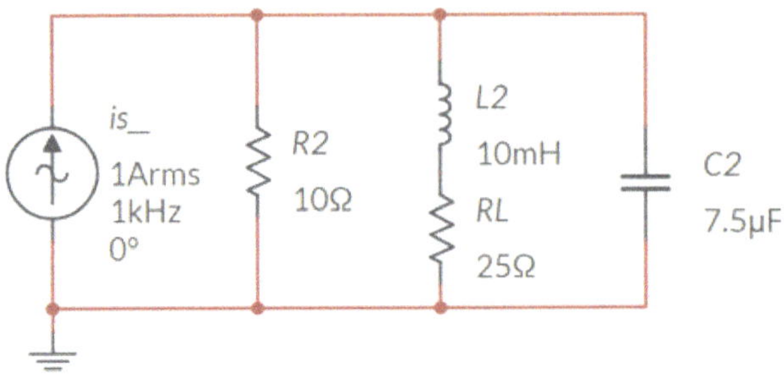

Figure 31.9 Parallel *RLC* circuit with a practical inductor and sinusoidal current generator.

degrees is given by

$$\Delta\varphi = \frac{360°}{T} \cdot \Delta t = \frac{360°}{1.72 \cdot 10^{-3}} \cdot 860.6 \cdot 10^{-6} = 180°. \quad (31.29)$$

31.2.3 Problem 2

In the given circuit depicted in Figure 31.9, a sinusoidal current generator $i\,(t) = \sqrt{2} \cdot \sin\,(\omega t)$ A is connected to a parallel *RLC* circuit with a practical inductor. To analyze the circuit, perform the following tasks:

1. Calculate the resonant frequency f_r and determine the parameters R_{Lp} and L_p of the equivalent parallel circuit of the practical inductor. Confirm your calculation with a Multisim simulation.
2. Determine the cutoff frequencies f_1 and f_2, the bandwidth *BW* and the quality factor *Q*. Confirm your calculation with a Multisim simulation.
3. At resonance, determine the *rms* value of the voltage $v(t)$ across the parallel. Compare the theoretical result with a Multisim simulation and verify that the generator current $i_s(t)$ is in phase with $v(t)$.

31.2.3.1 Solutions to Problem 2

1. The resonant frequency f_r can be calculated as follows:

$$f_r = \frac{1}{2\pi L_2}\sqrt{\frac{L_2}{C} - R_L^2} = 423.6 \text{ Hz}. \quad (31.30)$$

The parameters R_{Lp} and L_p of the equivalent parallel circuit of the practical inductor can be calculated as follows:

$$R_{Lp} = \frac{R_L^2 + (\omega L)^2}{R_L} = 53.3\ \Omega, \quad (31.31)$$

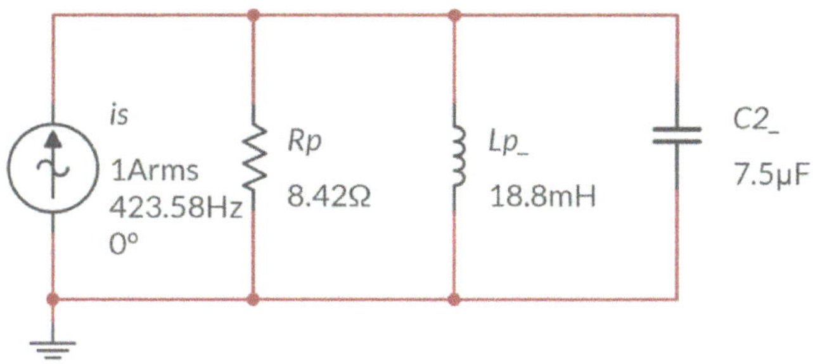

Figure 31.10 Equivalent resonant parallel circuit.

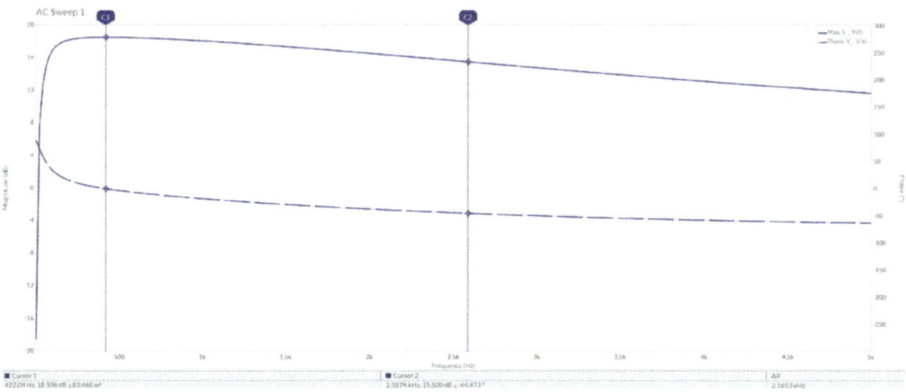

Figure 31.11 Multisim plot of the magnitude and phase angle of the voltage $v(t)$ confirms calculated value of f_r.

$$L_p = \frac{R_L^2 + (\omega L)^2}{\omega^2 L} = 18.8 \text{ mH}. \tag{31.32}$$

The equivalent resonant parallel circuit is provided in Figure 31.10.

R_p is the parallel between the original resistor R_2 and R_{Lp}.

The plot of the magnitude and phase angle of the voltage $v(t)$ across the parallel combination as function of frequency, obtained with Multisim AC Sweep, is given in Figure 31.11.

The plot confirms that the voltage $v(t)$ reaches its maximum magnitude at the resonant frequency, as expected in a resonant parallel *RLC* circuit. Additionally, the phase angle of $v(t)$ is zero at the resonant frequency, which indicates that the current is in phase with the supply voltage. This observation aligns with the theory, which states that at resonance, the circuit becomes purely resistive, causing the current and voltage to be in phase.

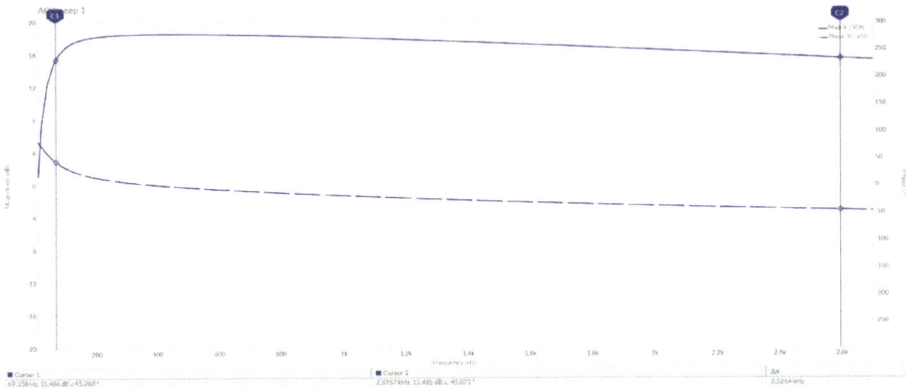

Figure 31.12 Plot of magnitude of the voltage $v(t)$ as a function of the frequency showing the cutoff frequencies.

2. The quality factor Q_r can be calculated as follows:

$$Q_r = \frac{R||R_{Lp}}{X_{Lp}} = 0.17. \tag{31.33}$$

The cutoff frequencies can be calculated as follows:

$$f_1 = \frac{f_r}{2}\left(-\frac{1}{Q_r} + \sqrt{\frac{1}{Q_r^2} + 4}\right) = 69.3 \text{ Hz}, \tag{31.34}$$

$$f_2 = \frac{f_r}{2}\left(\frac{1}{Q_r} + \sqrt{\frac{1}{Q_r^2} + 4}\right) = 2.59 \text{ kHz}. \tag{31.35}$$

The bandwidth is given by

$$\text{BW} = f_2 - f_1 = 2.52 \text{ kHz}. \tag{31.36}$$

The Multisim plot of Figure 31.12 confirms the analytical results by showing the cutoff frequencies at which the voltage magnitude decreases by 3 dB compared to its maximum resonant value.

Furthermore, the close match between the simulated and theoretical values for the bandwidth confirms the accuracy of the calculations.

3. At resonance, the impedance of the parallel combination is just its resistance. Thus, the phasor $\overline{V}$ of the voltage across the parallel can be calculated as follows:

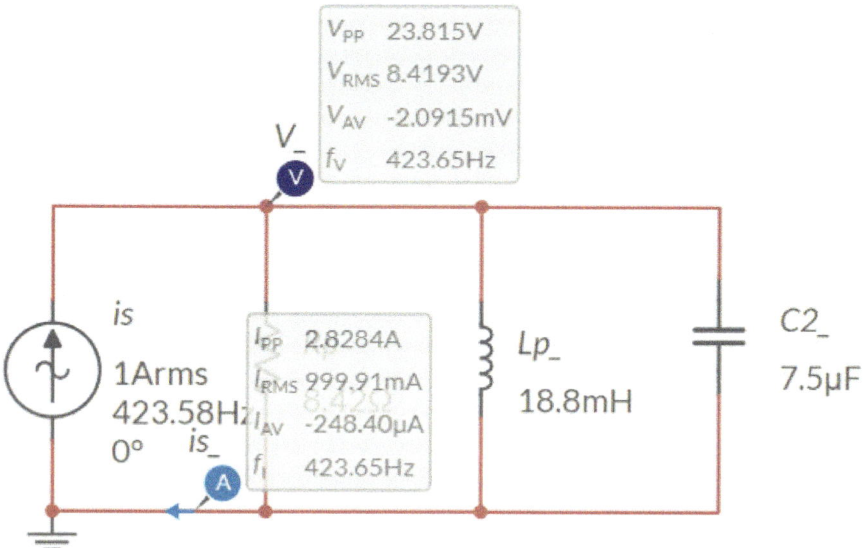

Figure 31.13 Multisim simulation confirming the calculated *rms* value for $v(t)$.

$$\overline{V} = I\angle 0° \cdot R_P = 8.42\ \angle 0°\ \text{V}. \tag{31.37}$$

Multisim simulation confirms the calculated *rms* value for $v(t)$ (Figure 31.13).

The Multisim simulation also confirms that at resonance the waveforms $v(t)$ and $i_s(t)$ are in phase (Figure 31.14).

This observation confirms the theoretical prediction that at resonance, the parallel circuit behaves purely resistive, resulting in the current being in phase with the applied voltage.

31.3 Lab Procedure 1

In the given circuit depicted in Figure 31.15, a sinusoidal current generator $i_s\,(t) = \sqrt{2} \cdot 2\sin\,(\omega t)$ A is connected to a parallel *RLC* circuit. To analyze the circuit, perform the following tasks:

1. Calculate the resonant frequency f_r, the cutoff frequencies f_1 and f_2, the bandwidth BW and the quality factor Q. Record results in Table 31.1.
2. At resonance, determine the *rms* value of the voltage $v_{R1}(t)$ across the resistor R_1. Compare the result with a Multisim simulation and verify

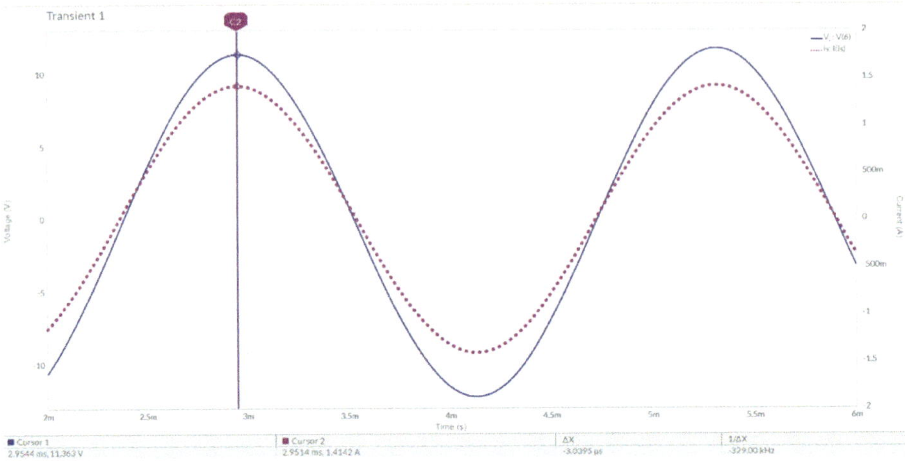

Figure 31.14 Multisim simulation confirms that in resonance $\overline{V}$ and $\overline{I}_s$ are in phase.

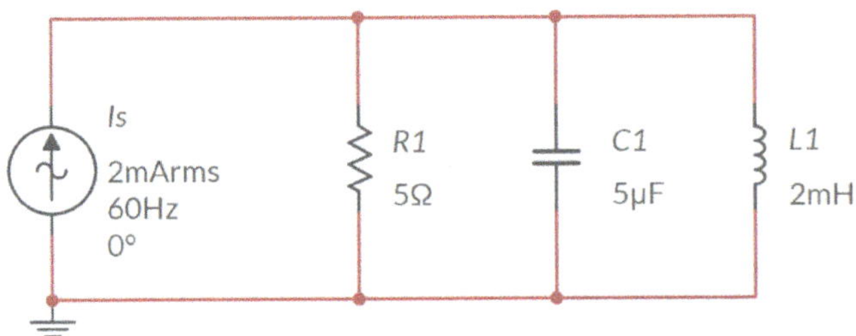

Figure 31.15 Parallel *RLC* circuit with a current generator.

that the generator current $i_s\,(t)$ is in phase with $v_{R1}(t)$. Record results in Table 31.1.

3. Using a Multisim simulation (*AC Sweep*), plot the magnitude of the voltage $v_{R1}(t)$ as function of frequency and confirm the occurrence of its maximum at resonance. Additionally, confirm the calculated values of the cutoff frequencies.
4. Verify with Multisim simulation that at resonance the current $i_{C1}(t)$ through the capacitor and the current $i_{L1}(t)$ through the inductor are equal.

Table 31.1 Summary of calculations

f_r	f_1	f_2	BW	Q	$\overline{V}_{R1}= V_R\angle\varphi$

31.4 Lab Procedure 2

In the given circuit depicted in Figure 31.16, a sinusoidal current generator $i_s(t) = \sqrt{2} \cdot 2 \cdot \sin(\omega t)$ A is connected to a parallel *RLC* circuit with a practical inductor. To analyze the circuit, perform the following tasks:

1. Calculate the resonant frequency f_r and determine the parameters R_{Lp} and L_p of the equivalent parallel circuit of the practical inductor. Records results in Table 31.2. Confirm your calculations with a Multisim simulation.
2. Calculate the resonant frequency f_r, the cutoff frequencies f_1 and f_2, the bandwidth *BW* and *Q*. Confirm your calculation with a Multisim simulation.
3. At resonance, determine the *rms* value of the voltage $v_{R2}(t)$ across the resistor R_2. Compare the theoretical result with a Multisim simulation and verify that the generator current $i_{s2}(t)$ is in phase with $v_{R2}(t)$.

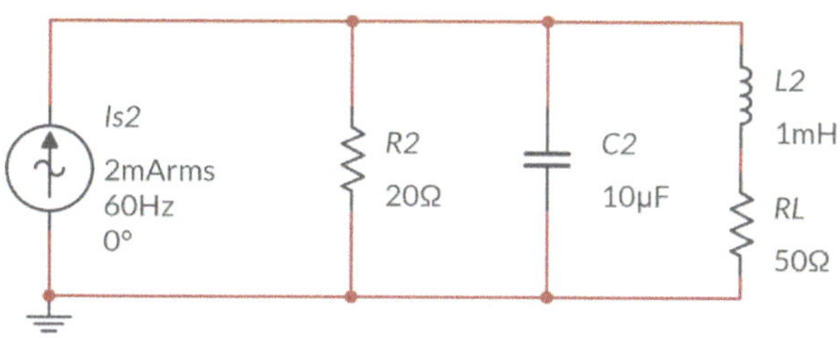

Figure 31.16 Parallel *RLC* circuit with a practical inductor and sinusoidal current generator.

Table 31.2 Summary of calculations

R_{Lp}	L_p	f_r	f_1	f_2	BW	Q_r	$\overline{V}(f_r) = V\angle\varphi_v$

32

Low-pass Filters

32.1 Student Learning Outcomes

- Analyze the frequency response of *RC* and *RL* low-pass passive filters and understand their characteristics.
- Evaluate the cutoff frequency of *RC* and *RL* low-pass passive filters.
- Validate calculated values of parameters of *RC* and *RL* low-pass passive filters using Multisim simulation.

32.2 Prelab

A low-pass filter is an essential component in electronic circuits used to control the flow of signals based on their frequency. It selectively allows signals with frequencies below a certain *cutoff* frequency to pass through while attenuating or rejecting signals with higher frequencies. This type of filter finds applications in various fields, including audio processing, communication systems, and instrumentation. Figure 32.1 illustrates the fundamental concept where $\overline{V}_i$ represents the phasor of the input voltage and $\overline{V}_o$ represents the phasor of the output voltage, which represents the circuit's response to the input signal. The *voltage gain* $\overline{A}$ of the circuit is defined as the ratio of $\overline{V}_o$ to $\overline{V}_i$ and varies with the angular frequency ω.

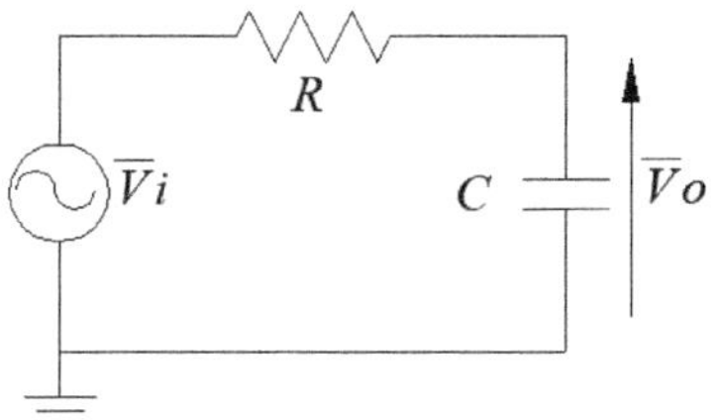

(a) RC low-pass filter

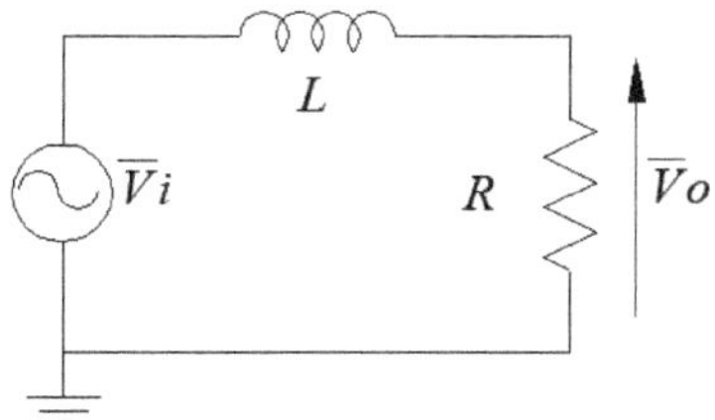

(b) RL low-pass filter

Figure 32.1 *RC* and *RL* low-pass filters.

32.2.1 Low-pass RC filter

For the *RC* filter depicted in Figure 32.1(a), the voltage gain $\overline{A}$ can be expressed as a complex function:

$$\overline{A} = \frac{\overline{V}_o}{\overline{V}_i} = \frac{1}{1 + j\omega RC} = \left|\overline{A}\right| \angle \varphi_{RC}. \tag{32.1}$$

This equation can be further separated into two distinct equations representing the magnitude $\left|\overline{A}\right|$ and the phase φ_{RC} of the voltage gain:

$$\left|\overline{A}\right| = \frac{\left|\overline{V}_o\right|}{\left|\overline{V}_i\right|} = \frac{1}{\sqrt{1 + (\omega RC)^2}}, \tag{32.2}$$

$$\varphi_{RC} = -\tan^{-1}(\omega RC). \tag{32.3}$$

Eqn (32.2) reveals that as the frequency increases, the magnitude $\left|\overline{A}\right|$ of the voltage gain decreases, resulting in the attenuation of higher-frequency components.

The arctangent function $\tan^{-1}$, also known as the inverse tangent function or atan function, is a mathematical function that calculates the angle whose tangent equals a given value. The phase angle φ_{RC} indicates the phase shift of the output signal relative to the input signal.

32.2.2 Low-pass RL filter

For the *RL* filter depicted in Figure 32.1(b), the voltage gain $\overline{A}$ can be expressed as a complex function:

$$\overline{A} = \frac{\overline{V}_o}{\overline{V}_i} = \frac{R}{R + j\omega L} = \left|\overline{A}\right| \angle \varphi_{RL}. \tag{32.4}$$

This equation can be separated into two distinct equations representing the magnitude $\left|\overline{A}\right|$ and the phase φ_{RL} of the voltage gain:

$$\left|\overline{A}\right| = \frac{\left|\overline{V}_o\right|}{\left|\overline{V}_i\right|} = \frac{R}{\sqrt{R^2 + (\omega L)^2}} = \frac{1}{\sqrt{1 + \left(\frac{\omega L}{R}\right)^2}}, \tag{32.5}$$

$$\varphi_{RL} = -\tan^{-1}\left(\frac{\omega L}{R}\right). \tag{32.6}$$

Eqn (32.5) reveals that as the frequency increases, the magnitude of the voltage gain decreases, resulting in the attenuation of higher-frequency components.

The phase angle φ_{RL} indicates the phase shift of the output signal relative to the input signal.

32.2.3 Cutoff frequency f_b

The *cutoff* frequency, denoted as f_b, is a crucial parameter in the analysis of low-pass filters. It represents the frequency at which the magnitude of the voltage gain is reduced to $\frac{1}{\sqrt{2}}$ or approximately 0.707.

For the *RC* filter, the cutoff frequency f_b is determined by equating the magnitude of the voltage gain to $\frac{1}{\sqrt{2}}$.

$$|\overline{A}| = \frac{1}{\sqrt{1 + (2\pi f_b RC)^2}} = \frac{1}{\sqrt{2}}. \tag{32.7}$$

Solving for f_b yields the following expression:

$$f_b = \frac{1}{2\pi RC}. \tag{32.8}$$

Similarly, for the *RL* filter, the cutoff frequency f_b is determined by equating the magnitude of the voltage gain to $\frac{1}{\sqrt{2}}$.

$$|\overline{A}| = \frac{1}{\sqrt{1 + \left(\frac{2\pi f_b L}{R}\right)^2}} = \frac{1}{\sqrt{2}}. \tag{32.9}$$

Solving for f_b yields the following expression:

$$f_b = \frac{R}{2\pi L}. \tag{32.10}$$

At the cutoff frequency, the phase of the complex voltage gain is –45° for both *RC* and *RL* filters. This means that the output voltage is shifted in phase by –45° with respect to the input voltage.

$$\left.\frac{\overline{V}_o}{\overline{V}_i}\right|_{f_b} = 0.707\angle -45^\circ. \tag{32.11}$$

The cutoff frequency allows us to determine the frequency range over which a low-pass filter will attenuate, or reject, high-frequency signals.

32.2.4 Decibel (dB)

The magnitudes of the voltage gain in the low-pass filters can also be expressed in decibels (dB). Decibels are a logarithmic unit used to represent the relative difference between two values. The decibel representation of the voltage gain, denoted as A_{dB}, is calculated using the following formula:

$$A_{\text{dB}} = 20 \cdot \log\left(\left|\overline{A}\right|\right) . \tag{32.12}$$

At the cutoff frequency, the magnitude of the voltage gain in decibels is –3 dB.

$$A_{\text{dB}}\big|_{f_b} = -3 \text{ dB}. \tag{32.13}$$

This indicates that the output voltage is reduced to $1/\sqrt{2}$ of the input voltage level.

In terms of frequency response of the low-pass filter, a *flat response* corresponds to a voltage gain of 0 dB per decade. This means that the magnitude of the voltage gain remains constant as the frequency changes logarithmically by a factor of 10. However, as the frequency increases beyond the cutoff frequency, the voltage gain starts to decrease, with a slope of –20 dB per decade. This indicates that the filter attenuates the amplitudes of higher-frequency components in the input signal.

32.2.5 Problem 1

In the given circuit depicted in Figure 32.2, a sinusoidal voltage generator $v_{i1}(t) = \sqrt{2} \cdot 10 \cdot \sin(\omega t)$ V is connected to an *RC* low-pass filter. To analyze the circuit, perform the following tasks:

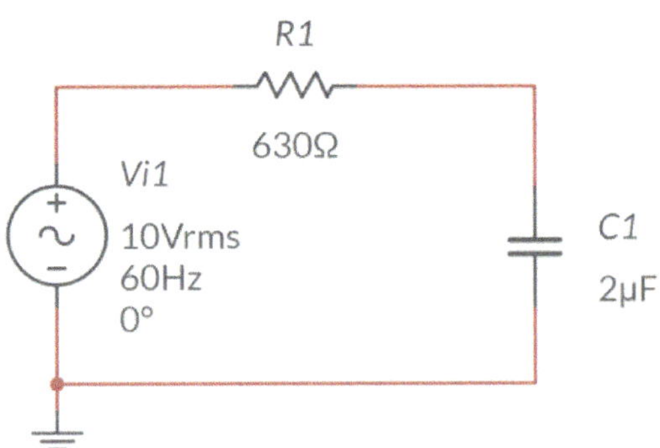

Figure 32.2 *RC* low-pass filter with a voltage generator.

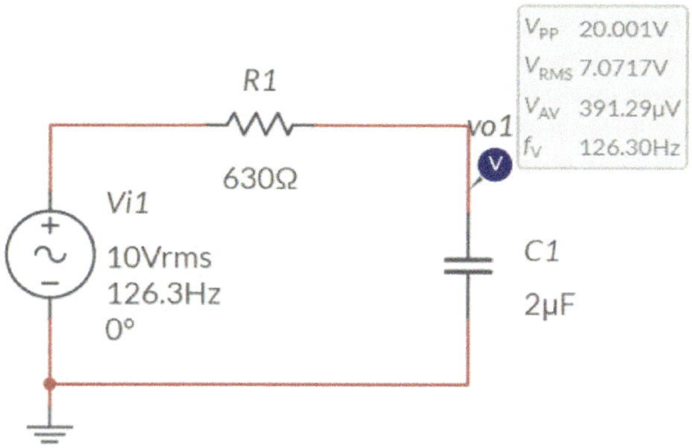

Figure 32.3 Multisim simulation confirming the *rms* value of $v_{01}(t)$ at cutoff frequency.

1. Determine the cutoff frequency f_b of the *RC* low-pass filter. Verify the calculation with Multisim simulation.
2. Calculate the *rms* values of the output voltage $v_{01}(t)$ across the capacitor C_1 for various frequencies: f_b, $10 \cdot f_b$, $10^2 \cdot f_b$, $10^3 \cdot f_b$, and $10^4 \cdot f_b$. Verify your calculations using Multisim simulation (*AC Sweep*[1]) on the simulated circuit.

32.2.5.1 Solutions to problem 1

1. The cutoff frequency f_b can be calculated using the following formula:

$$f_b = \frac{1}{2\pi RC} = 126.3 \text{ Hz}. \tag{32.14}$$

To verify the calculation with Multisim simulation, let us determine the phasor $\overline{V}_{o1}$ representing the output voltage at the cutoff frequency:

$$\overline{V}_{o1}\Big|_{f_b} = 0.707\angle - 45° \cdot \overline{V}_{i1} = 7\,\angle - 45° \text{ V}. \tag{32.15}$$

The calculated *rms* value of $v_{01}(t)$ is confirmed by the Multisim simulation, as shown in Figure 32.3. This agreement between the calculated and simulated results validates the accuracy of the calculation.

The Multisim plot of Figure 32.4 illustrates the waveform $v_{i1}(t)$ and $v_{o1}(t)$.

[1]In Multisim, the *AC* Sweep simulation calculates the frequency response of linear circuits.

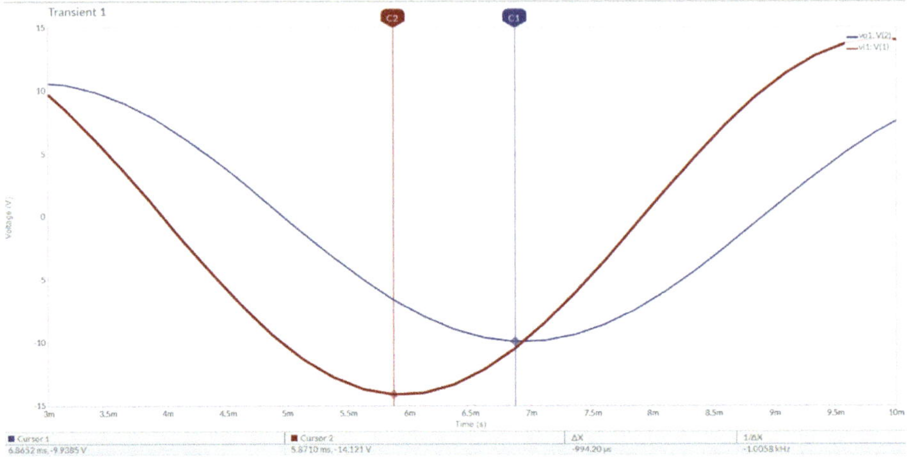

Figure 32.4 Multisim simulation confirming the calculated phase angle of the output voltage $v_{01}\,(t)$.

To estimate the phase difference $\Delta\varphi$ in degrees between the waveforms, we can calculate the time difference Δt between corresponding points on the plots. The formula to convert the time difference to phase difference in degrees is given by

$$\Delta\varphi = \frac{360^\circ}{T} \cdot \Delta t = \frac{360^\circ}{7.9 \cdot 10^{-3}} \cdot \left(-994.2 \cdot 10^{-6}\right) = -45.3^\circ. \qquad (32.16)$$

This estimate confirms the theoretical result for the phase angle.

2. The *rms* value of the output voltage $\overline{V}_{o1}$ at different frequencies can be calculated with the following formula:

$$\left|\overline{V}_{o1}\right| = \frac{\left|\overline{V}_{i1}\right|}{\sqrt{1 + (2\pi f RC)^2}}. \qquad (32.17)$$

Results are shown in Table 32.1.

The calculations demonstrate that as the frequency increases by multiples of 10 of the cutoff frequency, the magnitude of the output voltage ($\left|\overline{V}_{o1}\right|$) decreases by a factor of 10. This relationship is in line with the behavior of a low-pass filter, where higher frequencies are attenuated more significantly compared to lower frequencies.

Table 32.1 Summary of calculations

f_b = 126.3 Hz	$\lvert\overline{V}_{o1}\rvert$ (V)
f_b	7
$10{\cdot}f_b$	1
$10^2{\cdot}f_b$	10^{-1}
$10^3{\cdot}f_b$	10^{-2}
$10^4{\cdot}f_b$	10^{-3}

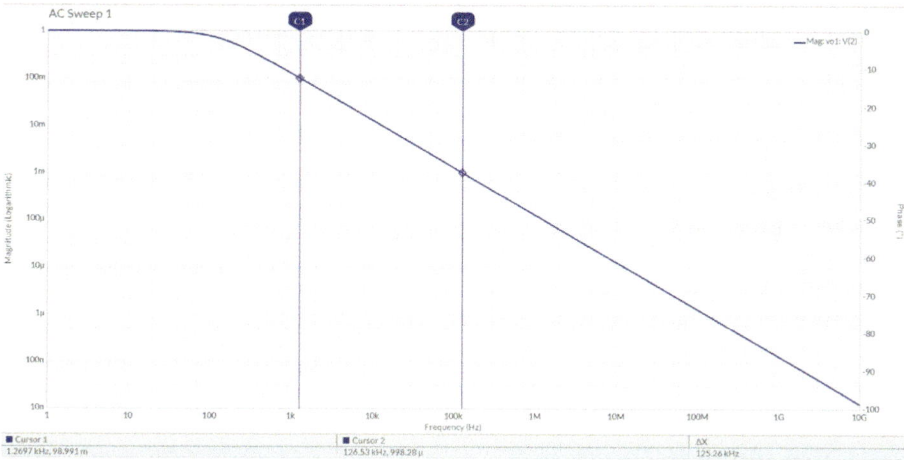

Figure 32.5 Multisim simulation confirming the validity of the calculations for the magnitude of the output voltage $\lvert\overline{V}_{o1}\rvert$.

The Multisim simulation, as illustrated in Figure 32.5, confirms the validity of the calculations. The simulation results align precisely with the expected behavior, exhibiting a decrease in the output voltage magnitude by a factor of 10 as the frequency increases by a factor of 10. This agreement between the calculated values and the simulated results provides strong evidence of the accuracy of the analysis.

The simulated output voltage magnitudes align with the calculated values, further confirming the expected decrease in magnitude as the frequency increases.

32.2.6 Problem 2

In the given circuit depicted in Figure 32.6, a sinusoidal voltage generator $v_{i2}\,(t) = \sqrt{2}{\cdot}10 \cdot \sin\,(\omega t)$ V is connected to an *LC* low-pass filter. To analyze the circuit, perform the following tasks:

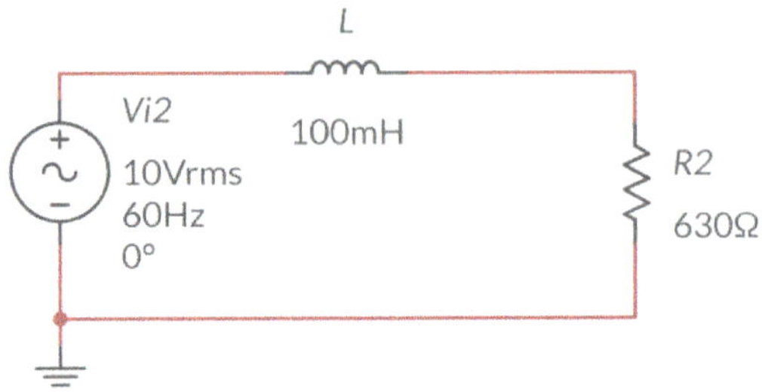

Figure 32.6 *RL* low-pass filter with a voltage generator.

1. Determine the cutoff frequency f_b of the *RL* low-pass filter. Verify the calculation with Multisim simulation.
2. Calculate the *rms* values of the output voltage $v_{02}(t)$ across the resistor R_2 for various frequencies: f_b, $10 \cdot f_b$, $10^2 \cdot f_b$, $10^3 \cdot f_b$, and $10^4 \cdot f_b$. Verify your calculations using Multisim simulation (*AC Sweep*) on the simulated circuit.
3. Calculate the voltage gain A_{dB} expressed in decibels for f_1 = 10 kHz and f_2 = 100 kHz. Verify your calculations using Multisim simulation (*AC Sweep*) on the simulated circuit.

32.2.6.1 Solutions to Problem 2

1. The cutoff frequency f_b can be calculated using the following formula:

$$f_b = \frac{R}{2\pi L} = 1 \text{ kHz}. \tag{32.18}$$

To verify the calculation with Multisim simulation, let us determine the phasor $\overline{V}_{o2}$ representing the output voltage at the cutoff frequency:

$$\overline{V}_{o2}\Big|_{f_b} = 0.707\angle - 45^\circ \cdot \left|\overline{V}_{i2}\right|_{f_b} \angle 0^\circ = 7 \angle - 45^\circ \text{ V}. \tag{32.19}$$

The calculated *rms* value of $\overline{V}_{o2}$ is confirmed by the Multisim simulation, as shown in Figure 32.7. This agreement between the calculated and simulated results validates the accuracy of the calculation.

The Multisim plot of Figure 32.8 illustrates the waveform $v_{i2}(t)$ and $v_{o2}(t)$.

To estimate the phase difference $\Delta\varphi$ in degrees between the waveforms, we can calculate the time difference Δt between corresponding points on the plots. The formula to convert the time difference to phase difference in

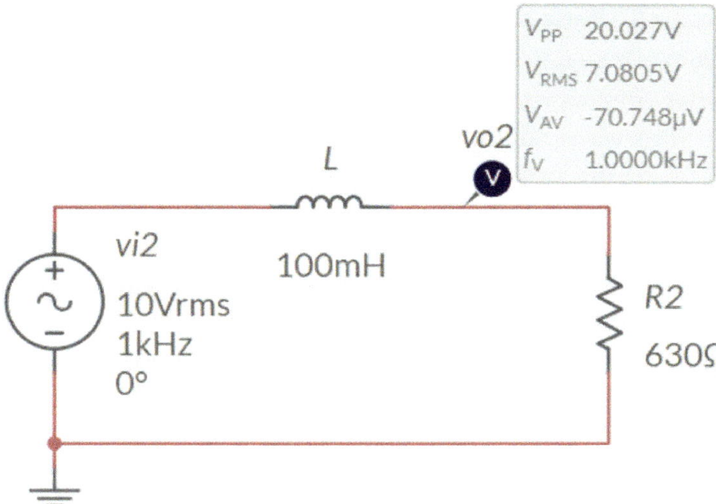

Figure 32.7 Multisim simulation confirming the *rms* value of $v_{02}(t)$.

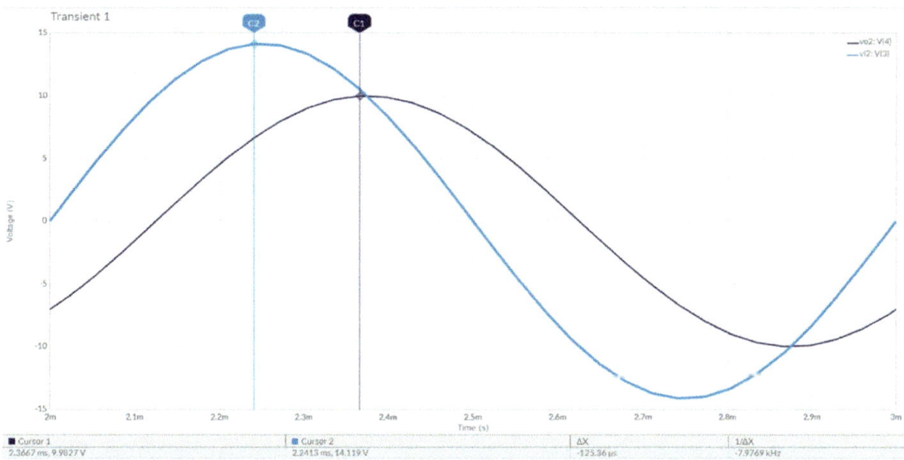

Figure 32.8 Multisim simulation confirming the calculated phase angle of $\overline{V}_{o2}$.

degrees is given by

$$\Delta\varphi = \frac{360^\circ}{T} \cdot \Delta t = \frac{360^\circ}{10^{-3}} \cdot \left(-125.36 \cdot 10^{-6}\right) = -45.1^\circ. \tag{32.20}$$

This estimate confirms the theoretical result.

Table 32.2 Summary of calculations

f_b = 1 kHz	$\left\|\overline{V}_{o2}\right\|$ (V)
f_b	7
$10{\cdot}f_b$	1
$10^2{\cdot}f_b$	10^{-1}
$10^3{\cdot}f_b$	10^{-2}
$10^4{\cdot}f_b$	10^{-3}

2. The *rms* value of the output voltage $\overline{V}_{o2}$ at different frequencies can be calculated with the following formula:

$$\left|\overline{V}_{o2}\right| = \frac{\left|\overline{V}_{i2}\right|}{\sqrt{1+\left(\frac{2\pi fL}{R}\right)^2}}. \tag{32.21}$$

Results are shown in Table 32.2.

The calculations demonstrate that as the frequency increases by multiples of 10 of the cutoff frequency, the magnitude of the output voltage, denoted as $\left|\overline{V}_{o2}\right|$, decreases by a factor of 10. This relationship is in line with the behavior of a low-pass filter, where higher frequencies are attenuated more significantly compared to lower frequencies.

The Multisim simulation, as illustrated in Figure 32.9, confirms the validity of the calculations.

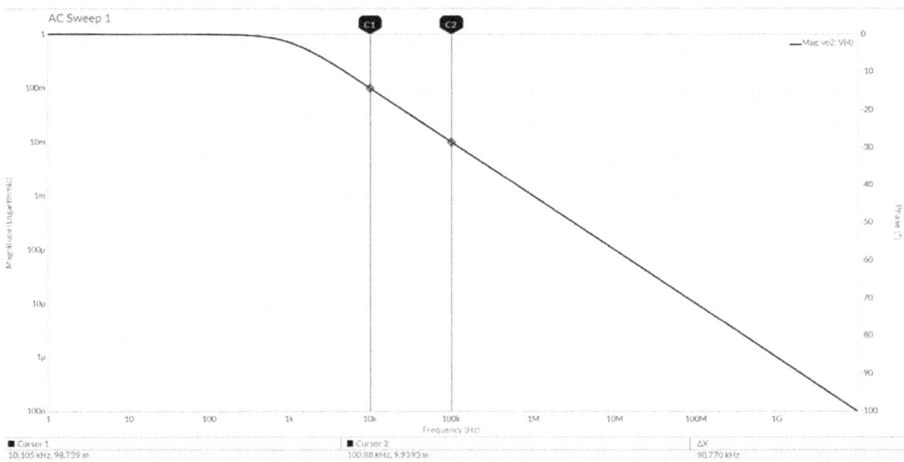

Figure 32.9 Multisim simulation confirming the calculations for the output voltage $\left|\overline{V}_{o2}\right|$.

The simulation results align precisely with the expected behavior, exhibiting a decrease in the output voltage magnitude by a factor of 10 as the frequency increases by a factor of 10. This agreement between the calculated values and the simulated results provides strong evidence of the accuracy of the analysis.

3. The voltage gain A_{dB} expressed in decibels at f_1 = 10 kHz and f_2 = 100 kHz can be calculated as follows:

$$\left|\overline{A}\right|_{\mathrm{dB}} = 20\log\left(\frac{1}{\sqrt{1+\left(\frac{2\pi fL}{R}\right)^2}}\right). \tag{32.22}$$

Substituting the respective frequencies into the equation, we obtain

$$A_{\mathrm{dB}}\big|_{10\ \mathrm{kHz}} = -20\ \mathrm{dB}, \tag{32.23}$$

$$A_{\mathrm{dB}}\big|_{100\ \mathrm{kHz}} = -40\ \mathrm{dB}. \tag{32.24}$$

The Multisim simulation results, as depicted in Figure 32.10, confirm the accuracy of these calculations.

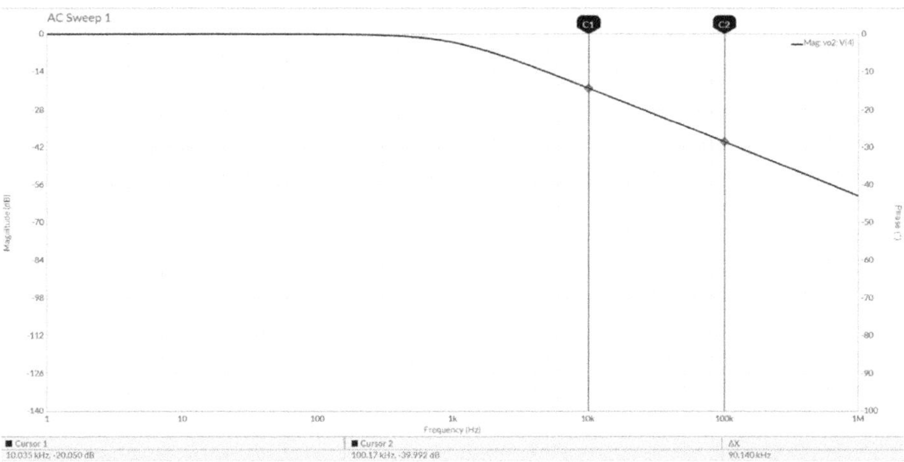

Figure 32.10 Multisim simulation confirming the validity of the voltage gain A_{dB} at the frequencies 10 kHz and 100 kHz.

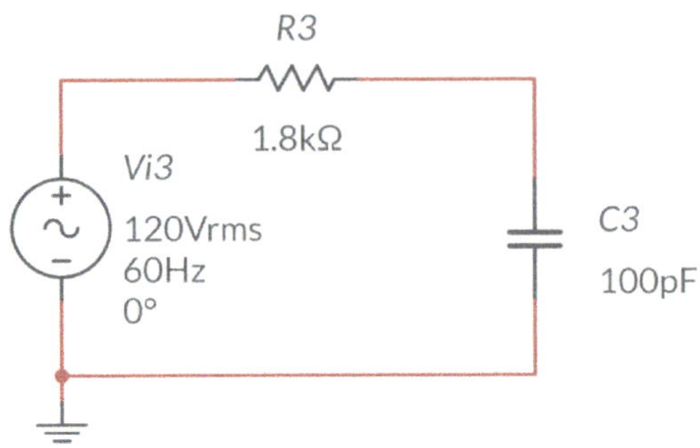

Figure 32.11 *RC* low-pass filter with a voltage generator.

Table 32.3 Summary of calculations

f_b =	$\left\|V_{o3}\right\|$ (V)
f_b	
$10{\cdot}f_b$	
$10^2{\cdot}f_b$	
$10^3{\cdot}f_b$	
$10^4{\cdot}f_b$	

32.3 Lab Procedure 1

In the given circuit depicted in Figure 32.11, a sinusoidal voltage generator $v_{i3}\left(t\right) = \sqrt{2} \cdot 120 \cdot \sin\left(\omega t\right)$ V is connected to an *RC* low-pass filter. To analyze the circuit, perform the following tasks:

1. Determine the cutoff frequency f_b of the low-pass *RC* filter. Verify the calculation with Multisim simulation. Record result in Table 32.3.
2. Calculate the *rms* values of the output voltage $v_{03}(t)$ across the capacitor C_3 for various frequencies: f_b, $10{\cdot}f_b$, $10^2 \cdot f_b$, $10^3 \cdot f_b$, and $10^4 \cdot f_b$. Verify your calculations using Multisim simulation (*AC Sweep*) on the simulated circuit. Record results in Table 32.3.

32.4 Lab Procedure 2

In the given circuit depicted in Figure 32.12, a sinusoidal voltage generator $v_{i4}\left(t\right) = \sqrt{2} \cdot 120 \cdot \sin\left(\omega t\right)$ V is connected to an *LC* low-pass filter. To analyze the circuit, perform the following tasks:

1. Determine the cutoff frequency f_b of the *RL* low-pass filter. Verify the calculation with Multisim simulation. Record results in Table 32.4.

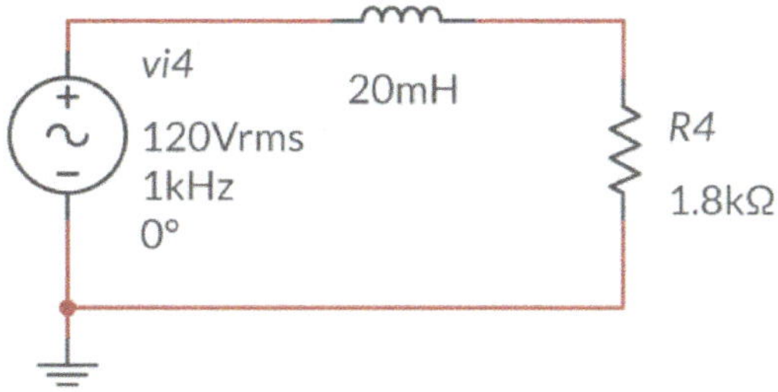

Figure 32.12 *RL* low-pass filter with a voltage generator.

Table 32.4 Summary of calculations

f_b =	$\overline{V_{o4}}$ (V)
f_b	
$10{\cdot}f_b$	
$10^2{\cdot}f_b$	
$10^3{\cdot}f_b$	
$10^4{\cdot}f_b$	

2. Calculate the *rms* values of the output voltage $v_{04}(t)$ across the resistor R_4 for various frequencies: f_b, $10{\cdot}f_b$,$10^2 \cdot f_b$, $10^3 \cdot f_b$, and $10^4 \cdot f_b$. Verify your calculations using Multisim simulation (*AC Sweep*) on the simulated circuit. Record results in Table 32.4.
3. Calculate the voltage gain A_{dB} expressed in decibels for f_1 = 10 kHz and f_2 = 100 kHz. Verify your calculations using Multisim simulation (*AC Sweep*) on the simulated circuit.

33

High-pass Filters

33.1 Student Learning Outcomes

- Analyze the frequency response of *RC* and *RL* high-pass passive filters and understand their characteristics.
- Evaluate the cutoff frequency of *RC* and *RL* high-pass passive filters.
- Validate calculated values of parameters of *RC* and *RL* low-pass passive filters using Multisim simulation.

33.2 Prelab

A high-pass filter operates differently from a low-pass filter as it allows signals with frequencies higher than a specific cutoff value to pass through while attenuating signals with frequencies lower than the *cutoff* frequency. Unlike low-pass filters, high-pass filters do not pass DC signals due to the capacitor blocking the DC component.

In Figure 33.1, the phasor $\overline{V}_i$ represents the input voltage and the phasor $\overline{V}_o$ the output voltage across the resistor *R*, which is the circuit's response to the input. The *voltage gain* $\overline{A}$ is defined as the ratio of $\overline{V}_o$ to $\overline{V}_i$ and depends on the angular frequency ω.

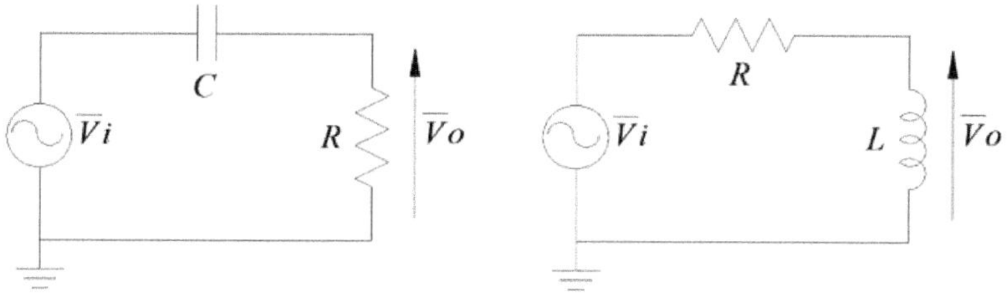

(a) RC high-pass filters (b) RL high-pass filters

Figure 33.1 *RC* and *RL* high-pass filters.

33.2.1 High-pass RC filter

For the *RC* filter shown in Figure 33.1(a), the voltage gain $\overline{A}$ is given by a complex function involving the angular frequency ω and the *RC* time constant.

$$\overline{A} = \frac{\overline{V}_o}{\overline{V}_i} = \frac{j\omega RC}{1 + j\omega RC} = \left|\overline{A}\right| \angle\varphi_{RC}. \tag{33.1}$$

It is worth noting that $\overline{A}\,(0) = 0$ and $\overline{A}\,(\infty) = 1$[1].

This equation can be further separated into two distinct equations representing the magnitude $\left|\overline{A}\right|$ and the phase φ_{RC} of the voltage gain:

$$\left|\overline{A}\right| = \frac{\left|\overline{V}_o\right|}{\left|\overline{V}_i\right|} = \frac{\omega RC}{\sqrt{1 + (\omega RC)^2}} \tag{33.2}$$

$$\varphi_{RC} = 90^\circ - \tan^{-1}(\omega RC). \tag{33.3}$$

Eqn (33.2) reveals that as the frequency decreases, the magnitude $\left|\overline{A}\right|$ of the voltage gain decreases, resulting in the attenuation of lower-frequency components.

The arctangent function $\tan^{-1}$, also known as the inverse tangent function or atan function, is a mathematical function that calculates the angle whose tangent equals a given value. The phase angle φ_{RC} indicates the phase shift of the output signal relative to the input signal.

The *cutoff frequency* f_b is the frequency at which the magnitude of the voltage gain is $1/\sqrt{2}$.

$$\left|\overline{A}\right| = \frac{2\pi f RC}{\sqrt{1 + (2\pi f RC)^2}} = \frac{1}{\sqrt{2}}. \tag{33.4}$$

Solving for f_b yields the following expression:

$$f_b = \frac{1}{2\pi RC}. \tag{33.5}$$

33.2.2 High-pass RL filter

The high-pass *RL* filter shown in Figure 33.1(b) exhibits a voltage gain $\overline{A}$ that can be expressed as a complex function:

$$\overline{A} = \frac{\overline{V}_o}{\overline{V}_i} = \frac{j\omega L}{R + j\omega L} = \left|\overline{A}\right| \angle\varphi_{RL}. \tag{33.6}$$

---

[1] $\lim\limits_{\omega\to\infty} \frac{1}{\frac{1}{\omega}+1} = 1.$

This equation can be further separated into two distinct equations representing the magnitude $|\overline{A}|$ and the phase φ_{RL} of the voltage gain:

$$|\overline{A}| = \frac{|\overline{V}_o|}{|\overline{V}_i|} = \frac{\omega L}{\sqrt{R^2 + (\omega L)^2}} \tag{33.7}$$

$$\varphi_{RL} = 90° - \tan^{-1}\left(\frac{\omega L}{R}\right). \tag{33.8}$$

The *cutoff frequency* f_b for the *RL* filter, representing the frequency at which the magnitude of the voltage gain is $1/\sqrt{2}$, is determined by the following equation:

$$|\overline{A}| = \frac{2\pi f L}{\sqrt{R^2 + (2\pi f L)^2}} = \frac{1}{\sqrt{2}}. \tag{33.9}$$

Solving for f_b yields the following expression:

$$f_b = \frac{R}{2\pi L}. \tag{33.10}$$

As the resistance increases or the inductance decreases, the cutoff frequency shifts to higher values, indicating that the *RL* high-pass filter allows higher-frequency signals to pass while attenuating lower-frequency signals.

For both *RC* and *RL* filters, when the frequency of the input signal reaches the cutoff frequency, the phase angle of the voltage gain $\overline{A}$ is 45°. This can be represented as follows:

$$\overline{A}\Big|_{f_b} = 0.707\ \angle 45°. \tag{33.11}$$

The phase angle of 45° signifies that at the cutoff frequency, the output signal has phase shift of 45° with respect to the input signal.

33.2.3 Decibel (dB)

The magnitudes of the voltage gain for the high-pass filters can also be expressed in decibels (dB). Decibels are a logarithmic unit used to represent the relative difference between two values. The decibel representation of the voltage gain, denoted as A_{dB}, is calculated using the following formula:

$$A_{\text{dB}} = 20 \cdot \log\left(|\overline{A}|\right). \tag{33.12}$$

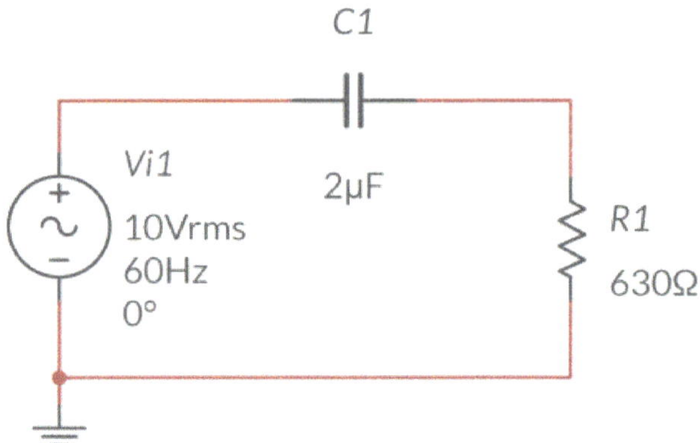

Figure 33.2 *RC* high-pass filter with a voltage generator.

At the cutoff frequency, the magnitude of the voltage gain in decibels is -3 dB.

$$A_{\mathrm{dB}}\big|_{f_b} = -3 \text{ dB}. \tag{33.13}$$

This indicates that the output voltage is reduced to $1/\sqrt{2}$ of the input voltage level.

In terms of frequency response of the high-pass filter, the voltage gain A_{dB} as a function of the frequency ramps up with a slope of 20 dB per decade until it reaches the -3 dB in correspondence with the cutoff frequency.

A *flat response* corresponds to a voltage gain of 0 dB per decade. This means that the magnitude of the voltage gain remains constant as the frequency increases logarithmically by a factor of 10. This indicates that the filter attenuates the amplitudes of lower-frequency components in the input signal.

33.2.4 Problem 1

In the given circuit depicted in Figure 33.2, a sinusoidal voltage generator $v_{i1}(t) = \sqrt{2} \cdot 10 \cdot \sin(\omega t)$ V is connected to a high-pass *RC* filter. To analyze the circuit, perform the following tasks:

1. Determine the cutoff frequency f_b of the RC high-pass filter. Verify the calculation with Multisim simulation.
2. Calculate the *rms* values of the output voltage $v_{01}(t)$ across the resistor R_1 for various frequencies: f_b, $10^{-1} \cdot f_b$, $10^{-2} \cdot f_b$, $10^{-3} \cdot f_b$, and $10^{-4} \cdot f_b$. Verify your calculations using Multisim simulation (*AC Sweep*) on the simulated circuit.

3. Calculate the voltage gain A_{dB} expressed in decibels for f_1 = 1 Hz and f_2 = 10 Hz. Verify your calculations using Multisim simulation (*AC Sweep*) on the simulated circuit.

33.2.4.1 Solutions to Problem 1

1. The cutoff frequency f_b can be calculated using the following formula:

$$f_b = \frac{1}{2\pi RC} = 126.3 \text{ Hz}. \tag{33.14}$$

To verify the calculation with Multisim simulation, let us determine the phasor $\overline{V}_{o1}$ representing the output voltage at the cutoff frequency:

$$\overline{V}_{o1}\Big|_{f_b} = 0.707\angle 45° \cdot \left|\overline{V}_{i1}\right|_{f_b} \angle 0° = 7.07\angle 45° \text{ V}. \tag{33.15}$$

The calculated *rms* value of $v_{01}(t)$ is confirmed by the Multisim simulation conducted at the cutoff frequency, as shown in Figure 33.3.

The Multisim plot of Figure 33.4 illustrates the waveforms $v_{i1}(t)$ and $v_{o1}(t)$.

To estimate the phase difference $\Delta\varphi$ in degrees between the waveforms, we can calculate the time difference Δt between corresponding points on

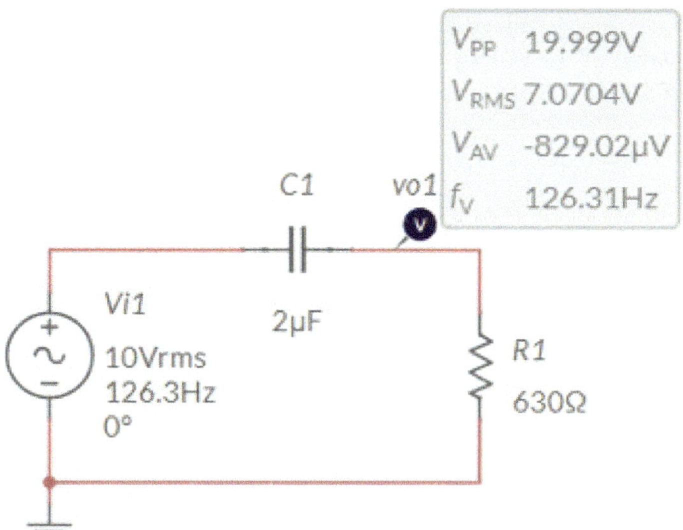

Figure 33.3 Multisim simulation confirming the *rms* value of $v_{01}(t)$ at cutoff frequency.

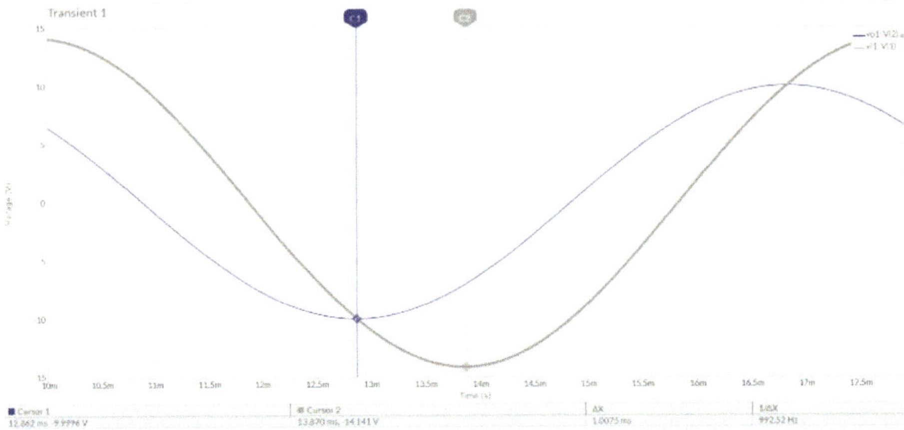

Figure 33.4 Multisim plot illustrating the waveforms $v_{i1}(t)$ and $v_{o1}(t)$.

the plots. The formula to convert the time difference to phase difference in degrees is given by

$$\Delta\varphi = \frac{360°}{T} \cdot \Delta t = \frac{360°}{7.9 \cdot 10^{-3}} \cdot 10^{-3} = 45.5°. \tag{33.16}$$

This estimate confirms the theoretical result for the phase angle of $v_{o1}(t)$ at resonance.

2. The *rms* value of the output voltage $\overline{V}_{o1}$ at different frequencies can be calculated with the following formula:

$$\left|\overline{V}_{o1}\right| = \left|\overline{V}_{i1}\right| \frac{2\pi f RC}{\sqrt{1+(2\pi f RC)^2}}. \tag{33.17}$$

Results are shown in Table 33.1.

The calculations demonstrate that as the frequency decreases by multiples of 10 of the cutoff frequency, the magnitude of the output voltage denoted

Table 33.1 Summary of calculations

f_b = 126.3 Hz	$\left|\overline{V}_{o1}\right|$ (V)
f_b	7.01
$10^{-1} \cdot f_b$	1
$10^{-2} \cdot f_b$	10^{-1}
$10^{-3} \cdot f_b$	10^{-2}
$10^{-4} \cdot f_b$	10^{-3}

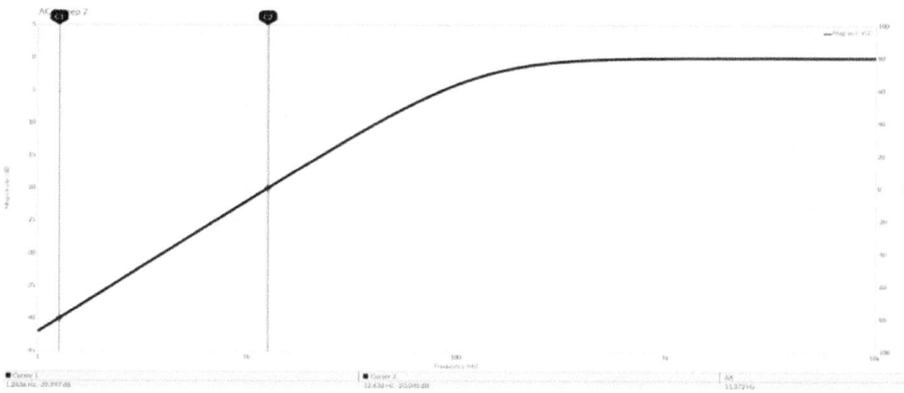

Figure 33.5 Multisim simulation confirming the validity of the calculations for the magnitude of the output voltage $|\overline{V}_{o1}|$.

as $|\overline{V}_{o1}|$ decreases by a factor of 10. This relationship is in line with the behavior of a high-pass filter, where lower frequencies are attenuated more significantly compared to higher frequencies.

The Multisim simulation results in Figure 33.5 align precisely with the calculation, exhibiting a decrease in the output voltage magnitude by a factor of 10 as the frequency decreases by a factor of 10.

3. The voltage gain A_{dB} expressed in decibels at f_1 = 1 Hz and f_2 = 10Hz can be calculated as follows:

$$|\overline{A}|_{\text{dB}} = 20\log\left(\frac{2\pi fRC}{\sqrt{1+(2\pi fRC)^2}}\right). \tag{33.18}$$

Substituting the respective frequencies into the equation, we obtain:

$$A_{\text{dB}}|_{1\ \text{Hz}} = -42\ \text{dB} \tag{33.19}$$

$$A_{\text{dB}}|_{10\ \text{Hz}} = -22.1\ \text{dB}. \tag{33.20}$$

The Multisim simulation results, as depicted in Figure 33.6, confirm the accuracy of these calculations.

33.2.5 Problem 2

In the given circuit depicted in Figure 33.7, an *RL* high-pass filter is connected to a sinusoidal voltage generator $v_{i2}(t) = \sqrt{2}\cdot 10\cdot \sin(\omega t)$ V. To analyze the circuit, perform the following tasks:

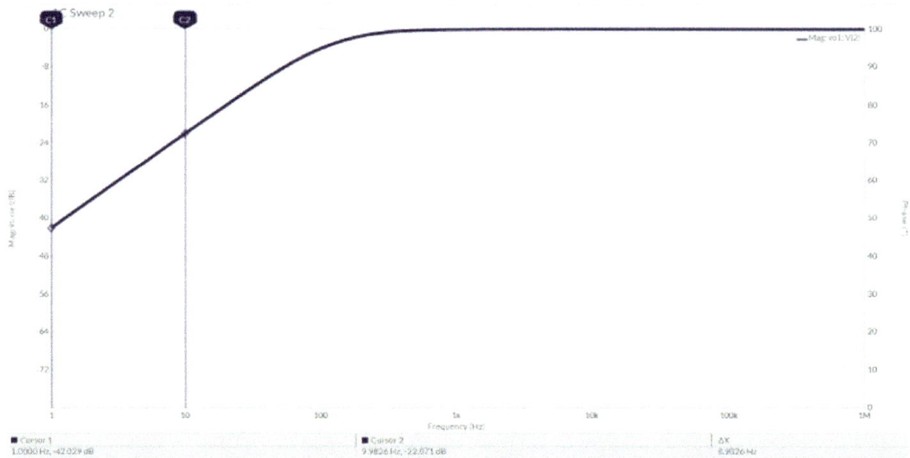

Figure 33.6 Multisim simulation confirming the validity of the voltage gain A_{dB} at the frequencies 1 and 10 Hz.

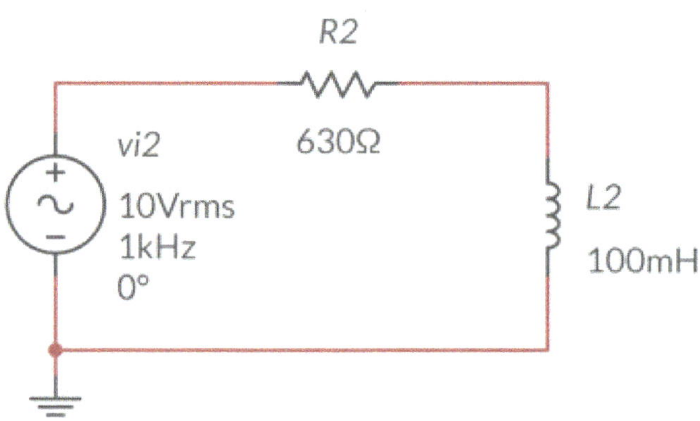

Figure 33.7 *RL* high-pass filter with a AC voltage generator.

1. Determine the cutoff frequency f_b of the *RL* low-pass filter. Verify the calculation with Multisim simulation.
2. Calculate the *rms* values of the output voltage $v_{02}(t)$ across the inductor L_2 for various frequencies: f_b, $10^{-1}\cdot f_b$, $10^{-2}\cdot f_b$, $10^{-3}\cdot f_b$ and $10^{-4}\cdot f_b$. Verify your calculations using Multisim simulation (*AC Sweep*) on the simulated circuit.
3. Calculate the voltage gain A_{dB} expressed in decibels for f_1 = 10 Hz and f_2 = 100 Hz. Verify your calculations using Multisim simulation (*AC Sweep*) on the simulated circuit.

4. Calculate the power P_2 dissipated across the resistor R_2 when the voltage gain is $A_{\text{dB}} = -20$ dB. Verify your calculations with measurements on the simulated circuit.

33.2.5.1 Solutions to Problem 2

1. The cutoff frequency f_b can be calculated using the following formula:

$$f_b = \frac{R}{2\pi L} = 1 \text{ kHz}. \tag{33.21}$$

To verify the calculation with Multisim simulation, let us determine the phasor $\overline{V}_{o2}$ representing the output voltage at the cutoff frequency:

$$\overline{V}_{o2}\Big|_{f_b} = 0.707 \angle 45° \cdot \left|\overline{V}_{i2}\right|_{\omega_b} \angle 0° = 7\angle 45° \text{ V}. \tag{33.22}$$

The calculated *rms* value of $v_{o2}(t)$ is confirmed by the Multisim simulation conducted at the cutoff frequency, as shown in Figure 33.8.

The Multisim plot of Figure 33.9 illustrates the waveform $v_{i2}(t)$ and $v_{o2}(t)$.

To estimate the phase difference $\Delta\varphi$ in degrees between the waveforms, we can calculate the time difference Δt between corresponding points on the plots. The formula to convert the time difference to phase difference in

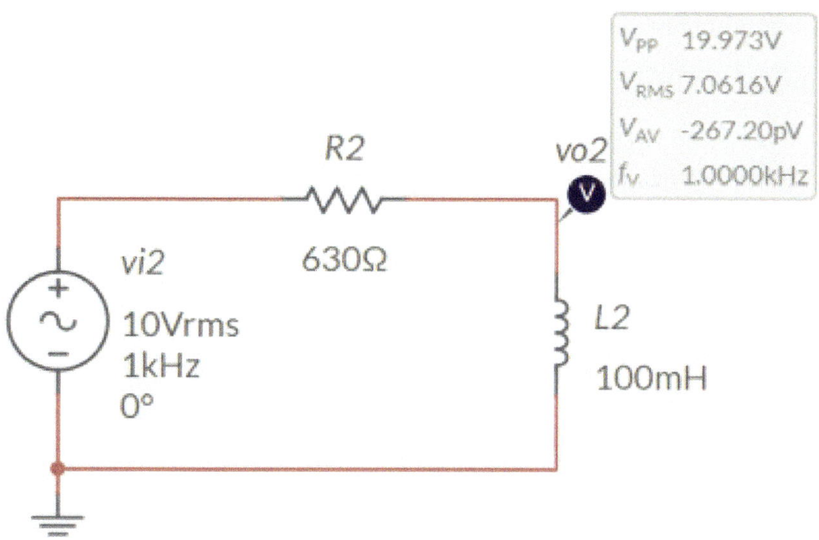

Figure 33.8 Multisim simulation confirming the *rms* value of $v_{o2}(t)$.

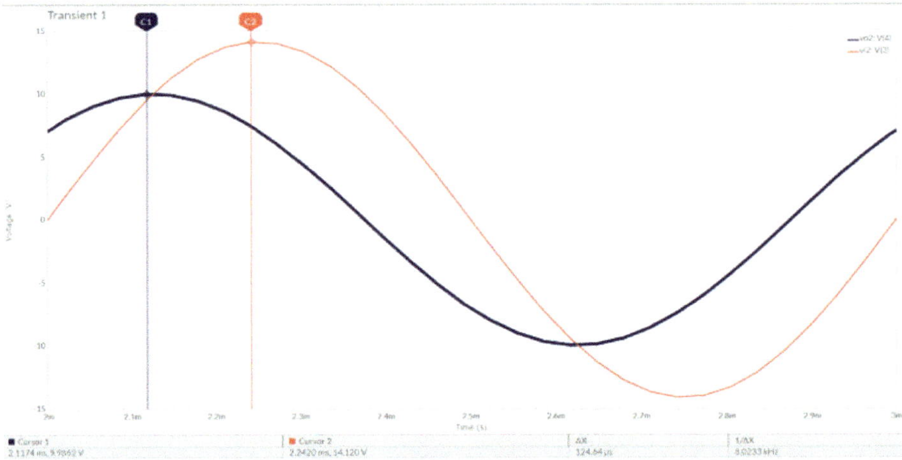

Figure 33.9 Multisim plot illustrating the waveforms $v_{i2}(t)$ and $v_{o2}(t)$.

degrees is given by

$$\Delta\varphi = \frac{360^\circ}{T}\cdot\Delta t = \frac{360^\circ}{10^{-3}}\cdot\left(124.64\cdot10^{-6}\right) = 44.87^\circ. \tag{33.23}$$

This estimate confirms the theoretical result for the phase angle of $v_{o2}(t)$ at resonance.

2. The *rms* value of the output voltage $v_{o2}(t)$ at different frequencies can be calculated with the following formula:

$$\left|\overline{V}_{o2}\right| = \left|\overline{V}_{i2}\right|\frac{2\pi L}{\sqrt{R^2+(2\pi L)^2}}. \tag{33.24}$$

Results are shown in Table 33.2.

Table 33.2 Summary of calculations

$f_b = 1$ kHz	$\left\|\overline{V}_{o2}\right\|$ (V)
f_b	7
$10^{-1}\cdot f_b$	1
$10^{-2}\cdot f_b$	10^{-1}
$10^{-3}\cdot f_b$	10^{-2}
$10^{-4}\cdot f_b$	10^{-3}

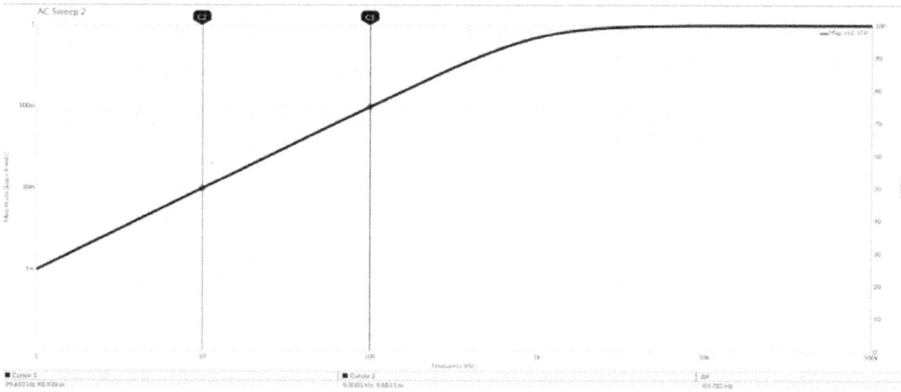

Figure 33.10 Multisim simulation confirming the validity of the calculations for the magnitude of the output voltage $\left|\overline{V}_{o2}\right|$.

The calculations demonstrate that as the frequency decreases by multiples of 10 of the cutoff frequency, the magnitude of the output voltage denoted as $\left|\overline{V}_{o2}\right|$ decreases by a factor of 10. This relationship is in line with the behavior of a high-pass filter, where lower frequencies are attenuated more significantly compared to higher frequencies.

The Multisim simulation results in Figure 33.10 align precisely with the calculation, exhibiting a decrease in the output voltage magnitude by a factor of 10 as the frequency decreases by a factor of 10.

3. The voltage gain A_{dB} expressed in decibels at f_1 = 10 Hz and f_2 = 100 Hz can be calculated as follows:

$$\left|\overline{A}\right|_{\mathrm{dB}} = 20\log\left(\frac{2\pi f L}{\sqrt{R^2 + (2\pi f L)^2}}\right). \tag{33.25}$$

Substituting the respective frequencies into the equation, we obtain

$$A_{\mathrm{dB}}\big|_{10\ \mathrm{Hz}} = -40\ \mathrm{dB} \tag{33.26}$$

$$A_{\mathrm{dB}}\big|_{100\ \mathrm{Hz}} = -20\ \mathrm{dB}. \tag{33.27}$$

The Multisim simulation results, as depicted in Figure 33.11, confirm the accuracy of these calculations.

4. To calculate the power P_2 dissipated across the resistor R_2, we first determine the phasor $\overline{V}_{R2}$ representing the voltage across R_2. $\overline{V}_{R2}$ must

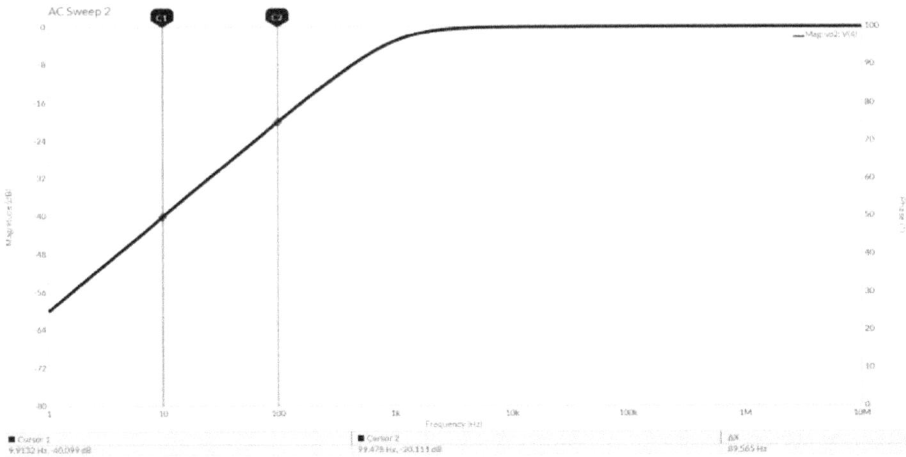

Figure 33.11 Multisim simulation confirming the validity of the voltage gain A_{dB} at the frequencies 10 and 100 Hz.

be determined at the specific frequency, f = 100 Hz, when the voltage gain is -20 dB. $\overline{V}_{R2}$ can be determined by utilizing the voltage divider formula:

$$\begin{aligned}\overline{V}_{R2} &= \overline{V}_{i2}\frac{R_2}{R_2 + j2\pi f L} = 10\angle 0^{\circ}\frac{630}{630 + j2\pi \cdot 100 \cdot 100 \cdot 10^{-3}} \\ &= 9.95\angle{-5.7^{\circ}}\ \text{V}. \end{aligned} \tag{33.28}$$

To measure with Multisim the *rms* value of the voltage $v_{R2}(t)$, we can conduct a simulation of the circuit (Figure 33.12).

The theoretical *rms* value of the voltage $v_{R2}(t)$ matches the result of the Multisim simulation.

The plot of the waveforms $v_{i2}\,(t)$ and $v_{R2}\,(t)$ at f = 100 Hz, obtained with Multisim simulation, is presented in Figure 33.13.

To estimate the phase difference $\Delta\varphi$ in degrees between the waveforms, we can calculate the time difference Δt between corresponding points on the plots. The formula to convert the time difference to phase difference in degrees is given by

$$\Delta\varphi = \frac{360^{\circ}}{T} \cdot \Delta t = \frac{360^{\circ}}{10^{-2}} \cdot \left(-160.77 \cdot 10^{-6}\right) = -5.8^{\circ}. \tag{33.29}$$

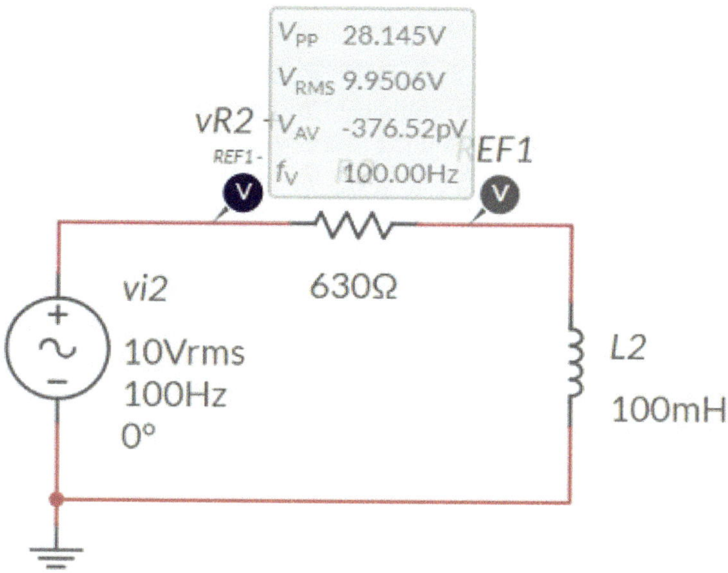

Figure 33.12 Multisim simulation confirming the *rms* value of $v_{R_2}(t)$ at the frequency f = 100 Hz.

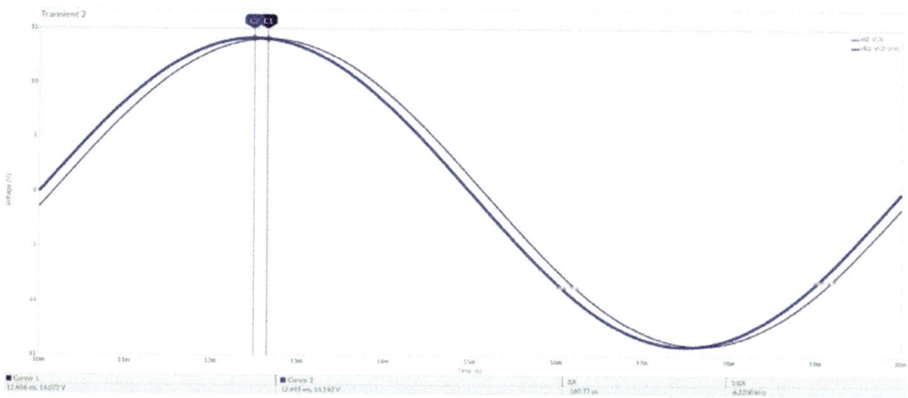

Figure 33.13 Multisim simulation confirming the calculated phase angle of $v_{R2}\ (t)$.

The plot confirms the theoretical result for the phase angle of $v_{R2}(t)$ at f = 100 Hz.

The power P_2 dissipated across the resistor R_2 can be calculated as follows:

$$P_2 = \frac{V_{R2}^2}{R_2} = \frac{9.95^2}{630} = 157\ \text{mW}. \qquad (33.30)$$

33.3 Lab Procedure 1

In the given circuit depicted in Figure 33.14, a sinusoidal voltage generator $v_{i3}(t) = \sqrt{2} \cdot 120 \cdot \sin(\omega t)$ V is connected to an *RC* high-pass filter. To analyze the circuit, perform the following tasks:

1. Determine the cutoff frequency f_b of the *RC* high-pass filter. Verify the calculation with Multisim simulation. Record results in Table 33.3.
2. Calculate the *rms* values of the output voltage $v_{03}(t)$ across the resistor R_3 for various frequencies: f_b, $10^{-1} \cdot f_b$, $10^{-2} \cdot f_b$, and $10^{-3} \cdot f_b$. Verify your calculations using Multisim simulation (*AC Sweep*). Record results in Table 33.3.
3. Calculate the voltage gain A_{dB} expressed in decibels for f_1 = 10 Hz and f_2 = 100 Hz. Verify your calculations using Multisim simulation (*AC Sweep*). Record results in Table 33.3.

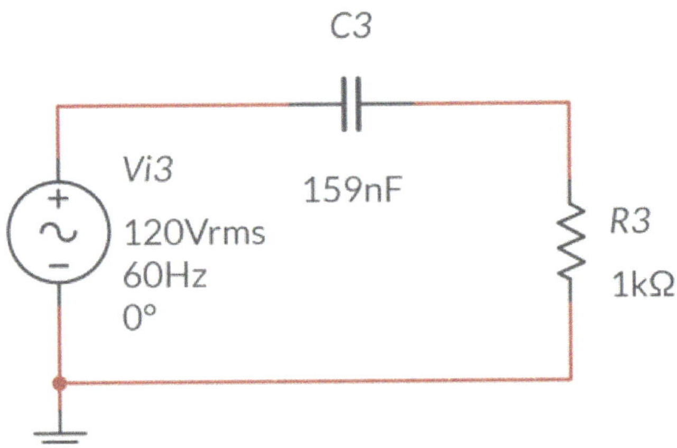

Figure 33.14 *RC* high-pass filter circuit with a sinusoidal voltage generator.

Table 33.3 Summary of calculations

f_b =	$\overline{V}_{o3}$ (V)	A_{dB}
f_b		
$0.1 \cdot f_b$		
$0.01 \cdot f_b$		
$0.001 \cdot f_b$		
f = 10 Hz		
f = 100 Hz		

33.4 Lab Procedure 2

In the given circuit depicted in Figure 33.15, a sinusoidal voltage generator $v_{i4}(t) = \sqrt{2} \cdot 120 \cdot \sin(\omega t)$ V is connected to an *RL* high-pass filter. To analyze the circuit, perform the following tasks:

1. Determine the cutoff frequency f_b of the *RL* low-pass filter. Verify the calculation with Multisim simulation. Record results in Table 33.4.
2. Calculate the *rms* values of the output voltage $v_{04}(t)$ across the inductor L_4 for various frequencies: f_b, $10^{-10} \cdot f_b$, $10^{-2} \cdot f_b$, and $10^{-3} \cdot f_b$. Verify your calculations using Multisim simulation (*AC Sweep*). Record results in Table 33.4.
3. Calculate the voltage gain A_{dB} expressed in decibels for f_1 = 10 Hz and f_2 = 100 Hz. Verify your calculations using Multisim simulation (*AC Sweep*). Record results in Table 16.4.
4. Calculate the power P_4 dissipated across the resistor R_4 when the voltage gain is $A_{\mathrm{dB}} = -20$ dB. Verify your calculations with measurements on the simulated circuit. Record results in Table 33.4.

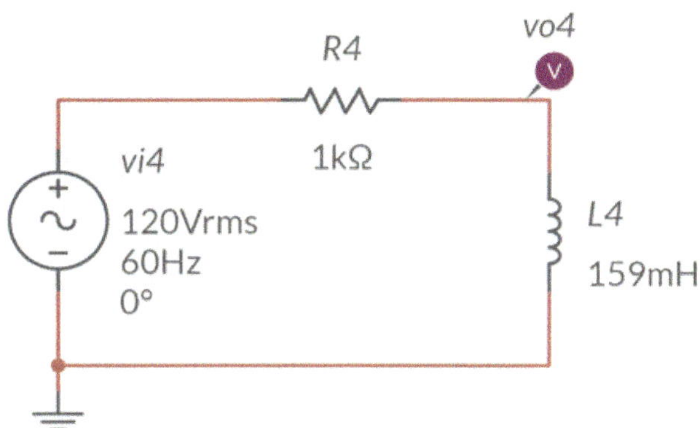

Figure 33.15 *RL* high-pass filter circuit with a sinusoidal voltage generator.

Table 33.4 Summary of calculations

f_b =	$\overline{V}_{o4}$ (V)	A_{dB}	P_4
f_b			
$0.1 \cdot f_b$			
$0.01 \cdot f_b$			
$0.001 \cdot f_b$			
f = 10 Hz			
f = 100 Hz			

34

The Transformer and the Impedance Matching

34.1 Student Learning Outcomes

- Measure and analyze the magnitude and phase relationship between the primary and secondary quantities of transformers using appropriate Multisim simulations.
- Analyze and measure with Multisim simulation transformer voltages and currents in relation to the turns ratio, understanding the concept of voltage transformation and the effect of turns ratio on the magnitude of voltages and currents.
- Apply the concept of matching the internal impedance of a source and a load to achieve maximum power transfer, with specific reference to transformer impedance on either side.

34.2 Prelab

A transformer is a fundamental device in electrical engineering, consisting of two magnetically coupled windings, the primary and secondary, that enable the transfer of energy from one side to the other without the need for a physical connection. The primary and secondary windings are indicated by solid dots, which provide information about the phase relationship between the voltages.

When both solid dots are shown on the top of the transformer symbol (Figure 34.1), it signifies that the primary and secondary voltages are in phase (i.e., with a phase difference of 0°). Conversely, if one solid dot is on the top and the other is on the bottom, it indicates a 180° phase difference between the primary and secondary voltages. This phase relationship is crucial for understanding the behavior and operation of transformers.

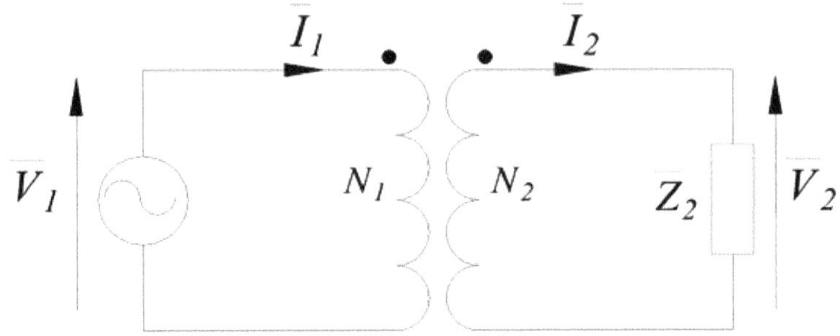

Figure 34.1 Transformer circuit.

The *turns ratio*, denoted as $a = \frac{N_1}{N_2}$, and calculated as the ratio of the number of primary turns ($N1$) to the number of secondary turns ($N2$), is an essential parameter in transformer analysis. In an ideal transformer, where there are no losses, the primary and secondary electric powers are equal, as expressed by the following equation:

$$p = v_1 i_1 = v_2 i_2. \tag{34.1}$$

In phasor form, the relationship between the primary and secondary voltages and currents can be expressed through the *turns ratio*, denoted as a:

$$a = \frac{N_1}{N_2} = \frac{\overline{V}_1}{\overline{V}_2} = \frac{\overline{I}_2}{\overline{I}_1}. \tag{34.2}$$

If the turns ratio a is greater than 1, indicating that N_1 is greater than N_2, the primary voltage is higher than the secondary voltage. In this configuration, the transformer functions as a *step-down* transformer, stepping down the voltage from the primary to the secondary side.

In practical applications, it is often convenient to *refer*, or *reflect*, the impedances connected to either side of the transformer to a single side. This allows for easier analysis and matching of source and load impedances to achieve maximum power transfer.

The impedance seen looking into the primary side of the transformer, denoted as $\overline{Z}_1$, can be determined using Ohm's law and the turns ratio:

$$\overline{Z}_1 = \frac{\overline{V}_1}{\overline{I}_1} = a^2 \frac{\overline{V}_2}{\overline{I}_2} = a^2 \overline{Z}_2. \tag{34.3}$$

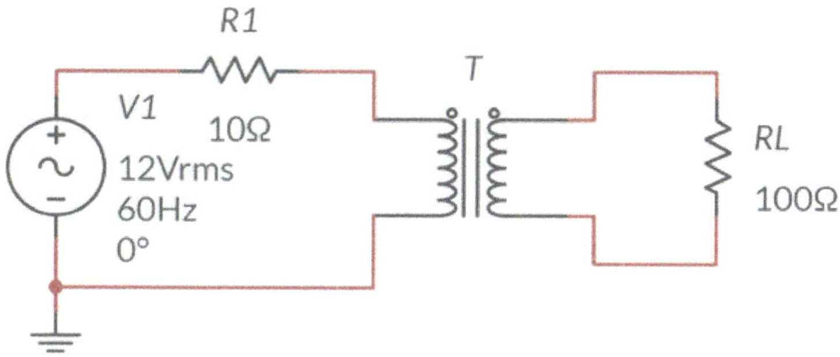

Figure 34.2 Ideal transformer connected to a resistive load.

Similarly, the impedance seen looking into the secondary side of the transformer, denoted as $\overline{Z}_2$, can be calculated as follows:

$$\overline{Z}_2 = \frac{\overline{V}_2}{\overline{I}_2} = \frac{1}{a^2}\frac{\overline{V}_1}{\overline{I}_1} = \frac{1}{a^2}\overline{Z}_1. \tag{34.4}$$

The process of referring the transformer impedance to either side allows to match the internal impedance of a source and that of a load to realize the maximum power transfer.

This process of impedance *reflection* facilitates impedance matching, enabling the optimal transfer of power between a source and the load. By adjusting the impedance on one side of the transformer, it is possible to achieve maximum power transfer and minimize losses within the system.

34.2.1 Problem 1

In the given circuit depicted in Figure 34.2, the primary of an ideal transformer with N_1 =100 and N_2 = 50 is supplied by a voltage $v_1(t) = \sqrt{2} \cdot 12 \cdot \sin(376.8t)$ V. To analyze the circuit, perform the following tasks:

1. Calculate the phasors representing the primary and secondary currents, denoted as $\overline{I}_1$ and I_2. Confirm your calculations with measurements on a Multisim simulated circuit.
2. Calculate the *rms* values of voltages across primary and secondary windings, denoted as V_1 and V_2. Confirm your calculations with measurements on a Multisim simulated circuit.

34.2.1.1 Solutions to Problem 1

1. To determine the phasor $\overline{I}_1$ representing the primary current, we begin by reflecting the resistance of the load, R_L, to the primary side of the transformer:

$$R'_L = a^2 R_L = 400\ \Omega. \tag{34.5}$$

We can calculate $\overline{I}_1$ using Ohm's law by considering the primary voltage, $\overline{V}_1$, and the total resistance seen from the primary side, which is the sum of the primary resistance, R_1, and the reflected load resistance, R'_L. We find

$$\overline{I}_1 = \frac{V_1\ \angle 0°}{R_1 + R'_L} = 29.3\ \angle 0° \text{mA}. \tag{34.6}$$

The *rms* value of the secondary current, I_2, can be determined by multiplying the *rms* value of the primary current, I_1, by the turns ratio a:

$$\overline{I}_2 = a\overline{I}_1 = 58.6\ \angle 0° \text{mA}. \tag{34.7}$$

These calculations are confirmed by performing measurements on a Multisim simulated circuit, ensuring the accuracy and validity of our results (Figure 34.3).

2. The *rms* value V_2 of the voltage across the secondary winding can be calculated using Ohm's law:

$$V_2 = I_2 R_L = 5.8\ \text{V}. \tag{34.8}$$

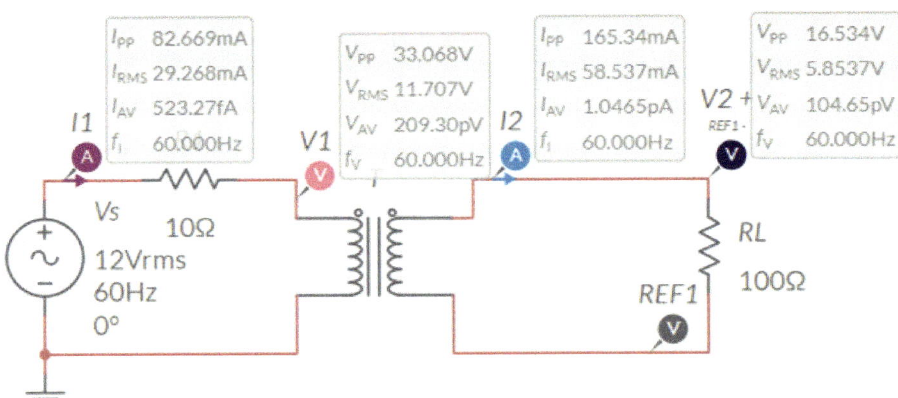

Figure 34.3 Multisim simulation confirming the calculated values of I_1, I_2, V_1, and V_2.

The *rms* value V_1 of the voltage across the primary winding can be calculated by multiplying V_2 by the turns ratio:

$$V_1 = aV_2 = 11.6 \text{ V}. \quad (34.9)$$

These calculations are further validated through Multisim simulation, as shown in Figure 34.3.

34.2.2 Problem 2

In the given circuit depicted in Figure 34.4, the primary of an ideal transformer with $N_1 = N_2 = 100$ is supplied by a voltage $v_{s1}(t) = \sqrt{2} \cdot 24 \cdot \sin(376.8t)$ V. To analyze the circuit, perform the following tasks:

1. Using Multisim, plot the waveforms of the primary and secondary voltages, $v_1(t)$ and $v_2(t)$, as functions of time. Evaluate the phase difference $\Delta\varphi$ between the primary and secondary voltages. From the plot, calculate the ratio of the peak voltage of the primary to the peak voltage of the secondary. Compare the calculated ratio with the theoretical value.
2. Calculate the phasor $\overline{V}_2$ representing the voltage across the load R_L and the phasor $\overline{I}_2$ representing the current through the load R_L.
3. Determine the turns ratio a_M that allows for maximum power transfer to the secondary load. Calculate the value P_{LM} of the maximum power transferred to the load under this condition.

34.2.2.1 Solutions to Problem 2

1. The Multisim plot of Figure 34.5 illustrates the waveforms $v_1(t)$ and $v_2(t)$.

The plot confirms that $v_1(t)$ and $v_2(t)$ perfectly overlap, indicating that there is no phase difference between them (i.e., $\Delta\varphi = 0 \Delta\varphi = 0$). Furthermore, the

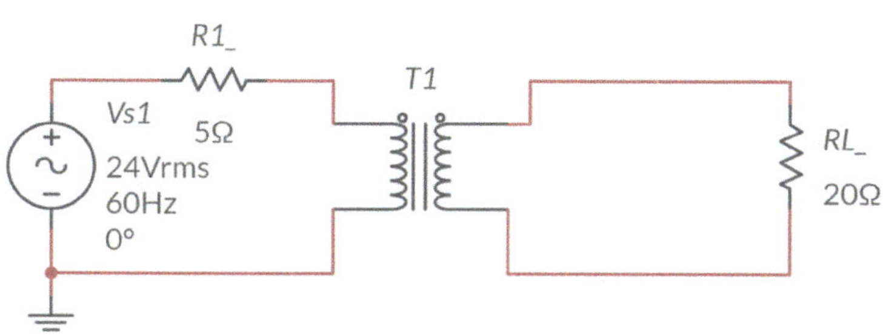

Figure 34.4 Transformer with a =1.

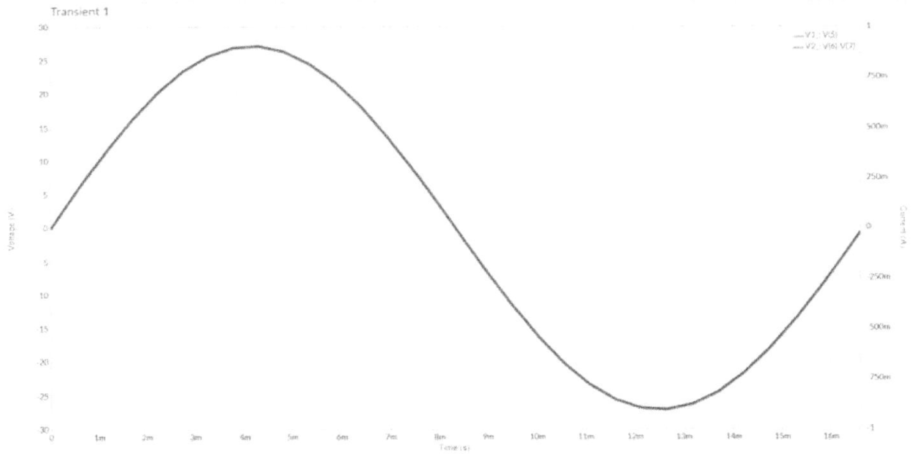

Figure 34.5 Multisim plot illustrating the waveforms $v_1(t)$ and $v_2(t)$.

ratio of the peak voltages of the primary and secondary waveforms is found to be unity, which is consistent with the theoretical turns ratio of the transformer, $a = 1$.

2. If $a = 1$, the load resistance R_L can be reflected to the primary side without modifications. The phasor primary current $\overline{I}_1$ is therefore,

$$\overline{I}_1 = \frac{V_1\angle 0°}{R_L + R_1} = 0.96\ \angle 0° \,\text{A}. \tag{34.10}$$

When $a = 1$, $\overline{I}_2 = \overline{I}_1$. Thus, the phasor $\overline{V}_2$ of the voltage across R_L can be calculated as follows:

$$\overline{V}_2 = \overline{I}_1 R_L = 19.2\ \angle 0°\ \text{V}. \tag{34.11}$$

The values of the above calculated currents are confirmed by the Multisim simulation (Figure 34.6).

3. The condition of maximum power transfer is achieved when the reflected load resistance R'_L, as seen from the primary winding, is equal to the resistance R_1 of the primary winding:

$$R'_L = a_M^2 R_L = R_1 = 5\ \Omega. \tag{34.12}$$

The corresponding turns ratio a_M can be calculated as follows:

$$a_M = \sqrt{\frac{R_1}{R_L}} = 0.5. \tag{34.13}$$

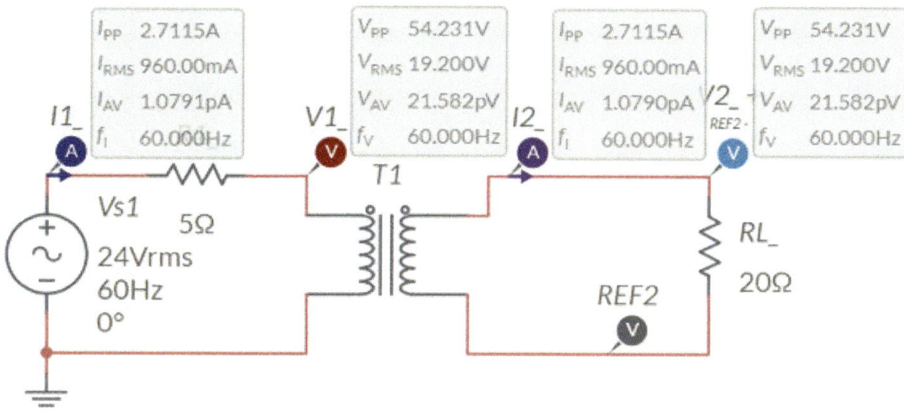

Figure 34.6 Multisim simulation confirming the calculated values of currents and voltages when $a = 1$.

Under these conditions, the phasor $\overline{I}_1$ representing the primary current can be determined as follows:

$$\overline{I}_1 = \frac{\overline{V}_{s1}}{2R_1} = 2.4\ \angle 0° \,\text{A}. \tag{34.14}$$

The phasor $\overline{I}_2$ representing the secondary current can be determined as follows:

$$\overline{I}_2 = a_M \overline{I}_1 = 1.2\ \angle 0° \,\text{A}. \tag{34.15}$$

The maximum power transferred to R_L, denoted as P_{LM}, is given by the following formula:

$$P_{LM} = I_2^2 R_L = 28.8\ \text{W}. \tag{34.16}$$

These calculated values of the currents are confirmed by the Multisim simulation, as shown in Figure 34.7.

34.2.3 Problem 3

An ideal transformer with a primary to secondary turns ratio of $N_1/N_2 = 100/50$ is excited by a primary voltage $v_{s2}(t) = \sqrt{2} \cdot 30 \cdot \sin(376.8t)$ V, as depicted in Figure 34.8. Perform the following tasks:

1. Determine the load impedance $\overline{Z}'_L$ as reflected to the primary circuit and calculate the primary current I_p. Validate these calculations by conducting measurements on a Multisim simulated circuit.

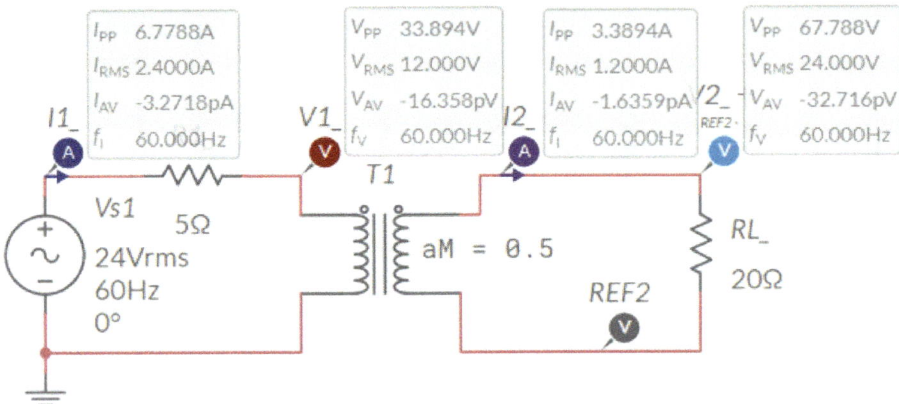

Figure 34.7 Multisim simulation confirming the calculated values of the currents.

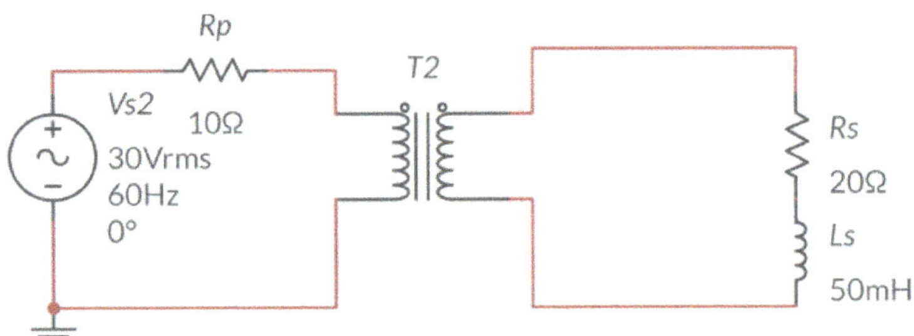

Figure 34.8 Transformer with a secondary ohmic-inductive load.

2. Calculate the phasor $\overline{V}_s$ representing the secondary voltage across the load $\overline{Z}_L$ and the phasor $\overline{I}_s$ representing the current flowing through it. Confirm the accuracy of these calculations through measurements conducted on a Multisim simulated circuit.

34.2.3.1 Solutions to Problem 3

1. The load impedance $\overline{Z}_L$ is characterized by a resistive component of and a reactive component, resulting in a complex impedance:

$$\overline{Z}_L = R_s + jX_s = 20 + j18.8\,\Omega = 27.4\,\angle 43.2°\Omega. \tag{34.17}$$

The load impedance reflected to the primary circuit, denoted as $\overline{Z}'_L$, is calculated as follows:

$$\overline{Z}'_L = a^2\overline{Z}_L = 80 + j75.2\,\Omega = 109.8\,\angle 43.2°\Omega. \tag{34.18}$$

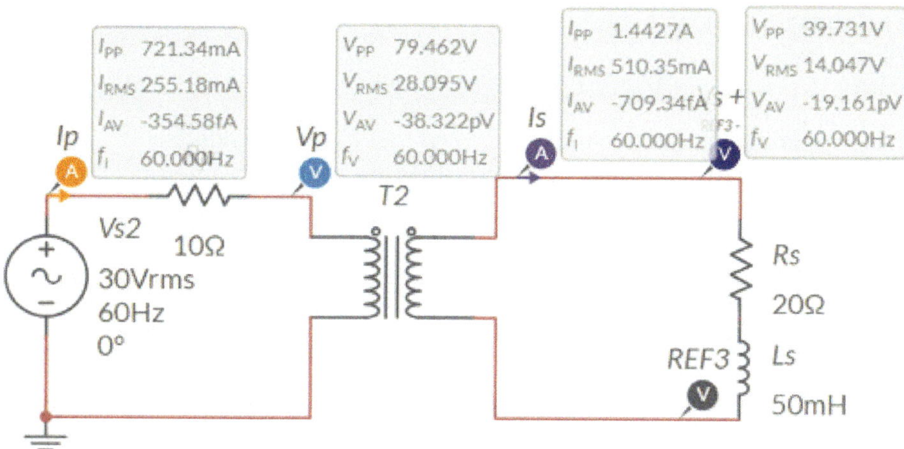

Figure 34.9 Multisim simulation confirming the calculated values of currents and voltages.

The total primary impedance $\overline{Z}_p$ is the sum of the primary resistance and the reflected load impedance, yielding a value:

$$\overline{Z}_p = R_p + \overline{Z}'_L = 90 + j75.2\ \Omega = 117.3\ \angle 39.9°\Omega\ . \tag{34.19}$$

The phasor $\overline{I}_p$ representing primary current is determined as follows:

$$\overline{I}_p = \frac{\overline{V}_{s2}}{\overline{Z}_p} = 255.7\ \angle{-39.9°}\text{mA} = 196.2 - j164\ \text{mA}. \tag{34.20}$$

The Multisim simulation confirms the calculated *rms* value of $\overline{I}_p$ as depicted in Figure 34.9.

Measurements also confirms the magnitude of the calculated total primary impedance: Measurements conducted on the simulated circuit validate the magnitude of the calculated total primary impedance $\overline{Z}_p$:

$$\overline{Z}_p = \frac{V_{s2\ \text{rms}}}{I_{p\ \text{rms}}} = 117.6\ \Omega. \tag{34.21}$$

2. The phasor $\overline{I}_s$ representing the secondary current flowing through the load $\overline{Z}_L$ is obtained by multiplying the primary current $\overline{I}_p$ by the turns ratio a:

$$\overline{I}_s = a\overline{I}_p = 392.4 - j328\ \text{mA} = 511.4\ \angle{-39.9°}\text{mA}. \tag{34.22}$$

The voltage $\overline{V}_s$ across the secondary load $\overline{Z}_L$ is determined by applying the Ohm's law:

$$\overline{V}_s = \overline{I}_s \overline{Z}_L = 14 \angle 3.2° \text{ V}. \tag{34.23}$$

The Multisim simulation confirms the *rms* values of the calculated $\overline{V}_s$ and $\overline{I}_s$ (Figure 34.9). This result is also confirmed by the Multisim simulation, which validates the accuracy of the calculated *rms* values for $\overline{V}_s$ and $\overline{I}_s$.

34.3 Lab Procedure 1

In the given circuit depicted in Figure 34.10, the primary of an ideal transformer with N_1 = 200 and N_2 = 100 is supplied by a voltage $v_{s3}(t) = \sqrt{2} \cdot 120 \cdot \sin(376.8t)$ V. To analyze the circuit, perform the following tasks:

1. Determine the phasors representing the primary and secondary currents, denoted as $\overline{I}_p$ and $\overline{I}_s$. Confirm your calculations with measurements on a Multisim simulated circuit. Record values in Table 34.1.
2. Calculate the *rms* values of voltages across primary and secondary windings, denoted as V_p and V_s. Confirm your calculations with measurements on a Multisim simulated circuit. Record values in Table 34.1.

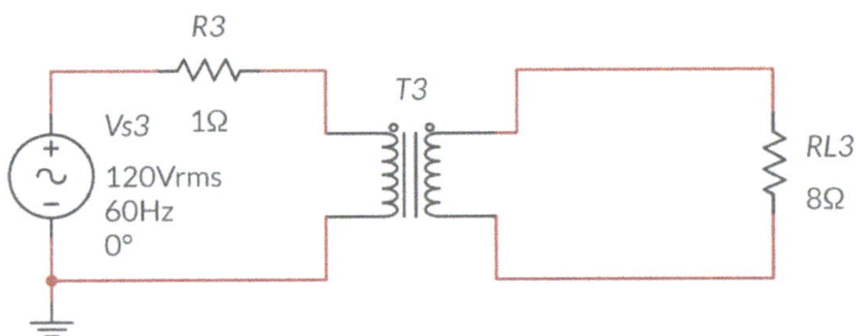

Figure 34.10 Ideal transformer connected to a resistive load.

Table 34.1 Summary of calculations

$\overline{I}_p$	$\overline{I}_s$	V_p	V_s

34.4 Lab Procedure 2

In the given circuit depicted in Figure 34.11, the primary of an ideal transformer with $N_1 = N_2 = 10$ is supplied by a voltage $v_{s4}(t) = \sqrt{2} \cdot 45 \cdot \sin(376.8t)$ V. To analyze the circuit, perform the following tasks:

1. Using Multisim, plot the waveforms of the primary and secondary voltages, $v_1(t)$ and $v_2(t)$, as functions of time. Evaluate the phase difference $\Delta\varphi$ between the primary and secondary voltages. From the plot, calculate the ratio of the peak voltage of the primary to the peak voltage of the secondary. Compare the calculated ratio with the theoretical value.
2. Calculate the phasor $\overline{V}_2$ representing the voltage across the load R_L and the phasor $\overline{I}_2$ representing the current through the load R_L. Record values in Table 34.2.
3. Determine the turns ratio a_M that allows for maximum power transfer to the secondary load. Calculate the value P_{LM} of the maximum power transferred to the load under this condition. Record values in Table 34.2.

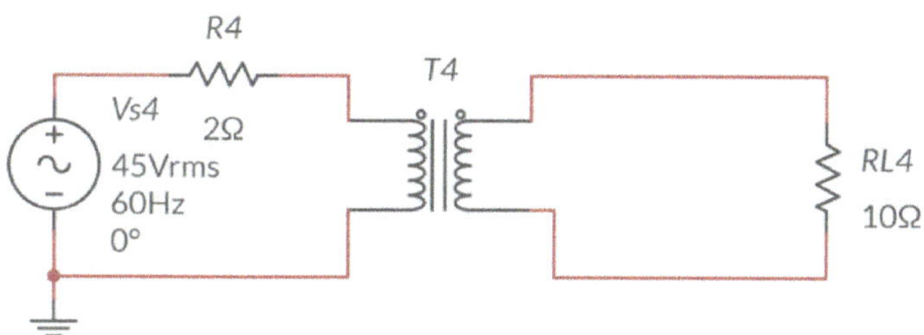

Figure 34.11 Ideal transformer with a =1, connected to a resistive load.

Table 34.2 Summary of calculations

$\overline{V}_2$	$\overline{I}_2$	a_M	P_{LM}

Appendix A
Complex Numbers, Sinewaves, and Phasors

A.1 Student Learning Outcomes

- Analyze and perform calculations involving complex numbers in the context of electric circuits, including addition, subtraction, multiplication, and division of complex numbers.
- Apply Euler's formula, which relates complex numbers to trigonometric functions, to simplify and solve complex circuit problems.
- Utilize phasor representation to analyze and calculate voltages, currents, and impedances in AC circuits, taking into account magnitude and phase relationships.

A.2 Complex Numbers

Complex numbers are a fundamental concept in mathematics and play a crucial role in various fields, including electrical engineering. They extend the set of real numbers by introducing an imaginary unit, denoted as *j*, which is defined as the square root of -1. The use of complex numbers enables the representation and analysis of quantities that involve both real and imaginary components.

In Cartesian notation, a complex number is expressed as follows:

$$\overline{Z} = a + jb, \tag{A.1}$$

where *a* and *b* are real numbers representing the *real* and *imaginary* parts of the complex number, respectively. Geometrically, a complex number can be represented as an oriented segment in a two-dimensional coordinate plane known as the Gauss plane, with the real part along the horizontal axis and the imaginary part along the vertical axis (Figure A.1). The oriented segment has a length $|\overline{Z}|$ equal to the module of the complex number.

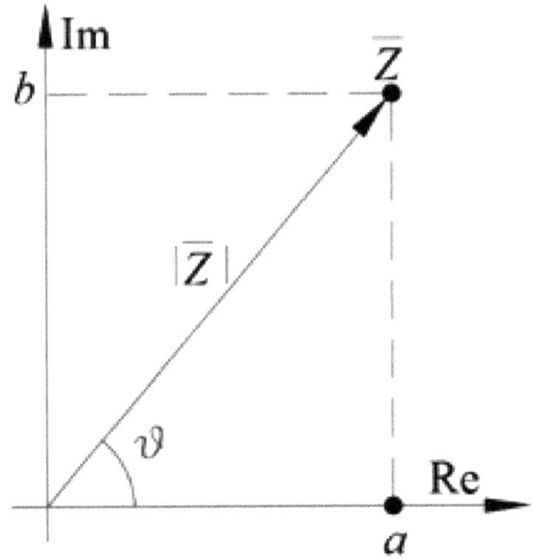

Figure A.1 Complex number in the Gauss plane.

Based on the representation in the Gauss plane, the real and imaginary parts of a complex number can be related to its magnitude $|\overline{Z}|$ and argument ϑ as follows:

$$a = |\overline{Z}| \cos\vartheta \tag{A.2}$$

$$b = |\overline{Z}| \sin\vartheta \,. \tag{A.3}$$

The module or magnitude of the complex number $|\overline{Z}|$ can be calculated with the Pythagorean's theorem:

$$|\overline{Z}| = \sqrt{a^2 + b^2}. \tag{A.4}$$

The argument ϑ can be determined using the arctangent function applied to the ratio of the imaginary part b to the real part a.

$$\vartheta = \tan^{-1}\frac{b}{a} \,. \tag{A.5}$$

Complex numbers can also be expressed in *polar* form, where the magnitude $|\overline{Z}|$ and the argument ϑ are directly written:

$$\overline{Z} = |\overline{Z}| \; \angle\vartheta. \tag{A.6}$$

This form is particularly useful when dealing with calculations involving multiplication, division, and exponentiation of complex numbers.

Complex numbers can also be represented in terms of complex exponentials, which is an effective notation that allows the application of the properties of the exponentials to calculations.

The Euler relationship relates trigonometric functions to complex exponentials.

Another useful representation of complex numbers is in terms of *complex exponentials*. The Euler relationship states that:

$$e^{\pm j\vartheta} = \left|\overline{Z}\right| (\cos\vartheta \pm j\sin\vartheta\ ). \quad \text{(A.7)}$$

This relationship connects complex numbers to exponential functions and facilitates the application of properties and operations involving exponential functions in complex number calculations.

A.3 Symbolic Method

The symbolic method is a powerful tool that establishes a one-to-one correspondence between sinusoidal functions and complex functions of time, denoted as $\overline{X}\ (t)$. This correspondence allows us to represent sinusoids using complex numbers, enabling more convenient and efficient calculations.

This correspondence is expressed as follows:

$$x\ (t) = \sqrt{2}X\cos\left(\omega t + \varphi\right) \leftrightarrow \overline{X}(t), \quad \text{(A.8)}$$

where, for any given time *t*, a complex number is returned.

The complex function $\overline{X}\ (t)$ is defined as follows:

$$\begin{aligned}\overline{X}\ (t) &= \sqrt{2}Xe^{j(\omega t+\varphi)} = \sqrt{2}X\ e^{j\omega t}e^{j\varphi} = \sqrt{2}X\cos\left(\omega t + \varphi\right) \\ &\quad + j\sqrt{2}X\sin\left(\omega t + \varphi\right), \end{aligned} \quad \text{(A.9)}$$

where *X* is the *rms* value of the sinewave.

The complex quantity $\overline{X} = \sqrt{2}Xe^{j\varphi}$ is referred to as the *phasor*. It captures both the peak value and phase angle of the sinusoid, effectively describing the sinusoidal waveform without explicitly indicating the time dependence. In the complex plane, the phasor can be represented as an oriented segment, allowing for a geometric interpretation in relation to the sinusoidal waveform.

Using the phasor representation, sinusoids can be obtained by extracting either the real part or the imaginary part of the complex function $\overline{X}\ (t)$:

$$\text{Re}\ \left[\overline{X}e^{j\omega t}\right] = \sqrt{2}X\cos\left(\omega t + \varphi\right) \quad \text{(A.10)}$$

$$\text{Im}\,\left[\overline{X}e^{j\omega t}\right] = \sqrt{2}X\sin\left(\omega t + \varphi\right). \tag{A.11}$$

These relationships enable us to work in the *phasor domain*, where calculations are often more convenient, and then "reconstruct" the sinusoidal waveform in the *time domain* using the formulae provided.

A.4 Linear Combination of Sinusoids in the Phasor Domain

The linear combination, specifically the addition and subtraction, of multiple sinusoidal signals with identical frequencies, but arbitrary magnitudes and phase angles, results in a new sinusoid of the same frequency. This principle can be leveraged by representing the sinusoids as phasors and applying the simpler rules of complex number addition.

For example, the summation $s\left(t\right)$ of two sinusoidal signals $a\left(t\right) = a_p\cos\left(\omega t + \alpha\right)$ and $b\left(t\right) = b_p\cos\left(\omega t + \beta\right)$ can be obtained by manipulating their corresponding complex functions:

$$\overline{S}\left(t\right) = \overline{A}\left(t\right) + \overline{B}\left(t\right) = \overline{A}e^{j\omega t} + \overline{B}e^{j\omega t} \Leftrightarrow s\left(t\right) = \text{Re}\left[\overline{A}e^{j\omega t} + \overline{B}e^{j\omega t}\right] \tag{A.12}$$

Since the operator Re[·] is linear, we can express the sum $s\left(t\right)$ as follows:

$$s\left(t\right) = \text{Re}\left[\left(\overline{A} + \overline{B}\right)e^{j\omega t}\right], \tag{A.13}$$

which implies that the phasor of the sum is equal to the sum of the phasors.

This approach can be extended to the sum of any number *n* of sinusoids.

$$\begin{aligned} s\left(t\right) &= \sum_{i}^{n} X_{pi}\cos\left(\omega t + \varphi_i\right) = \sum_{i}^{n} \text{Re}\left[X_{pi}e^{j\varphi_i}e^{j\omega t}\right] \\ &= \text{Re}\left[\left(\sum_{i}^{n} X_{pi}e^{j\varphi_i}\right)e^{j\omega t}\right] = \text{Re}\left[S_p e^{j\theta}e^{j\omega t}\right] = S_p\cos\left(\omega t + \theta\right). \end{aligned} \tag{A.14}$$

Therefore, the resulting sinusoid $s\left(t\right)$ has an amplitude S_p and a phase angle θ, which can be determined by performing the summation of the respective phasors.

A.4.1 Problem 1

Sum the following sinusoids: $x_1\left(t\right) = 2\cos\left(\omega t + 45°\right)$ and $x_2\left(t\right) = 4\cos\left(\omega t\right)$.

A.4.1.1 Solution to Problem 1

To sum the sinusoids $x_1(t)$ and $x_2(t)$, we can calculate the sum $\overline{S}$ of their corresponding phasors:

$$\overline{S} = 2e^{j45^\circ} + 4e^{j0^\circ} = 1.41 + j1.41 + 4 = 5.6e^{j14.6^\circ} = 5.6\angle 14.6^\circ. \quad \text{(A.15)}$$

The sum in the time domain can be obtained by extracting the real part of the complex function:

$$s(t) = \text{Re}\left[\overline{S}e^{j\omega t}\right] = 5.6 \cdot \cos(\omega t + 14.6^\circ). \quad \text{(A.16)}$$

A.5 Derivation and Integration of Sinusoids in the Phasor Domain

Differentiation of sinusoidal signals can be conveniently performed in the phasor domain. Consider a sinusoidal signal $x(t) = X_p \cos(\omega t + \varphi)$[1]. Its time derivative is

$$\frac{dx(t)}{dt} = \omega X_p \sin(\omega t + \varphi + 180^\circ) = \omega X_p \cos(\omega t + \varphi + 90^\circ). \quad \text{(A.17)}$$

This expression can be rewritten in terms of a complex function:

$$\frac{dx(t)}{dt} = \text{Re}\left[\omega X_p e^{j90^\circ} e^{j\varphi} e^{j\omega t}\right] = \text{Re}\left[j\omega X_p e^{j\varphi} e^{j\omega t}\right] = \text{Re}\left[j\omega \overline{X} e^{j\omega t}\right], \quad \text{(A.18)}$$

where $\overline{X} = X_p e^{j\varphi}$ is the phasor associated with $x(t)$. We can establish the correspondence between the time domain and the phasor domain as follows:

$$\frac{dx(t)}{dt} \leftrightarrow j\omega \overline{X}. \quad \text{(A.19)}$$

In the phasor domain, the time derivative of a sinusoid $x(t)$ corresponds to multiplying its associated phasor $\overline{X}$ by $j\omega$.

Similarly, in the phasor domain, the integral of a sinusoid corresponds to dividing its associated phasor by $j\omega$:

$$\int x(t)\, dt \leftrightarrow \frac{\overline{X}}{j\omega}. \quad \text{(A.20)}$$

---

[1] To transform a sinusoid from the sine form to the cosine form or vice versa, we can make use of the following trigonometric identities: $-\cos(\omega t) = \cos(\omega t \pm 180^\circ)$; $-\sin(\omega t) = \sin(\omega t \pm 180^\circ)$; $\cos(\omega t) = \sin(\omega t + 90^\circ)$; $\sin \omega t = \cos(\omega t - 90^\circ)$.

A.6 Phasor Relationships in a Circuit

The voltage and current phasors in circuit elements have specific phase relationships:

A.6.1 Resistors

For a resistor *R*, the sinusoidal voltage $v_R(t)$ across it and the corresponding sinusoidal current $i_R(t)$ through it are related by Ohm's law:

$$v_R\,(t) = \; i_R\,(t)\,R. \tag{A.21}$$

In terms of phasors, we have

$$\overline{V}_R = \overline{I}_R R. \tag{A.22}$$

The phasors of voltage and current in a resistor are in phase with each other.

A.6.2 Inductors

For an inductor, the sinusoidal voltage $v_L(t)$ across it and the corresponding sinusoidal current $i_L(t)$ through it are related by Faraday's law of induction:

$$v_L\,(t) = L\frac{di_L(t)}{dt}. \tag{A.23}$$

In terms of phasors, we have

$$v_L\,(t) = \text{Re}\left[\overline{V}_L e^{j\omega t}\right] = L\cdot\frac{d}{dt}\left(\text{Re}\left[\overline{I}_L e^{j\omega t}\right]\right) = \text{Re}\left[j\omega L\overline{I}_L e^{j\omega t}\right]. \tag{A.24}$$

Thus, the relationship between the voltage and current phasors in an inductor is given by

$$\overline{V}_L = j\omega L\overline{I}_L, \tag{A.25}$$

Here, the voltage phasor leads the current phasor by 90° due to the presence of the imaginary unit $j = e^{j90^\circ}$.

A.6.3 Capacitors

For a capacitor, the sinusoidal current $i_C\,(t)$ through it and the corresponding sinusoidal voltage $v_C\,(t)$ across it are related by the following equation:

$$i_C\,(t) = C\frac{dv_C\,(t)}{dt}. \tag{A.26}$$

In terms of phasors, we have

$$i_C(t) = \mathrm{Re}\left[\overline{I}_C e^{j\omega t}\right] = C\frac{d}{dt}\left(\mathrm{Re}\left[\overline{V}_C e^{j\omega t}\right]\right) = \mathrm{Re}\left[j\omega C\overline{V}_C e^{j\omega t}\right]. \quad \text{(A.27)}$$

Hence, the relationship between the current and voltage phasors in a capacitor is given by

$$\overline{I}_C = j\omega C\overline{V}_C. \quad \text{(A.28)}$$

Here, the current phasor leads the voltage phasor by 90° due to the presence of the imaginary unit $j = e^{j90°}$.

A.7 Complex Impedance of a Circuit Element

The impedance $\overline{Z}(j\omega)$ of a circuit element is defined as the ratio of the phasor of the voltage across the element to the phasor of the current through it. Mathematically, it can be expressed as follows:

$$\overline{Z}(j\omega) = \frac{\overline{V}}{\overline{I}}. \quad \text{(A.29)}$$

For a resistor *R*, the impedance is equal to the resistance itself:

$$\overline{Z}(j\omega) = \frac{\overline{V}_R}{\overline{I}_R} = R. \quad \text{(A.30)}$$

The voltage and current phasors are in phase.

For an inductor *L*, the impedance is given by

$$\overline{Z}(j\omega) = \frac{\overline{V}_L}{\overline{I}_L} = j\omega L. \quad \text{(A.31)}$$

The voltage phasor leads the current phasor by 90°.

For a capacitor *C*, the impedance is given by

$$\overline{Z}(j\omega) = \frac{\overline{V}_C}{\overline{I}_C} = \frac{\overline{V}_C}{j\omega C\overline{V}_C} = -j\frac{1}{\omega C}. \quad \text{(A.32)}$$

The current phasor leads the voltage phasor by 90°.

The complex impedances of circuit elements reflect their frequency-dependent oppositions to current flow, encompassing both resistive and reactive properties. The impedance concept allows all elements to be treated equally using phasors.

A.8 MATLAB Code

MATLAB is a powerful programming language and software package that includes built-in functions for complex numbers, making it straightforward to perform calculations with them. Below are some examples of how to work with complex numbers.

```
>> % define a complex number
z = 3 + 4i;

% display the real and imaginary parts of z
disp(['Real part: ' num2str(real(z))]);
disp(['Imaginary part: ' num2str(imag(z))]);

% calculate the magnitude (or absolute value) of z
mag_z = abs(z);
disp(['Magnitude of z: ' num2str(mag_z)]);

% calculate the complex conjugate of z
conj_z = conj(z);
disp(['Complex conjugate of z: ' num2str(conj_z)]);

% perform some arithmetic operations with z and another complex number
w = 2 - i;
sum_zw = z + w;
diff_zw = z - w;
prod_zw = z * w;
quotient_zw = z / w;

% display the results of the arithmetic operations
disp(['Sum of z and w: ' num2str(sum_zw)]);
disp(['Difference of z and w: ' num2str(diff_zw)]);
disp(['Product of z and w: ' num2str(prod_zw)]);
disp(['Quotient of z and w: ' num2str(quotient_zw)]);
Real part: 3
Imaginary part: 4
Magnitude of z: 5
Complex conjugate of z: 3-4i
Sum of z and w: 5+3i
Difference of z and w: 1+5i
Product of z and w: 10+5i
Quotient of z and w: 0.4+2.2i
>>
```

The complex variables z and w are defined representing the number $3 + 4i$ and $2 - i$, respectively. The *real()* and *imag()* functions extract the real and imaginary parts of z, respectively.

The *abs*() function calculates the magnitude, or modulus, of a complex number. For *z*, this yields 5 due to the Pythagorean theorem.

The complex conjugate is obtained using *conj*(), which negates the imaginary part. This gives 3 – 4*i* for *z*.

Arithmetic operations with complex numbers are done naturally in MATLAB.

*z***w* multiplies the two complex numbers, giving 10 + 5*i*. *z*/*w* divides them, resulting in 0.4 + 2.2*i*.

The results of these operations are displayed using the *disp ()* function.

MATLAB provides convenient built-in functions for working with complex numbers and phasors, including polar and rectangular conversions, arithmetic operations, computing modulus and angle, graphing, and more. This makes MATLAB a useful tool for programming and visualizing solutions to problems involving complex numbers and phasors.

Index

About the Author

Massimo Mitolo, a distinguished scholar and scientist, has been bestowed the Knighthood in the *Order of Merit of the Italian Republic* in acknowledgment of his exceptional contributions to scientific endeavors that have brought great honor to the nation. He is renowned for his remarkable achievements in the field of electrical engineering.

Sir Massimo earned his doctoral degree in Electrical Engineering from the University of Napoli "Federico II" in Italy. His unwavering dedication and significant impact on the field have led to his recognition as a Fellow of IEEE "for contributions to the electrical safety of low-voltage systems". Furthermore, he holds the distinguished title of Fellow from the Institution of Engineering and Technology (IET) in London, United Kingdom, and is a member of the *IEEE-HKN* Honor Society. Additionally, he is a registered Professional Engineer in both the state of California and Italy.

Presently, Dr. Mitolo serves as a Full Professor of Electrical Engineering at Irvine Valley College in California. In addition to his academic responsibilities, he is a Senior Consultant specializing in the domains of failure analysis and electrical safety engineering. His extensive research and industrial experience revolve around the comprehensive analysis and grounding of power systems.

Dr. Mitolo's expertise is reflected in his publication record, encompassing more than 180 journal papers, as well as the authorship of several influential books. Noteworthy titles authored by him include "Electrical Safety of Low-Voltage Systems" published by McGraw-Hill, "Laboratory Manual for Introduction to Electronics: A Basic Approach" published by Pearson, "Analysis of Grounding and Bonding Systems" published by CRC Press, "Electrical Safety Engineering of Renewable Energy Systems" published by IEEE Wiley, and "Smart and Power Grid Systems: Design Challenges and Paradigms" published by River Publishers.

His scholarly endeavors have garnered significant recognition, culminating in his inclusion in the World's Top 2% Most-cited Scientists List, as compiled by Stanford University.

Within the Industrial and Commercial Power Systems Department of the IEEE Industry Applications Society (IAS), Dr. Mitolo actively engages in various committees and working groups, demonstrating his commitment to advancing the field and fostering collaborative efforts.

Acknowledging his achievements, Dr. Mitolo has been the recipient of numerous prestigious accolades throughout his career. Notably, he has been honored with the IEEE Region 6 Outstanding Engineer Award and has garnered nine Best Paper Awards for his exceptional scholarly contributions. Furthermore, he has received recognitions such as the IEEE Ralph H. Lee I&CPS Department Prize Award, the IEEE I&CPS Department Achievement Award, and the James E. Ballinger Engineer of the Year Award from the Orange County Engineering Council in California.